The Facts On File

DICTIONARY of PHYSICS

The Facts On File

DICTIONARY of PHYSICS

Fourth Edition

Edited by
John Daintith
Richard Rennie

Checkmark Books®
An imprint of Facts On File, Inc.

The Facts On File Dictionary of Physics
Fourth Edition

Checkmark Books
An imprint of Facts On File, Inc.
132 West 31st Street
New York NY 10001

For Library of Congress Cataloging-in-Publication Data,
please contact Facts On File, Inc.
ISBN 0-8160-5654-4

You can find Facts On File on the World Wide Web at
http://www.factsonfile.com

Compiled and typeset by Market House Books Ltd, Aylesbury, UK

Printed in the United States of America

MP PKG 10 9 8 7 6 5 4 3 2 1

This book is printed on acid-free paper.

PREFACE

This dictionary is one of a series designed for use in schools. It is intended for students of physics, but we hope that it will also be helpful to other science students and to anyone interested in science. Facts On File also publishes dictionaries in a variety of disciplines, including biology, chemistry, forensic science, marine science, mathematics, space and astronomy, and weather and climate.

The Facts On File Dictionary of Physics was first published in 1980 and the third edition was published in 1999. This fourth edition of the dictionary has been extensively revised and extended. The dictionary now contains over 2,800 headwords covering the terminology of modern physics. A totally new feature of this edition is the inclusion of over 1,000 pronunciations for terms that are not in everyday use. A number of appendixes have been included at the end of the book containing useful information, including a list of chemical elements, a periodic table, a list of symbols, a number of useful conversion tables, and a Greek alphabet. There is also a list of Web sites and a bibliography. A guide to using the dictionary has also been added to this latest version of the book.

We would like to thank all the people who have cooperated in producing this book. A list of contributors is given on the acknowledgments page. We are also grateful to the many people who have given additional help and advice.

ACKNOWLEDGMENTS

Consultant Editors (First and Second Editions)

Eric Deeson M.Sc., F.C.P., F.R.A.S.
J. W. Warren B.Sc.

Contributors

Roger Adams B.Sc.
Jane Craig B.Sc., A.R.C.S.
Sue Flint B.Sc.
B. W. Lowthian B.Sc., Ph.D.
Carol Russell B.Sc.
Michael S. Slater B.Sc.
James M. Struthers B.Sc.
Thomas A. Shields B.A., C.S.A.
M. Welton B.Sc.

Pronunciations

William Gould B.A.

Note

Unless otherwise stated, the melting and boiling points given in the dictionary are at standard pressure. Relative densities of liquids are at standard pressure with the liquid at 20°C relative to water at 4°C. Relative densities of gases are relative to air, both gases being at standard temperature and pressure.

The following abbreviations are used in the text:

p.n.	proton number (atomic number)
r.a.m.	relative atomic mass (atomic weight)
m.p.	melting point
b.p.	boiling point

CONTENTS

GUIDE TO USING THE DICTIONARY

The main features of dictionary entries are as follows.

Headwords

The main term being defined is in bold type:

> **aberration** A defect in an optical system such that the image is not a true picture of the object.

Plurals

Irregular plurals are given in brackets after the headword.

> **spectrum** (*pl.* **spectra**) ... A range of electromagnetic radiation emitted or absorbed by a substance under particular circumstances.

Variants

Sometimes a word has a synonym or alternative spelling. This is placed in brackets after the headword, and is also in bold type:

> **eyepiece** (**ocular**) The lens or combination of lenses nearest the eye in an optical instrument.

Here, 'ocular' is another word for eyepiece. Generally, the entry for the synonym consists of a simple cross-reference:

> **ocular** *See* eyepiece.

Abbreviations

Abbreviations for terms are treated in the same way as variants:

> **electron spin resonance** (**ESR**) A branch of microwave spectroscopy ....

The entry for the synonym consists of a simple cross-reference:

> **ESR** *See* electron spin resonance.

Multiple definitions

Some terms have two or more distinct senses. These are numbered in bold type

> **abundance** **1.** The relative amount of a given element among others; for example, the abundance of oxygen in the Earth's crust is approximately 50% by weight.
> **2.** The amount of a nuclide (stable or radioactive) relative to other nuclides of the same element in a given sample.

Cross-references

These are references within an entry to other entries that may give additional useful information. Cross-references are indicated in two ways. When the word appears in the definition, it is printed in small capitals:

> **accommodation** The action of the EYE in changing its focal power. ...

In this case the cross-reference is to the entry for 'eye'.

Alternatively, a cross-reference may be indicated by 'See', 'See also', or 'Compare', usually at the end of an entry:

> **angular momentum** Symbol: *L* The product of the moment of inertia of a body and its angular velocity. *See also* rotational motion.

Hidden entries

Sometimes it is convenient to define one term within the entry for another term:

> **charcoal** An amorphous form of carbon made by... *Activated charcoal* is charcoal heated to...

Here, 'activated charcoal' is a hidden entry under charcoal, and is indicated by italic type. The entry for 'activated charcoal' consists of a simple cross-reference:

> **activated charcoal** *See* charcoal.

Pronunciations

Where appropriate pronunciations are indicated immediately after the headword, enclosed in forward slashes:

> **ablation** /ab-**lay**-shŏn/ The burning away of the outer surface of ....

Note that simple words in everyday language are not given pronunciations. Also headwords that are two-word phrases do not have pronunciations if the component words are pronounced elsewhere in the dictionary.

Pronunciation Key

Bold type indicates a stressed syllable. In pronunciations, a consonant is sometimes doubled to prevent accidental mispronunciation of a syllable resembling a familiar word; for example, /**ass**-id/ (acid), rather than /**as**-id/ and /ul-tră- **sonn**-iks/ (ultrasonics), rather than /ul-tră-**son**-iks/. An apostrophe is used: (a) between two consonants forming a syllable, as in /**den**-t'l/ (dental), and (b) between two letters when the syllable might otherwise be mispronounced through resembling a familiar word, as in /**th'e**-ră-pee/ (therapy) and /tal'k/ (talc). The symbols used are:

/a/ *as in* back /bak/, active /**ak**-tiv/
/ă/ *as in* abduct /ăb-**dukt**/, gamma /**gam**-ă/
/ah/ *as in* palm /pahm/, father /**fah**-*th*er/,
/air/ *as in* care /kair/, aerospace /**air**-ŏ-spays/
/ar/ *as in* tar /tar/, starfish /**star**-fish/, heart /hart/
/aw/ *as in* jaw /jaw/, gall /gawl/, taut /tawt/
/ay/ *as in* mania /**may**-niă/ ,grey /gray/
/b/ *as in* bed /bed/
/ch/ *as in* chin /chin/
/d/ *as in* day /day/
/e/ *as in* red /red/
/ĕ/ *as in* bowel /**bow**-ĕl/
/ee/ *as in* see /see/, haem /heem/, caffeine /**ka**f-een/, baby /**bay**-bee/
/eer/ *as in* fear /feer/, serum /**see**r-ŭm/
/er/ *as in* dermal /**der**-măl/, labour /**lay**-ber/
/ew/ *as in* dew /dew/, nucleus /**new**-klee-ŭs/
/ewr/ *as in* epidural /ep-i-**dewr**-ăl/
/f/ *as in* fat /fat/, phobia /**foh**-biă/, rough /ruf/
/g/ *as in* gag /gag/
/h/ *as in* hip /hip/
/i/ *as in* fit /fit/, reduction /ri-**duk**-shăn/
/j/ *as in* jaw /jaw/, gene /jeen/, ridge /rij/
/k/ *as in* kidney /**kid**-nee/, chlorine /**klor**-een/, crisis /**krÿ**-sis/
/ks/ *as in* toxic /**toks**-ik/
/kw/ *as in* quadrate /**kwod**-rayt/
/l/ *as in* liver /**liv**-er/, seal /seel/
/m/ *as in* milk /milk/
/n/ *as in* nit /nit/
/ng/ *as in* sing /sing/
/nk/ *as in* rank /rank/, bronchus /**bronk**-ŭs/
/o/ *as in* pot /pot/
/ô/ *as in* dog /dôg/
/ŏ/ *as in* buttock /**but**-ŏk/
/oh/ *as in* home /hohm/, post /pohst/
/oi/ *as in* boil /boil/
/oo/ *as in* food /food/, croup /kroop/, fluke /flook/
/oor/ *as in* pruritus /proor-**ÿ**-tis/
/or/ *as in* organ /**or**-găn/, wart /wort/
/ow/ *as in* powder /**pow**-der/, pouch /powch/
/p/ *as in* pill /pil/
/r/ *as in* rib /rib/
/s/ *as in* skin /skin/, cell /sel/
/sh/ *as in* shock /shok/, action /**ak**-shŏn/
/t/ *as in* tone /tohn/
/th/ *as in* thin /thin/, stealth /stelth/
/*th*/ as in then /*th*en/, bathe /bay*th*/
/u/ *as in* pulp /pulp/, blood /blud/
/ŭ/ *as in* typhus /**tÿ**-fŭs/
/û/ *as in* pull /pûl/, hook /hûk/
/v/ *as in* vein /vayn/
/w/ *as in* wind /wind/
/y/ *as in* yeast /yeest/
/ÿ/ *as in* bite /bÿt/, high /hÿ/, hyperfine /**hÿ**-per-fÿn/
/yoo/ *as in* unit /**yoo**-nit/, formula /**form**-yoo-lă/
/yoor/ *as in* pure /pyoor/, ureter /yoor-**ee**-ter/
/ÿr/ *as in* fire /fÿr/

A

ab- A prefix used with a practical electrical unit to name the corresponding electromagnetic unit. For example, the electromagnetic unit of charge is called the *abcoulomb*. *Compare* stat-.

aberration A defect in an optical system such that the image is not a true picture of the object. For instance, colored fringes may appear, the image may not be focused, or the shape may show distortion. Techniques of aberration correction exist; these can, however, be complex and costly.

Chromatic (color) *aberration* is found with a single lens; mirrors do not suffer from chromatic aberration. Because dispersion always accompanies refractive deviation, the 'red' image will be farther from the lens than the 'blue'. Consequently, the image is surrounded by colored fringes. Chromatic aberration is corrected by forming a compound lens, whose elements have different refractive constants.

Spherical aberration always occurs with rays that are distant from the axis and incident on a spherical mirror or lens. It is the cause of the caustic curve. Spherical aberration is corrected by using parabolic reflecting and refracting surfaces.

Astigmatism affects rays neither close nor parallel to the axis. The cone of rays through a lens from an off-axis object does not focus at a point. Instead, two images in the form of short lines are formed at different distances from the lens. Between the two the image appears blurred. Mirrors forming images of off-axis points show a similar defect. The best method of minimizing astigmatism is to reduce the aperture with stops, thus allowing light only through the center of the lens.

Coma is rather similar in cause, effect, and correction to astigmatism. After refraction by a lens, a cone of rays from an off-axis object tends to have a tadpole-shaped section because of coma.

Distortion is the result of differences in a lens' magnifying power between different axes. Reduction of aperture is the normal solution to both coma and distortion.

ablation /ab-**lay**-shŏn/ The burning away of the outer surface of a satellite, shuttle, space probe or missile on re-entering the Earth's atmosphere. The term originally applied to meteors.

absolute expansion *See* expansivity.

absolute humidity The mass of water vapor per unit volume of air, usually measured in micrograms per cubic meter. *Compare* relative humidity. *See also* humidity.

absolute permeability *See* permeability.

absolute permittivity *See* permittivity.

absolute refractive index *See* refractive index.

absolute temperature Symbol: T A temperature expressed on the *thermodynamic* (ideal gas) scale, measured from absolute zero. If θ is the temperature on a Celsius scale calibrated against the *International Practical Temperature Scale*, then:

$$T = \theta + 273.15$$

See also temperature scale.

absolute unit A unit defined in terms of fundamental quantities (such as length, mass, time, and electric charge).

absolute zero The zero value of thermodynamic temperature; 0 kelvin or −273.15°C.

absorbance /ăb-**sor**-băns, **-zor-**/ (**optical density**) The logarithm of absorptance. *See* absorptance.

absorptance /ăb-**sorp**-tăns, **-zorp-**/ Symbol: α The ratio of the radiant or luminous flux absorbed by a body or material to the incident flux. It was formerly called the *absorptivity*.

absorption A process in which a gas is taken up by a liquid or solid, or in which a liquid is taken up by a solid. In absorption, the substance absorbed goes into the bulk of the material. Solids that absorb gases or liquids often have a porous structure. The absorption of gases in solids is sometimes called *sorption*. *Compare* adsorption.

absorption band *See* band spectrum

absorption coefficient *See* Lambert's law.

absorption of radiation No medium transmits radiation without some energy loss. This loss of energy is called absorption. The energy is converted to some other form within the medium. *See also* Lambert's law.

absorption spectrum *See* spectrum.

absorptivity /ab-sorp-**tiv**-ă-tee, -zorp-/ *See* absorptance.

abundance **1.** The relative amount of a given element among others; for example, the abundance of oxygen in the Earth's crust is approximately 50% by mass.
2. The amount of a nuclide (stable or radioactive) relative to other nuclides of the same element in a given sample. The *natural abundance* is the abundance of a nuclide as it occurs naturally. For instance, chlorine has two stable isotopes of masses 35 and 37. The abundance of ^{35}Cl is 75.5% and that of ^{37}Cl is 24.5%. For some elements the abundance of a particular nuclide depends on the source.

abundance ratio The ratio of the number of atoms of an isotope to the number of atoms of another isotope of the same element in a sample. *See* abundance.

a.c. *See* alternating current.

acceleration Symbol: a The SI unit of linear acceleration is the meter per second per second (m s^{-2}). The SI unit of angular acceleration is the radian per second per second (rad s^{-2}). **1.** When considering motion in one dimension, and in unscientific usage, acceleration means rate of increase of speed. This is a scalar quantity, which can be positive or negative. Negative values mean that the speed is decreasing and may be called deceleration or retardation.
2. In scientific study of motion in two or three dimensions acceleration means rate of change of velocity; $a = dv/dt$. This is a vector quantity having magnitude (which is always positive) and direction. Whenever speed changes (increasing or decreasing), or direction changes, or both speed and direction change, this is an acceleration.

By Newton's second law the net force F acting on a body of mass m gives it an acceleration a where $F = ma$.

See equations of motion; Newton's laws of motion.

acceleration due to gravity *See* acceleration of free fall.

acceleration of free fall (**acceleration due to gravity; gravitational acceleration**) Symbol: g The constant acceleration of a mass moving freely (without friction or any other force) close to the surface of the Earth. This acceleration is measured with respect to the nearby surface of the Earth so is not exactly equal to the acceleration toward the center, because of the centripetal acceleration of the reference point. On correcting for this small error g is a measure of the gravitational field strength, the force on unit mass. The force mg is sometimes called the weight.

The value of g decreases on going upward from the surface and increases on going down a mine. It increases slightly on going from the equator toward the poles, because of the effects of rotation and the slight flattening of the Earth at the poles. Certain geological features cause small local differences, the detection of which may be useful in prospecting for minerals, especially oil. The standard value assumed for g is 9.806 65 m s^{-2}. For rough calculations one can assume $g = 10$ m s^{-2}.

See Newton's law of universal gravitation, weight.

accelerator A device for accelerating charged particles to high energies so that they are able to penetrate to the nuclei of atoms in a target, causing nuclear reactions. The earliest accelerator was invented by Cockcroft and Walton and was first used to accelerate protons toward a target of lithium.

Two types are now in use. In LINEAR ACCELERATORS the particles are accelerated in a straight line. *Cyclic accelerators* use magnetic fields to keep the particles moving in circular or spiral paths. Examples of cyclic accelerators are the cyclotron, the synchrocyclotron, and the betatron.

acceptor /ak-**sep**-ter, -tor/ *See* semiconductor.

acceptor circuit *See* resonance.

accommodation The action of the EYE in changing its focal power. The normal eye has a high power (short focal distance) for viewing close objects; it relaxes to low power for very distant objects. Accommodation is accomplished by muscles in a ring round the lens of the eye, which are able to change the shape of the lens. The AMPLITUDE OF ACCOMMODATION decreases with age – the power range is around 11 diopters at age 10 and 1 diopter at age 70. Thus the distance between far point and near point decreases with age. This effect is presbyopia.

accumulator (**secondary cell; storage battery**) An electric cell or battery that can be charged by passing an electric current through it. The chemical reaction in the cell is reversible. When the cell begins to run down, current in the opposite direction will convert the reaction products back into their original forms. The most common example is the lead-acid accumulator, used in vehicle batteries.

achromat /**ak**-rŏ-mat/ An achromatic lens.

achromatic color /ak-rŏ-**mat**-ik/ A color that has no hue; i.e. black, white, or gray.

achromatic lens A compound lens whose elements differ in refractive constant in order to minimize chromatic aberration. Simple *achromatic doublets* are formed by combining two lenses of different glass. The condition for achromatism is:

$$\omega_1 P_1 + \omega_2 P_2 = 0$$

where ω_1 and ω_2 are the dispersive powers of the glasses of the lenses, and P_1 and P_2 are the powers of the lenses. Achromatic lenses are corrected for chromatic aberration at two different wavelengths. *See also* apochromatic lens.

aclinic line /ay-**klin**-ik/ (**magnetic equator**) *See* isoclinic line.

acoustics The study of the production and properties of sounds. The term is also used to describe the way in which sound is reproduced in practical situations.

acoustoelectronics /ă-koos-toh-i-lek-**tron**-iks/ The branch of electronics that deals with sound waves at the frequencies of microwaves.

actinic radiation /ak-**tin**-ik/ Radiation that can cause a chemical reaction; for example, ultraviolet radiation is actinic.

actinium /ak-**tin**-ee-ŭm/ A soft silvery-white radioactive metallic element that is the first member of the actinoid series. It occurs in minute quantities in uranium ores. It can be produced by neutron bombardment of radium and is used as a source of alpha particles. The metal glows in the dark.

Symbol: Ac; m.p. 1050±50°C; b.p. 3200±300°C; r.d. 10.06 (20°C); p.n. 89;

most stable isotope ^{227}Ac (half-life 21.77 years).

actinometer /ak-tă-**nom**-ĕ-ter/ An instrument for measuring the intensity of radiation.

actinon /**ak**-tă-non/ *See* emanation.

action 1. An out-dated term for force. *See* reaction.
2. The product of kinetic energy times time, or momentum times displacement, integrated along the path of a body. It has the dimensions of mass times the square of length over time, the same as for angular momentum. The Planck constant is sometimes called Planck's constant of action.

action at a distance A type of interaction between two bodies such that each influences the other through space. The descriptions of gravity by Newton's law and electrostatics by Coulomb's law are examples of action at a distance. Theories based upon action at a distance do not have a mechanism for explaining how the force gets across space from one body to the other, and it is usual to describe such phenomena using field theories.

activated charcoal *See* charcoal.

activation analysis A method used for analyzing materials in which a very small sample of the material is bombarded with neutrons. A nucleus of the material captures a neutron to form an excited nucleus with a nucleon number one higher than that of the original nucleus. The wavelength of gamma rays emitted by the excited nucleus returning to its ground state enables the element to be identified. The technique is very sensitive and can detect concentrations of a few parts per million.

activity Symbol: *A* For a radioactive substance, the average number of atoms disintegrating per unit time.

acuity, visual The ability of the eye to see separately two points close to each other. It is a measure of the resolving power of the eye's optical system and depends on the density of cells in the retina. The maximum acuity of the normal human eye is around 0.5 minutes of arc – points separated by this angle at the eye should be seen as separate. *See* resolution.

additive process A process of color mixing by addition. *See* color.

adhesion A force of attraction between atoms or molecules of different substances. For example, adhesion between water molecules and glass creates a MENISCUS. *See* capillary action.

adiabatic change /ad-ee-ă-**bat**-ik/ A change taking place in a system that has perfect thermal insulation, so that heat cannot enter or leave the system and energy can only be transferred by work. In practice, a close approximation to an adiabatic change can be achieved by the process being too rapid for significant heat transfer, or by the large scale of the system (e.g. a large volume of air in the atmosphere).

In an adiabatic expansion of a gas, mechanical work is done by the gas as its volume increases and the gas temperature falls. For an ideal gas undergoing a reversible adiabatic change it can be shown that

$$pV^{\gamma} = K_1$$
$$T^{\gamma}p^{1-\gamma} = K_2$$
$$\text{and } TV^{\gamma-1} = K_3$$

where K_1, K_2, and K_3 are constants and γ is the ratio of the principal specific heat capacities. *Compare* isothermal change.

adiabatic demagnetization A method of producing temperatures close to absolute zero. A sample of a paramagnetic salt is cooled in liquid helium in a strong magnetizing field. The sample is then thermally isolated by pumping away the helium, and the magnetic field is removed. The sample demagnetizes itself at the expense of its internal energy so that the temperature falls. Temperatures of the order of a millikelvin can be obtained.

admittance Symbol: *Y* The reciprocal of impedance, measured in siemens (S). It is a measure of the response of an electric circuit to an alternating signal. *See also* impedance.

adsorbent /ad-**sor**-bĕnt, **-zor-**/ The substance on whose surface ADSORPTION occurs.

adsorption /ad-**sorp**-shŏn, **-zorp-**/ A process in which a layer of atoms or molecules of one substance forms on the surface of a solid or liquid. All solid surfaces take up layers of gas from the surrounding atmosphere. The adsorbed layer may be held by chemical bonds (*chemisorption*) or by weaker van der Waals' forces (*physisorption*). *Compare* absorption.

advanced gas-cooled reactor *See* gas-cooled reactor.

advection /ad-**vek**-shŏn/ The transfer of matter such as water vapor or heat through the atmosphere as a result of horizontal movement of an air mass.

aerial *See* antenna.

aerodynamics The branch of physics concerned with the movement of gases and the motion of solid objects in gases (usually air). It is applied to the flight of birds and insects, and (particularly) to various kinds of aircraft. *See* drag; lift; Reynold's number; turbulent flow.

aerogenerator An electric generator driven by wind power.

aerosol A dispersion of small particles of solid or droplets of liquid in a gas.

aether /**ee**-th'er/ *See* ether.

after-image An image seen after the eye's retina has been exposed for a time to an intense or stationary light source. It may be negative or positive, or appear in complementary colors.

agate /**ag**-it, -ayt/ A crystalline form of silica used, because of its hardness, in making knife edges in balances, pendulums, etc.

age of the Earth An age of approximately 4.5 billion years. The Earth is thought to have formed, like the rest of the planets, soon after the solar system itself was formed. Early calculations in the 19th century based on rates of cooling gave much lower estimated ages. Present estimates are based on radioactive dating of rocks. The original estimates were too low because they did not allow for radioactivity in the Earth's core.

age of the universe An age of approximately 13.7 billion years since the BIG BANG. This figure has been arrived at by careful analysis of the cosmic microwave background radiation in the early years of this century.

agonic line /ă-**gon**-ik, ay-/ *See* isogonic line.

AGR Advanced gas-cooled reactor. *See* gas-cooled reactor.

air column A long column of air in which standing waves can be set up. This type of column is used in musical instruments based on pipes. If both ends of the pipe are open there are antinodes at both ends. If one end is open and the other end is closed there is a node at the closed end and an antinode at the open end.

airfoil A lifting wing on an aircraft, or animal, or the cross-sectional shape of such a wing.

air pressure The PRESSURE OF THE ATMOSPHERE at any point on Earth or in the Earth's atmosphere.

air wedge An arrangement producing localized interference patterns by reflection at the two sides of a wedge-shaped film of air (as between two glass slides at an angle). Newton's rings (variable wedge angle) and thin films (zero wedge angle) produce similar effects. In an air wedge the fringes (with monochromatic light, wavelength λ) are light and dark bands parallel to the thin edge of the wedge. A bright fringe occurs when $2t + \lambda/2 = m\lambda$, t being the thickness and m an integer. For a dark fringe $2t = m\lambda$. *See also* interference.

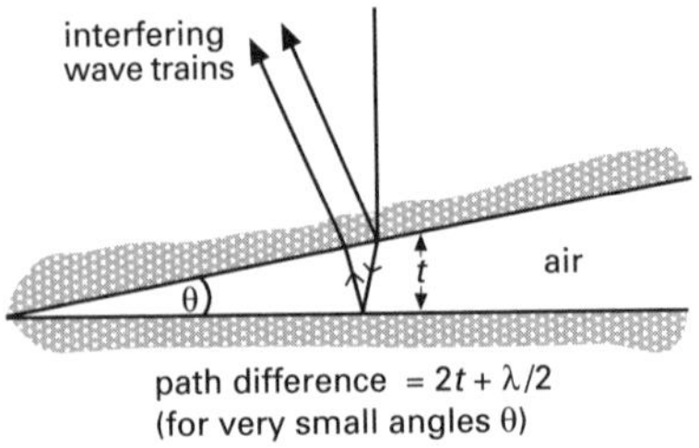

Air wedge

albedo /al-**bee**-doh/ The ratio of the amount of light reflected from a surface to the amount of incident light.

alcohol thermometer A liquid-in-glass thermometer that uses ethanol as its working substance. The ethanol commonly contains a red or blue dye to make the liquid more visible. *See also* thermometer.

allotropy /ă-**lot**-rŏ-pee/ The existence of a solid substance in different physical forms. Tin, for example, has metallic and non-metallic crystalline forms. Carbon has two crystalline allotropes: diamond and graphite.

alloy A mixture of two or more metals (e.g. bronze or brass) or a metal with small amounts of non-metals (e.g. steel). Alloys may be completely homogeneous mixtures or may contain small particles of one phase in the other phase.

alpha decay A type of radioactive decay in which the unstable nucleus emits a helium nucleus. The resulting nuclide has a mass number decreased by 4 and a proton number decreased by 2. An example is:

$$^{226}_{88}\text{Ra} \rightarrow {}^{222}_{86}\text{Rn} + {}_{2}^{4}\text{He}$$

The particles emitted in alpha decay are *alpha particles*. Streams of alpha particles are *alpha rays* or *alpha radiation*. They penetrate a few centimeters of air at STP or a metal foil of mass/area a few milligram/cm^2. *See also* beta decay.

alternating current (a.c.) Electric current that regularly reverses its direction. In the simplest case, the current varies with time (t) in a simple harmonic manner, represented by the equation $I = I_0\sin 2\pi ft$, f being the frequency. Alternating current can be described by its peak value I_0, or by its root-mean-square value I_{RMS} (= $I_0/\sqrt{2}$ for a sine wave). In the USA it is 220 V (RMS) at a frequency of 60 hertz. In the UK, the mains electricity supply is also alternating, about 250 V (RMS) at a frequency of 50 hertz. *Compare* direct current.

alternating-current circuit A circuit containing a resistance R, capacitance C, and inductance L, with an alternating voltage supply, is called an *LCR circuit*. The simplest type is one in which L, C, and R are all in series. The impedance of such a circuit is given by

$$Z = \sqrt{[(X_L - X_C)^2 + R^2]}$$

where X_L is the reactance of the inductor ($2\pi fL$), and X_C is the reactance of the capacitor ($\frac{1}{2}\pi C$). The current I is given by V/Z. There is a phase difference between the current in the circuit and the voltage. Current lags behind voltage by a phase angle ϕ:

$$\tan\phi = (X_L - X_C)/R$$

See also resonance.

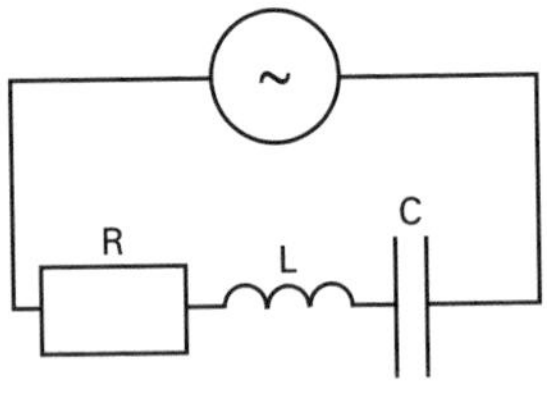

Alternating-current circuit

alternator A generator for producing an alternating electric current.

altimeter /al-**tim**-ĕ-ter/ An instrument for measuring altitude – that is, the height above a certain reference level (usually mean sea-level). Because atmospheric pressure varies with altitude, an aneroid barometer can be used as an altimeter. *See* barometer.

aluminum A soft moderately reactive metal. Aluminum has the electronic structure of neon plus three additional outer electrons. There are numerous minerals of aluminum; it is the most common metallic element in the Earth's crust (8.1% by mass) and the third in order of abundance.

Symbol: Al; m.p. 660.37°C; b.p. 2470°C; r.d. 2.698 (20°C); p.n. 13; r.a.m. 26.981539.

AM *See* amplitude modulation.

Amagat's experiments /ah-mah-**gah**/ *See* Andrews' experiments.

amalgam /ă-**mal**-găm/ An alloy of mercury with one or more other metals. Amalgams may be liquid or solid.

americium /am-ĕ-**rish**-ee-ŭm/ A highly toxic radioactive silvery element of the actinoid series of metals. A transuranic element, it is not found naturally on Earth but is synthesized from plutonium. ^{241}Am has been used in gamma-ray radiography.

Symbol: Am; m.p. 1172°C; b.p. 2607°C; r.d. 13.67 (20°C); p.n. 95; most stable isotope ^{243}Am (half-life 7.37×10^3 years).

ammeter /**am**-mee-ter/ A meter used to measure electric current. Ammeters have to have low resistance as they are connected in series in the circuit. Commonly, moving-coil instruments are used with shunt resistors to increase the current range. For alternating current a rectifier is necessary. Moving-iron instruments can be used both for d.c. and a.c. High-frequency currents may be measured with a hot-wire instrument.

amorphous /ă-**mor**-fŭs/ Denoting a solid that has no crystalline structure; i.e. there is no long-range ordering of atoms. Many substances that appear to be amorphous are in fact composed of many tiny crystals. Soot and GLASS are examples of truly amorphous materials.

amount of substance Symbol: n A physical quantity that is a measure of the number of entities present in a substance. *See* mole.

ampere /**am**-pair/ Symbol: A The SI base unit of electric current, defined as the constant current that, maintained in two straight parallel infinite conductors of negligible circular cross section placed one meter apart in vacuum, would produce a force between the conductors of 2×10^{-7} newton per meter. The ampere is named for the French physicist and mathematician André Marie Ampère (1775–1836).

ampere balance *See* current balance.

ampere-hour Symbol: Ah A unit of electric charge equal to the charge transferred through a conductor in one hour by a current of one ampere. It is equal to 3.6 coulombs.

Ampère–Laplace law /ahm-**pair** lah-**plahs**/ *See* Ampère's law.

Ampère's law **1.** (Ampère–Laplace law) The elemental force, dF, between two current elements, $I_1 dl_1$ and $I_2 dl_2$, parallel to each other at a distance r apart in free space is given by:

$$d = \mu_0 I_1 d_1 l_1 I_2 dl_2 \sin\theta / 4\pi r^2$$

Here μ_0 is the permeability of free space and θ is the angle between either element and the line joining them.

2. The principle that the sum or integral of the magnetic flux density B times the path length along a closed path round a current-carrying conductor is proportional to the current I. For a circular path of radius r round a long straight wire in a vacuum, $B.2\pi r = \mu_0 I$. (μ_0 is the magnetic permeability of free space.) Ampère's law enables the value of B inside a solenoid to be calculated using the equation $B = n\mu_0 I$, where n is the number of turns per unit length. *See also* Maxwell's equations.

ampere-turn /**am**-pair **tern**/ Symbol: At The SI unit of magnetomotive force (m.m.f.) equal to the magnetomotive force produced by a current of one ampere flowing through one turn of a conductor. *See also* magnetic circuit.

amplification factor *See* triode.

amplifier A device that increases an electrical signal applied to it as an input. If the input is an alternating voltage, the output voltage has a similar waveform with an increased amplitude.

The ratio of the output signal to the input signal (called the *gain*), will usually vary with the signal frequency. Amplifiers are usually designed to give a particular current, voltage, or power gain over the required frequency range. Some circuits containing a number of amplifying stages can cope with frequencies from 0 hertz (steady direct current) to radiofrequencies. In modern solid-state electronics, all of the amplifier circuit components, including many individual amplifying stages, are manufactured in a single integrated circuit.

amplitude The maximum value of a varying quantity from its mean or base value. In the case of a simple harmonic motion – a wave or vibration – it is half the maximum peak-to-peak value.

amplitude modulation (AM) A type of modulation in which the amplitude of a CARRIER WAVE is modulated by an imposed signal, usually at audio frequency.

In this way communication of a signal is made between two distant points using a radio transmission as carrier. When the carrier wave is received the audio component is extracted by the process of DEMODULATION, and the original sound may be reproduced.

amplitude of accommodation The eye's range of accommodation in terms of refractive power (often measured in diopters). It is given by $(1/u_1 - 1/u_2)$, where u_1 is the distance from the near point to the lens and u_2 is the distance from the far point to the lens.

amu /ay-em-**yoo**/ *See* atomic mass unit.

analyzer A device for determining the plane of *polarization* of plane-polarized radiation. Maximum intensity is transmitted if the plane is parallel with the analyzer's direction of polarization; the intensity is a minimum (theoretically zero) if the two are perpendicular. For visible radiation, analyzers are usually Polaroid sheets or Nicol prisms.

anastigmatic lens /an-ă-stig-**mat**-ik, an-as-tig-/ A lens designed so as to minimize its astigmatic aberration. Anastigmatic lenses have different curvatures in different directions; the surface of an anastigmatic lens is part of a toroid.

AND gate *See* logic gate.

Andrews' experiments Experiments performed (1863) on the effect of pressure and temperature on carbon dioxide and named for the Irish physical chemist Thomas Andrews (1813–85). Andrews used two thick-walled glass capillary tubes, one containing dry carbon dioxide and the other dry nitrogen. The top end of each tube was closed and the bottom end contained a plug of mercury to trap the gas. The bottom ends of the tubes were sealed into a case containing water, and pressure could be applied by means of a pair of screws. In this way Andrews achieved pres-

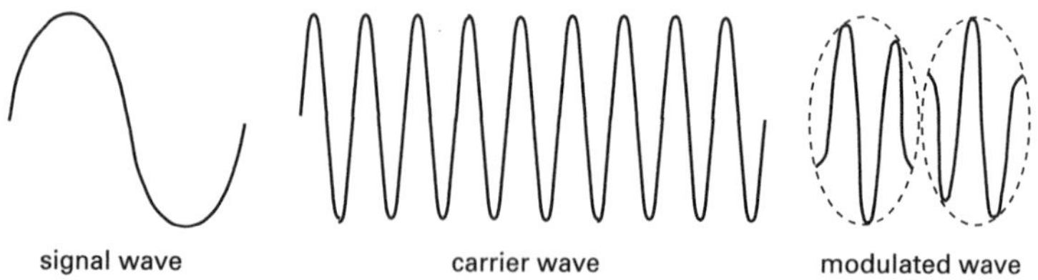

Amplitude modulation

sures up to above 10 MPa. The nitrogen was used to measure the pressure by assuming that it obeyed Boyle's law. The apparatus was surrounded by a constant temperature bath, so that isothermals (p–V curves) could be plotted at different temperatures.

In this way Andrews showed the behavior near the critical temperature, and the liquefaction of carbon dioxide by pressure below the critical temperature. Similar experiments were done on carbon dioxide and other gases by the French physicist Emile Hilaire Amagat (1841–1915).

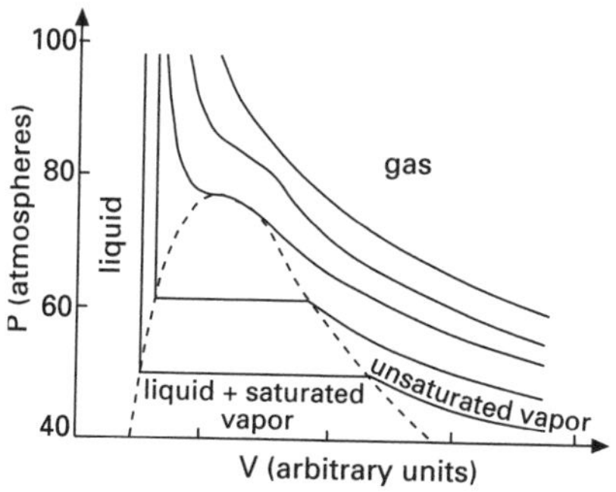

Andrew's isothermal for CO_2

anechoic chamber /an-ek-**oh**-ik/ (**dead room**) A room designed so that there is little or no reflection of sound from its internal walls. The walls are covered with pyramid shapes so that stationary waves are not produced between parallel surfaces. They are coated with absorbing material. Anechoic chambers are used for experiments in acoustics.

anemometer /an-ĕ-**mom**-ĕ-ter/ An instrument for measuring wind speed. A simple type consists of several cups or vanes attached to a vertical shaft that rotates when the cups/vanes are forced round by the wind. The shaft can be geared to a pointer to give a direct reading of wind speed on a dial.

aneroid (non-liquid) barometer /**an**-ĕ-roid/ *See* barometer.

angle of deviation *See* deviation.

angle of dip *See* inclination.

angle of incidence The angle between a ray incident on a surface and the normal to the surface at the point of incidence.

angle of polarization *See* Brewster angle.

angle of reflection The angle between a ray reflected by a surface and the normal to the surface at the point of reflection.

angle of refraction The angle between a ray refracted at the surface between two media and the normal to the surface at the point of refraction.

angstrom /**ang**-strŏm/ Symbol: Å A unit of length defined as 10^{-10} meter. The angstrom was formerly used for expressing wavelengths of light or ultraviolet radiation or for interatomic distances and the sizes of molecules. The angstrom is named for the Swedish physicist Anders Jonas Ångstrom (1814–74).

angular acceleration Symbol: α The rotational acceleration of an object about an axis:

$$\alpha = d\omega/dt \text{ or } \alpha = d^2\theta/dt^2$$

Here ω is angular velocity; θ is angular displacement. Angular acceleration is directly analogous to linear acceleration, a. *See also* equations of motion; rotational motion.

angular displacement Symbol: θ The rotational displacement of an object about an axis. If the object (or a point on it) moves from point P_1 to point P_2 in a plane perpendicular to the axis, θ is the angle P_1OP_2, where O is the point at which the perpendicular plane meets the axis. *See also* rotational motion.

angular frequency (**pulsatance**) Symbol: ω The number of complete rotations per unit time. A simple harmonic motion of frequency f can be represented by a point moving in a circular path at constant speed. The foot of a perpendicular from the point to a diameter of the circle moves backward and forward along the diameter with simple harmonic motion. The angular frequency of this motion is $2\pi f$, where f is the frequency. The unit is the radian per second.

angular magnification (**magnifying power**) Symbol: M The ratio of the angle subtended at the eye by an image to that subtended by the object: $M = \theta_1/\theta_0$. The object and image are considered to be at their actual positions, except in the case of microscopes. Here it is conventional to measure θ_0 for the object at the standard near-point distance (250 mm from the eye). The maximum useful magnifying power depends on the resolving power of the viewing system – i.e. the acuity of the eye or the grain of the photographic emulsion. *See also* magnification.

angular momentum Symbol: L The product of the moment of inertia of a body and its angular velocity. *See also* rotational motion.

angular velocity Symbol: ω The rate of change of angular displacement: $\omega = d\theta/dt$. *See also* rotational motion.

anharmonic oscillator /an-har-**mon**-ik/ A system whose vibration, while still periodic, cannot be described in terms of simple harmonic motions (i.e. sinusoidal motions). In such cases, the period of oscillation is not independent of the amplitude.

anion /**an**-ÿ-ŏn, -on/ A negatively charged ion, formed by addition of electrons to atoms or molecules. In electrolysis anions are attracted to the positive electrode (the anode). *Compare* cation.

anisotropy /an-ÿ-**sot**-ŏ-pee/ A medium is anisotropic if a certain physical quantity differs in value in different directions. Most crystals are anisotropic electrically; important polarization properties result from differences in transmission of electromagnetic radiation in different directions. *Compare* isotropy.

annealing /ă-**neel**-ing/ The process of heating a solid to a temperature below the melting point, and then cooling it slowly. Annealing removes crystal imperfections and strains in the solid.

annihilation A reaction between a particle and its antiparticle; for example, between an electron and a positron. The energy released is equal to the sum of the rest energies of the particles and their kinetic energies. In order that momentum be conserved two photons are formed, moving away in opposite directions. This radiation (*annihilation radiation*) is in the gamma-ray region of the electromagnetic spectrum. The quantum energy is about 0.51 MeV.

Annihilation also can occur between a nucleon and its antiparticle. In this case mesons can be produced.

annual variation The direction and strength of the Earth's magnetic field at any point changes with time. This must be allowed for by navigators. One such change is a variation with a period of a year, but there are others. The amplitude of the annual variation is greatest during maximum sun-spot activity. *See also* Earth's magnetism; magnetic variation.

annular eclipse *See* eclipse.

anode /**an**-ohd/ In electrolysis, the electrode that is at a positive potential with respect to the cathode. In any electrical system, such as a discharge tube or a solid-state electronic device, the anode is the terminal at which electrons flow out of the system.

anomalous dispersion The refractive index of a transparent medium normally increases as the wavelength is reduced. There is then a range of wavelengths (usually in the ultraviolet) in which the radiation is absorbed fairly strongly. Such little radiation as is transmitted in this region shows anomalous dispersion, that is the refractive index *decreases* as the wavelength is reduced. *See* dispersion.

anomalous expansion An increase in volume resulting from a decreased temperature. Most liquids increase in volume as their temperature rises. The density of the liquid falls with increased temperature. Water, however, shows anomalous behavior. Between 0 and 4°C the density increases with increasing temperature.

antenna (**aerial**) (*pl.* **antennae**) A device such as a wire, rod or dish used to transmit or receive radio waves. The simplest type is a rod of ferrite as used inside domestic radio sets; complex transmitting antennas may be mounted on a mast 100 meters tall. An *omnidirectional antenna* transmits or receives signals from all directions. A *directional antenna* is designed to operate preferentially in a single direction.

anthropic principle /an-**throp**-ik/ A theory in cosmology about the present state of the Universe. The *weak anthropic principle* concerns intelligent life on Earth and states that the Universe is the way that it is because if it were otherwise we would not be here to observe it. For example, if the value of the gravitational constant were significantly different, the Universe would not have evolved in a way that allowed intelligent life to form. The *strong anthropic principle* deals with ideas of other possible universes and with other lifeforms. The anthropic principle has been the subject of much controversy among physicists.

anticlastic curvature /an-tee-**klas**-tik/ The saddle-shaped curve on the upper surface of a horizontal bar that is being bent by a downward force at each end. The effect is easily demonstrated by bending an eraser.

antiferromagnetism /an-tee-fe-roh-**mag**-nĕ-ti-zăm/ A kind of MAGNETISM found in many solids at low temperatures. The molecular magnets form two arrays, aligned antiparallel. At the lowest temperatures there are equal numbers with equal magnetic moments in opposite directions, giving zero resultant magnetization. As the temperature is raised, the susceptibility increases up to the *Néel temperature* above which the substance is paramagnetic. The Néel temperature is named for the French physicist Louis Néel (1904–2000). *See also* ferrimagnetism.

antimatter Matter formed of antiparticles. Nuclei of antimatter would consist of antiprotons and antineutrons, and would be surrounded by orbiting positrons. When matter encounters antimatter annihilation occurs. *See* annihilation.

antimony /**an**-tă-moh-nee/ A metalloid element existing in three allotropic forms; the most stable is a brittle silvery metal. It is used in alloys – small amounts of antimony can harden other metals. It is also used in semiconductor devices.

Symbol: Sb; m.p. 630.74°C; b.p. 1635°C; r.d. 6.691; p.n. 51; r.a.m. 112.74.

antinode /**an**-tee-nohd/ *See* node.

antiparallel /an-tee-**pa**-ră-lel/ Having parallel lines of action that are directed in opposite directions.

antiparticle A particle of the same mass and spin, but opposite charge (and other properties) to its corresponding particle. For example, a proton and antiproton both have mass 1836 times that of an electron and spin ½ unit, but the charge on the proton is +1 unit, while that on the antiproton is –1 unit. For unstable particles, such as an isolated neutron, the particle and antiparticle have the same half-life. For uncharged particles the antiparticle is indicated by a bar above the symbol, such as ^{3} for the antineutron. For charged particles the distinction is indicated by the sign, for example, e^+ is the positron, the antiparticle of an electron. Antiparticles of fermions are subject to a conservation law according to which new particles can only be created in particle–antiparticle pairs, while particles can be destroyed only by annihilation with their antiparticles. This rule of number conservation does not apply to BOSONS. *See also* fermion.

aperture A measure of the effective diameter (d) of a mirror or lens compared with its focal distance (f): aperture = d/f. Thus a 50-mm camera lens may be used with an aperture diameter of 12.5 mm. Then, aperture = 12.5/50. This is usually described as an f-number. In this case the aperture diameter is $f/4$, often written as f4.

The transmitted light intensity depends on aperture diameter, so that I is proportional to d^2. However, large apertures lead to large aberrations although diffraction effects are more serious at small apertures. In many optical instruments, iris diaphragms vary the aperture to obtain the optimum results.

aplanatic lens /ap-lă-**nat**-ik/ A lens designed so as to minimize both its astigmatic and coma aberration.

apochromatic lens /ap-ŏ-krŏ-**mat**-ik/ A lens designed to correct for chromatic aberration at three different wavelengths. Apochromatic lenses are constructed of three or more kinds of glass. They thus have better correction than achromatic lenses, which correct at two different wavelengths (usually in the red and blue regions of the spectrum). *See* achromatic lens.

apparent depth Because radiation travels at different speeds in different media, the apparent depth or thickness of a transparent sample is not the same as its real depth or thickness. The effect is very obvious when one looks down into a glass of water or a clear pool. It is associated with the fact that a long object partly submerged in water seems bent at the water surface.

The refractive constant of the substance can be measured on this basis:

refractive index =
real depth/apparent depth

The relation is used in a number of methods for finding the refractive constant of a transparent medium. It applies to all wave radiations, not just to visible radiation.

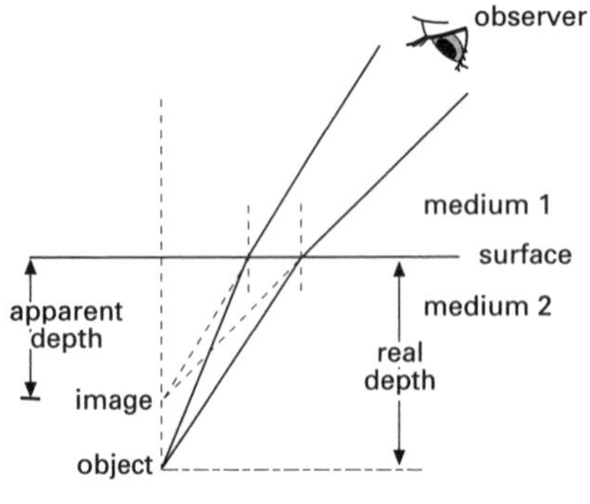

Apparent depth

apparent expansion *See* expansivity.

Appleton layer /**ap**-ăl-tŏn/ (**F-layer**) The upper of the main layers in the IONOSPHERE, at a height above about 150 km. It reflects radio waves. The Appleton layer is named for the British physicist Sir Edward Victor Appleton (1892–1965). *See* Heaviside layer.

aqueous humor The watery substance between the cornea and the lens in the EYE.

Arago's spot *See* Poisson's spot.

arc, electric *See* electric arc.

Archimedes' principle /ar-kă-**mee**-deez/ The upward force on an object totally or partly submerged in a fluid is equal to the weight of fluid displaced by the object. The upward force, often called *upthrust*, results from the fact that the pressure in a fluid (liquid or gas) increases with depth. If the object displaces a volume V of fluid of density ρ, then:

upthrust $u = V\rho g$

where g is the acceleration of free fall.

If the upthrust on the object equals the object's weight, the object will float. The principle is named for the Greek mathematician Archimedes (287 BC–212 BC). *See* flotation; law of.

arc lamp An intense light source in which the light is produced by an electric arc between two (usually carbon) electrodes. Arc lamps are used in lighthouses and as searchlights and spotlights.

argon /**ar**-gon/ An inert colorless odorless monatomic element of the rare-gas group. It forms 0.93% by volume of air. Argon is used to provide an inert atmosphere in electric and fluorescent lights, in welding, and in extracting titanium and silicon. The element forms no known compounds.

Symbol: Ar; m.p. −189.37°C; b.p. −185.86°C; d. 1.784 kg m^{-3} (0°C); p.n. 18; r.a.m. 39.95.

arithmetic series A series in which the difference between two consecutive members is a constant number. If the first term of an arithmetic series is a and the common difference between two consecutive terms is d then the nth term is given by $[a + (n-1)d]$ and the sum of the first n terms is given by $n[2a + (n-1)d]/2$.

armature **1.** The part of an electric motor or generator that carries the principal current. This is the rotating coil in a small motor but the stationary coil in a large motor or generator. Torque acting on the armature enables work to be done against the load. *See also* electric motor; rotor; stator.

2. The moving part of any electromechanical device, such as an electric bell or relay.

arsenic /**ar**-sĕ-nik, **ars**-nik; *adj.* ars-**sen**-ik/ A toxic metalloid element existing in several allotropic forms; the most stable is a brittle gray metal. It is used in semiconductor devices, alloys, and gun shot.

Symbol: As; m.p. 817°C (gray) at 3 MPa pressure; sublimes at 616°C (gray); r.d. 5.78 (gray at 20°C); p.n. 33; r.a.m. 74.92159.

artificial radioactivity Radioactivity produced by nuclear reactions. Various methods are available.

Bombardment by neutrons, particularly using a nuclear reactor, can give very many artificial radioactive substances, most of which are beta active, emitting electrons. For example:

$$^{59}_{27}\mathrm{Co} + ^{1}_{0}\mathrm{n} \rightarrow ^{60}_{27}\mathrm{Co} + \gamma \text{ rays}$$

$^{60}_{27}\mathrm{Co}$ is a beta active material with half-life 5.3 years and is very important as the beta emission is followed instantaneously by gamma rays of high quantum energy, which are used in radiography and cancer therapy.

Bombardment by light atomic nuclei accelerated in a cyclotron or similar machine often gives positron-emitting radionuclides. For example

$$^{24}_{12}\mathrm{Mg} + ^{2}_{1}\mathrm{H} \rightarrow ^{22}_{11}\mathrm{Na} + ^{4}_{2}\mathrm{He}$$

$^{22}_{11}\mathrm{Na}$ has half-life 2.6 years and emits positrons and gamma rays. Transuranic elements can be produced in this way.

The products of nuclear fission are often highly active; most emit electrons (beta rays) and several also emit gamma rays.

asdic /**az**-dik/ *See* sonar.

Aspect experiment *See* Bell's paradox.

astable circuit /ay-**stay**-băl/ (**pulse generator**) A multivibrator circuit that switches continually and regularly from one state to another. Unlike other forms of MULTIVIBRATOR, no trigger pulse is needed. It is used in computers as a source of clock pulses for counting, because the output is a rectangular voltage waveform.

In the astable multivibrator, two transistors are arranged with the base terminal of each connected to the collector terminal of the other through capacitors *C1* and *C2* respectively. There is a steady voltage supply. *C1* charges and *C2* discharges until the transistors switch from one state to another and the charging direction reverses. The value of the capacitances and resistances determines the switching frequency.

astatic coils /ay-**stat**-ik/ Two identical coils connected together in series and suspended on the same axis. When a current passes through them, any external magnetic field will result in the same turning force on each, but in opposite directions. Thus neither the Earth's magnetic field, nor any other external magnetic disturbance, will affect the rotation of the axis.

astatic pair Two identical magnetic needles suspended on the same vertical axis with their N- and S-poles pointing in opposite directions. The couples on the needles from an external magnetic field, such as the Earth's, are equal and opposite. Astatic pairs are used in very sensitive galvanometers in which the current-carrying coils are wound round each needle in opposite directions. The current therefore rotates them both in the same direction and external magnetic effects are canceled out.

astatine /**ass**-tă-teen, -tin/ A radioactive element belonging to the halogen group. It occurs in minute quantities in uranium ores. Many short-lived radioisotopes are known, all alpha-particle emitters.

Symbol: At; m.p. 302°C (est.); b.p. 337°C (est.); p.n. 85; most stable isotope ^{210}At (half-life 8.1 hours).

asteroid Any of a number of small objects that orbit round the Sun in a narrow belt of space (the *asteroid belt*) located between the orbit of Mars and the orbit of Jupiter. Asteroids have a range of sizes, with the largest having a radius of approximately 500 km. They are sometimes known as *minor planets* or *planetoids*.

astigmatism /ă-**stig**-nă-tiz-ăm/ **1.** A common eye defect in which the observer cannot focus clearly on objects at any distance. The cause is usually a non-spherical cornea. Visual astigmatism may be corrected with a lens with a suitable degree of cylindrical curvature. *See* anastigmatic lens.
2. *See* aberration.

astronomical telescope *See* Keplerian telescope; telescope.

astronomical unit (**au; AU**) The mean distance between the Sun and the Earth, used as a unit of distance in astronomy for measurements within the solar system. It is approximately 1.496×10^{11} meters.

astrophysics The science that deals with physical and chemical processes in astronomic phenomena, such as the formation and evolution of stars and galaxies.

atmolysis /at-**mol**-ă-sis/ The separation of a mixture of gases by using their different rates of diffusion.

atmosphere *See* standard pressure.

atmosphere of the Earth The layer of gas that surrounds the Earth. It consists mostly of nitrogen (about 78%) and oxygen (about 21%) with a little carbon dioxide and inert (noble) gases. The gas is held in place by the gravitational field of the Earth. The atmosphere does not have a sharp cut-off but becomes thinner as the distance from the surface of the Earth increases. *See also* pressure of the atmosphere.

atmospheric pressure *See* pressure of the atmosphere.

atmospheric window A region in the electromagnetic spectrum for which radiation of a particular range of wavelengths can reach the Earth from space without absorption by the atmosphere.

atom The smallest part of an element that can take part in a chemical reaction. Atoms consist of a small dense positively charged nucleus, made up of neutrons and protons, with electrons in a cloud around this nucleus. The chemical reactions of an element are determined by the number of electrons (which is equal to the number of protons in the nucleus). All atoms of a given element have the same number of protons (the proton number). A given element may have two or more isotopes, which differ in the number of neutrons in the nucleus.

The electrons surrounding the nucleus are grouped into *shells* – i.e. main orbits around the nucleus. Within these main orbits there may be sub-shells. These correspond to atomic orbitals. An electron in an atom is specified by four quantum numbers:
1. The *principal quantum number* (n) can have values 1, 2, etc. The corresponding shells are denoted by letters K, L, M, etc.,

the K shell ($n = 1$) being the nearest to the nucleus. The maximum number of electrons in a given shell is $2n^2$. This quantum number has the largest effect on the energies of the states; high values of n correspond to weakly bound (higher energy) electrons.
2. The *orbital quantum number* (l), which specifies the angular momentum. For a given value of n, l can have possible values of $n-1$, $n-2$, ... 2, 1, 0. For instance, the M shell ($n = 3$) has three sub-shells with different values of l (0, 1, and 2). Sub-shells with angular momentum, 0, 1, 2, and 3 are designated by letters s, p, d, and f. This quantum number has the second largest effect on the energies; higher values of l give moderately higher energy electrons.
3. The *magnetic quantum number* (m). This can have values $-l, -(l-1) \ldots 0 \ldots +(l+1), +l$. It determines the orientation of the electron orbital in a magnetic field. States with the same values of n and l but different values of m have the same energy in the absence of a magnetic field, but differ slightly when a field is applied.
4. The *spin quantum number* (m_s), which specifies the intrinsic angular momentum of the electron. It can have values +½ and –½. Quantum states in which the spin is parallel to the orbital angular momentum are at slightly higher energy than ones in which it is antiparallel. This results, for example, in the fact that the yellow light from a sodium lamp has two very close lines in its spectrum.

Each electron in the atom has four quantum numbers and, according to the Pauli exclusion principle, no two electrons can have the same set of quantum numbers. This explains the electronic structure of atoms. *See also* Bohr theory.

atom bomb A bomb in which the explosion is caused by a fast uncontrolled fission reaction. *See* nuclear weapon.

atomic clock An apparatus for measuring time by the frequency of radiation emitted or absorbed in transitions of atoms. *See* cesium clock.

atomic energy *See* nuclear energy.

atomic heat *See* Dulong and Petit's law.

atomicity /at-ŏ-**mis**-ă-tee/ The number of atoms per molecule of a compound. Methane, for instance has an atomicity of five (CH_4).

atomic mass Another name for RELATIVE ATOMIC MASS, often applied to isotopes.

atomic mass unit (**amu**) Symbol: u A unit of mass used for atoms and molecules, equal to 1/12 of the mass of an atom of carbon-12. It is equal to $1.660\,54 \times 10^{-27}$ kg.

atomic number *See* proton number.

atomic orbital *See* orbital.

atomic physics *See* nuclear physics.

atomic pile A nuclear reactor, particularly the early form constructed by piling up graphite blocks (the moderator) and uranium rods (the fuel).

atomic radius An imprecise measurement usually expressed as a half of the distance between neighboring atoms of the same kind in a crystal or molecule. Depending on the type of chemical bonding between the atoms, it may be qualified as covalent radius, ionic radius, or metallic radius.

atomic theory The theory that matter is made up of atoms that combine to form molecules. Each chemical element has a particular type of atom, which may join with like atoms to form molecules of the element, or with atoms of other elements to form molecules of a compound. The atom consists of a dense positively charged nucleus containing protons and neutrons, surrounded by electrons. The number of protons in the nucleus determines the number and distribution of the electrons, which are held by the positive charge of the nucleus. Because the outer electrons form the chemical bonds between atoms, the chemical properties of an element depend on the electronic structure of the atom, and therefore also on the number of protons. The number of neutrons in the nucleus may vary, forming different isotopes of an element. These cannot usually be separated by chemical means.

atomic volume The relative atomic mass of an element divided by its density.

atomic weight *See* relative atomic mass.

attenuation /ă-ten-yoo-**ay**-shŏn/ **1.** The reduction of intensity of a radiation as it passes through a medium. It includes reductions due to both absorption and scattering.
2. Reduction in current, voltage, or power of an electrical signal passing through a circuit.

atto- Symbol: a A prefix denoting 10^{-18}. For example, 1 attometer (am) = 10^{-18} meter (m).

attractor The point or set of points in phase space to which a changing system moves with time. The idea of an attractor for a system comes from CHAOS THEORY. The attractor of a system may be a single point (in which case the system reaches a fixed state that is independent of time). Alternatively, it may be a closed curve, known as a *limit cycle*. This is the type of behavior found in oscillating systems. In some systems, the attractor is a curve that is not closed and does not repeat itself. This, known as a *strange attractor*, is characteristic of chaotic systems. *See also* phase space.

AU (**au**) *See* astronomical unit.

audibility, limits of The frequencies beyond which sound cannot be heard by the human ear. The lowest audible frequency is about 20 hertz (a deep rumble), and the highest 15–20 kilohertz (a very high-pitched whistle). Because hearing deteriorates continuously with age, older people cannot detect sounds as high as children can. *See also* infrasound; ultrasonics.

audiofrequency /aw-dee-oh-**free**-kwĕn-see/ A frequency within the audible frequency range (about 20 hertz to about 20 kilohertz). Sound vibrations in this range can be detected by the human ear. Audiofrequency electrical signals are converted directly into sound in a loudspeaker.

audiometer /aw-dee-**om**-ĕ-ter/ A device for measuring the frequency range of the human ear and the minimum intensity of sound that can be detected at the different frequencies. It consists of a signal generator used to feed a tone of variable frequency and intensity through a set of earphones.

Auger effect /oh-**zhay**/ The ejection of an electron from an atom or ion without the emission of radiation (x-rays or gamma rays). It results from the de-excitation of an excited electron within the atom. It can be regarded as the internal conversion of the photon that would otherwise have been emitted. *See* internal conversion. The Auger effect is named for the French physicist Pierre Auger (1899–1994).

aurora /ô-**ror**-ă, **-roh**-ră/ (**polar lights**) (*pl.* **auroras** or **aurorae**) An atmospheric phenomenon in which colored luminous arcs and streamers appear in the night sky. It is caused by charged particles from the Sun interacting with atoms in the Earth's upper atmosphere and the effect is strongest near the magnetic poles, giving rise to the *aurora borealis* (*northern lights*) in the north and *aurora australis* (*southern lights*) in the south.

autoionization /aw-toh-ÿ-ŏ-ni-**zay**-shŏn/ The spontaneous IONIZATION of excited atoms, ions, or molecules, as in the AUGER EFFECT.

avalanche A process such as that in which a single ionization leads in stages to a large number of ions. The electrons and ions produced ionize more atoms, so that the number of ions multiplies quickly. *See* Geiger counter.

avalanche diode *See* Zener diode.

Avogadro constant /ah-vŏ-**gah**-droh/ Symbol: N_A The number of particles in one mole of a substance. Its value is $6.002\ 142 \times 10^{23}$. The constant is named for the Italian physicist and chemist Count Amedeo Avogadro (1776–1856).

Avogadro's law Equal volumes of all gases at the same temperature and pressure contain equal numbers of molecules. It is often called *Avogadro's hypothesis*. It is strictly true only for ideal gases and is readily explained by the kinetic theory of gases.

avoirdupois /av-er-dŭ-**poiz**/ A system of weights based on the pound, which is subdivided into 16 ounces or 7000 grains. In scientific use it has been superseded by SI UNITS.

axis *See* principal axis.

azeotropic mixture (**azeotrope**) A mixture of two liquids that boils without any change in composition. The proportions of components in vapor are the same as in the liquid. Azeotropic mixtures cannot be separated by distillation.

B

back e.m.f. An e.m.f. that opposes the normal flow of electric charge in a circuit or circuit element. **1.** In some electrolytic cells a back e.m.f. is caused by the layer of hydrogen bubbles that build up on the cathode as hydrogen ions pick up electrons and form gas molecules. *See also* polarization.
2. *See* self-induction.

background radiation Ever-present low-intensity ionizing radiation from natural sources, such as cosmic radiation from outer space and radioactivity from natural sources in the ground. In astronomy, the term background radiation refers to a cosmic background of microwave radiation thought to have originated with the big bang at the formation of the Universe. *See* big-bang theory.

ballistic galvanometer /bă-**lis**-tik/ An instrument that measures the total electric charge passing through it in a sudden pulse of current. It is a moving-coil instrument constructed and calibrated so that the maximum deflection of the pointer is proportional to the total charge that has passed. The coil suspension is lightly damped in a ballistic galvanometer. Provided that the discharge through it occurs in a much shorter time than the suspension's natural period of oscillation, the maximum deflection is proportional to the total charge.

ballistic pendulum A device for measuring the velocity of a projectile (e.g. a bullet). It consists of a heavy pendulum, which is struck by the projectile. The velocity can be calculated by measuring the displacement of the pendulum and using the law of conservation of momentum.

ballistics The study of the motion of projectiles.

Balmer series /**bahl**-mer/ A series of lines in the spectrum of radiation emitted by excited hydrogen atoms. The lines correspond to the atomic electrons falling into the second lowest energy level, emitting energy as radiation. The wavelengths (λ) of the radiation in the Balmer series are given by:

$$1/\lambda = R(1/2^2 - 1/n^2)$$

where n is an integer and R is the Rydberg constant. *See* Bohr theory. *See also* spectral series.

band, energy *See* energy bands.

band-pass filter An electrical or optical filter that transmits only frequencies within a single band.

band spectrum A spectrum that appears as a number of bands of emitted or absorbed radiation. Band spectra are characteristic of molecules. Often each band can be resolved into a number of closely spaced lines. The bands correspond to changes of electron orbit in the molecules. The close lines seen under higher resolution are the result of different vibrational states of the molecule. *See also* spectrum.

band theory (of solids) *See* energy bands.

band width An indication of the range of frequencies, or wavelengths (wave band) that:
1. an antenna can receive efficiently;
2. a radio receiver or amplifier can efficiently handle;
3. exist in a radio transmission above and below the carrier-wave frequency. *See also*

carrier wave.

Bandwidth is a measure of the amount of information that can be transmitted.

bar A unit of pressure defined as 10^5 pascals. The *millibar* (mbar) is more common; it is used for measuring atmospheric pressure in meteorology.

barium /bair-ee-ŭm/ A dense, low-melting reactive metal. The electronic configuration is that of xenon with two additional outer 6s electrons. Barium metal is used as a 'getter', i.e., a compound added to a system to seek out the last traces of oxygen; and as an alloy constituent for certain bearing metals. Metallic barium has the body-centred cubic structure.

Symbol: Ba; m.p. 729°C; b.p. 1640°C; r.d. 3.594 (20°C); p.n. 56; r.a.m. 137.327.

Barkhausen effect /bark-how-zĕn/ A phenomenon that demonstrates the domain theory of magnetism. When a ferromagnetic substance is being magnetized, changes of induction occur as domains reverse direction. The effect is demonstrated as shown in the diagram; a series of clicks is heard when the current is switched on and off. The Barkhausen effect is named for the German physicist Heinrich Georg Barkhausen (1881–1956).

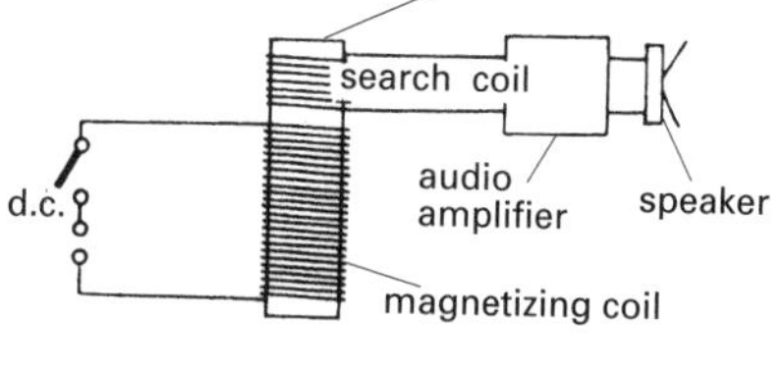

Barkhausen effect

barn Symbol: b A unit of area defined as 10^{-28} square meter. The barn is sometimes used to express the effective cross-sections of atoms or nuclei in the scattering or absorption of particles.

barometer A device that measures the pressure of the atmosphere: the standard value is around 100 kPa.

The *liquid barometer* has a column of liquid in a vertical tube. Various types of *mercury barometer* are commonly used. As the external atmospheric pressure rises and falls, the length of the liquid column rises and falls.

The *aneroid* (*non-liquid*) *barometer* employs a thin-metal evacuated box. Changes in atmospheric pressure move the sides of the box, and levers communicate this movement to a pointer. In general, it is not as accurate as a liquid barometer, but it is much easier to transport and use, and is much cheaper. It can also be used as an altimeter (*see* altimeter).

The liquid barometer provides an absolute measure; aneroid barometers must be calibrated.

See also barometric height.

barometric height The 'height' of a column in a liquid barometer. As the usual barometer liquid is mercury (because of its very high density), barometric height has historically been measured in millimeters of mercury (mmHg); 1 mmHg is about 133.322 pascals. The standard value of the

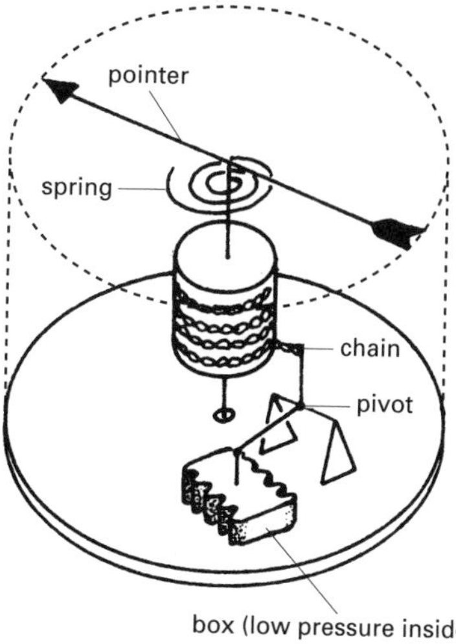

An aneroid barometer (not to scale)

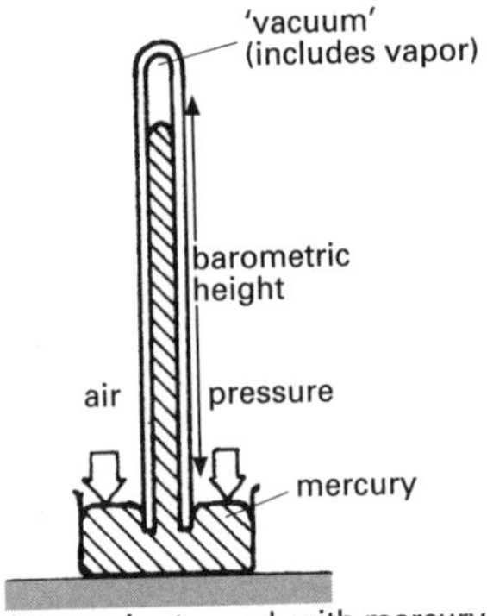

Caution It is not wise to work with mercury in this way.

A simple liquid barometer (not to scale)

pressure of the atmosphere is 760 mmHg, (101 325 Pa). *See also* STP.

baryon number /**ba**-ree-on/ A property of an elementary particle, equal to +1 for a baryon and -1 for an antibaryon. Gauge bosons, leptons, and mesons have a baryon number of 0. The baryon number is conserved in all observed types of particle–particle interaction. It has been suggested that baryon number might not be conserved in certain types of theories such as GRAND UNIFIED THEORIES. However, such baryon number violating processes have never been observed. *See* baryons; elementary particle.

baryons A group of heavy ELEMENTARY PARTICLES, which includes protons and neutrons. The baryons form a subclass of the hadrons. They are further subdivided into nucleons and hyperons.

base *See* transistor.

base unit A unit within a system of measurement from which other units may be derived by combining it with one or more other base units. With the exception of the kilogram, base SI UNITS are defined in terms of physical constants.

battery (*pl.* **batteries**) A number of similar units, such as electric cells, working together. Many dry 'batteries' used in radios, flashlights, etc., are in fact single cells. If a number of identical cells are connected in series, the total e.m.f. of the battery is the sum of the e.m.f.s of the individual cells. If the cells are in parallel, the e.m.f. of the battery is the same as that of one cell, but the current drawn from each is less (the total current is split among the cells).

Baumé scale /boh-**may**/ A scale of relative densities (specific gravities) of liquids, sometimes used on hydrometers, on which 0° is the density of water and 10° is the density of a 10% sodium chloride solution (both at a temperature of 12.5°C). The scale is named for the French chemist Antoine Baumé (1728–1805). *See* hydrometer.

beam A group of rays of light, or other forms of radiation, moving in the same direction. Strictly, a beam is the entire set of rays coming from a point or area of an object. A *pencil* is a narrow beam from a single point.

beat frequency *See* beats.

beats A regular increase and decrease in intensity of sound waves (or other waves) caused by two waves of slightly different frequencies being added together. The waves successively reinforce and cancel each other as they move in and out of phase. Sometimes radiofrequency waves produce audiofrequency beats in sound equipment. The frequency of the resulting signal (the *beat frequency*) is given by the difference in frequencies of the two signals; i.e. $f_1 - f_2$. If two waves of equal amplitude (a) produce beats, the resulting amplitude (A) is given by:

$$A = 2a\cos[2\pi(f_1 - f_2)t - \theta]/2$$

where θ is the phase angle between the original signals. *See also* heterodyne; interference.

beauty *See* quark.

becquerel /bek-ĕ-**rel**/ Symbol: Bq The SI unit of activity of radioactive nuclides. The activity in becquerels of a sample at a given time is the average number of disintegrations per second of its atoms at that time.

The unit is named for the French physicist Antoine Henri Becquerel (1852–1908). *See also* curie.

Beer's law The fraction of light of a given wavelength absorbed by a solution varies exponentially with the concentration of the absorbing substance and the thickness of its absorbing layer. The law is named for the German chemist, mathematician, and physicist August Beer (1825–63).

bel *See* decibel.

Bell's paradox A supposed paradox arising from theoretical work on quantum mechanics by the Irish physicist John Bell (1928–90). The work concerns the interpretation of quantum mechanics put forward by Niels Bohr, who argued that quantum mechanics depended on probabilities and that particles had an indeterminate existence until they were observed. Einstein never accepted this idea – he believed that there was some underlying deterministic mechanism governed by so-called *hidden variables*.

As an attack on Bohr's theories he (with others) postulated a thought experiment known as the *Einstein-Podolsky-Rosen experiment* (or *EPR experiment*). One simple form of it is to think of a particle of zero spin decaying into two particles with spin, which fly apart. Because spin is conserved, the particles must have opposite values; if one has a spin 'up' the other must have spin 'down'. In the experiment, one waits until the particles are several meters apart and then measures the spin of one particle. One instantly knows the spin of the other (because it must be opposite). But according to Bohr, the spin is neither 'up' nor 'down' until it is measured but is in an indeterminate state. Einstein argued that this could not be the case. Otherwise, one particle would have to communicate instantly across space. In Einstein's interpretation, the spins would be determined at the time of decay of the original particle and would be governed by hidden variables.

In the mid 1960s Bell proved a theorem, *Bell's theorem*, concerning measurements of spin in different directions for two particles. He showed that a certain set of inequalities (*Bell's inequalities*) would hold if hidden variables operated and Einstein was correct. If Bohr was correct, they would not hold.

The theorem opened the way for a real experimental test of the theories. In the early 1980s the French physicist Alain Aspect (1947–) and his team did such an experiment in Paris, making simultaneous measurements on photons separated by 12 meters. The results of the *Aspect experiment* supported Bohr's interpretation of quantum mechanics rather than Einstein's.

The consequence of this is very mysterious (hence the 'paradox'). It seems that two particles can be a large distance apart and still be part of a single system with neither one state nor another but a superposition of both. This phenomenon is known as *quantum entanglement*.

Bénard cell /bay-nard, bay-nar/ Any of a number of small convection cells that can appear in a liquid when it is heated from below under certain circumstances. At a certain temperature convection of the liquid suddenly occurs. The convection cells formed are an example of how a degree of order can be produced in a system when energy is supplied. They were studied by the French scientist Henri Bénard in about 1900.

berkelium /ber-klee-ŭm, ber-kee-lee-ŭm/ A silvery radioactive transuranic element of the actinoid series of metals, not found naturally on Earth. Several radioisotopes have been synthesized.

Symbol: Bk; m.p. 1050°C; p.n. 97; r.d. 14.79 (20°C); most stable isotope ^{247}Bk (half-life 1400 years).

Bernoulli effect /ber-noo-lee/ The relation between the pressure in a steadily flowing fluid, and its velocity. The pressure is less where the velocity is higher as, for example, where water flows through a narrower section in a pipe. The pressure that lifts an aircraft also depends on this effect. For horizontal flow, provided frictional resistance is negligible

$$p_1 - p_2 = \tfrac{1}{2}\rho(v_2^2 - v_1^2)$$

where p_1 is the pressure where the speed is v_1, and p_2 is the pressure where the speed is v_2. ρ is the density of the fluid. The principle is used in instruments for measuring the speed of flow, such as the Pitot tube. The Bernoulli effect is named for the Swiss mathematician Daniel Bernoulli (1700–82).

beryllium /bĕ-**ril**-ee-ŭm/ A light metallic element, similar to aluminum but somewhat harder. It has the electronic configuration of helium with two additional outer 2s electrons. It is used as an antioxidant and hardener in some alloys, such as copper and phosphor bronzes. Beryllium is extremely toxic.

Symbol: Be; m.p. 1278±5°C; b.p. 2970°C; r.d. 1.85 (20°C); p.n. 4; r.a.m. 9.012182.

beta decay /**bay**-tă, **bee**-tă/ A type of radioactive decay in which a nucleus emits, for instance, an electron. The result is a nuclide with the same mass number but a proton number one greater (electron emission) than the original nuclide. An example of beta decay is:

$$^{3}_{1}H \rightarrow ^{3}_{2}He + e^- + \bar{\nu}$$

The particles emitted in beta decay are *beta particles.* Streams of beta particles are *beta rays* or *beta radiation.* High-energy particles may penetrate metal sheets of mass/area a few gram/cm^2, or tens of meters of air at STP. The lowest energy particles may be absorbed in a few millimeters of air.

Beta particles may have a range of energies up to a maximum value characteristic of the nucleus concerned. The total energy is constant; it is carried by the beta particle and an antineutrino emitted at the same time. In another type of beta decay, positrons are emitted. In such cases the excess energy is carried by a neutrino. An example is:

$$^{13}_{7}N \rightarrow ^{13}_{6}C + e^+ + \bar{\nu}$$

See also alpha decay.

beta transformation The transformation of a nucleus by beta decay. Also the decay of a neutron to a proton, an electron, and an antineutrino:

$$n \rightarrow p + e^- + \bar{\nu}$$

betatron /**bay**-tă-tron, **bee**-/ A device for accelerating electrons to very high energies (300 MeV or more). Electrons produced from a source are injected into an evacuated donut-shaped ring between the poles of an electromagnet just as the magnetic field is being increased. As the magnetic field increases the electrons are accelerated, making as many as a quarter of a million circuits before the magnetic field reaches its maximum, at which time the orbit is changed by passing a current through auxiliary coils to deflect the electrons onto a target. A betatron can be compared to a transformer in which a cloud of electrons in the toroid constitutes the secondary circuit. Alternating current circulates in the primary coil (the magnetizing coil) but the electrons are extracted at the end of a quarter cycle before the decreasing primary current can cause deceleration.

BeV *See* GeV.

bevatron /**bev**-ă-tron/ The name given to the proton synchrotron at the University of California, which can accelerate protons to energies of about 10^{-9} joule (6 GeV).

bias A potential applied to an electrode in an electronic device to produce the desired characteristic.

biaxial crystal /bȳ-**aks**-ee-ăl/ A type of birefringent crystal having two axes, parallel to which the ordinary ray and the extraordinary ray travel at the same speed.

biconcave /bȳ-**kong**-kayv/ Describing a LENS with two concave faces. *Compare* biconvex.

biconvex /bȳ-**kon**-veks/ Describing a LENS with two convex faces. *Compare* biconcave.

big-bang theory The theory that the Universe originated in a very small hot dense state from which it has expanded. Evidence for the big-bang theory comes from the observed expansion of the Universe, the cosmic microwave background radiation, and the abundances of the light elements in the Universe.

bimetallic strip A strip of two metals joined side by side. When heated, the metals expand by different amounts, causing the strip to bend. Bimetallic strips are used in thermostats and circuit breakers.

binary star A system of two stars orbiting about their common center of mass. About half the stars in the Universe are thought to occur as binary stars. A particular type of binary-star system in which one of the stars is a pulsar has been used to give very accurate checks for general relativity theory.

binding energy (of a nucleus) The energy equivalent to the difference between the mass of the nucleus and the sum of the masses of its constituent nucleons. An example of calculating the binding energy of $^{7}_{3}$Li, with 4 neutrons and 3 protons is shown.

A useful measure is binding energy per nucleon. In the example the binding energy per nucleon is 39.2501/7 = approximately 5.6 MeV. For most nuclei, binding energy lies between about 7 and about 9 MeV per nucleon, reaching a maximum of about 9 MeV for nuclei of mass number about 60. The difference in mass in the example (i.e. the mass equivalent to the binding energy) is the *mass defect*.

CALCULATION OF MASS DEFECT AND BINDING ENERGY

mass of neutrons (4 × 1.008 983 amu)	4.035 932 amu
mass of protons (3 × 1.008 144 amu)	3.024 432 amu
total mass of constituents	7.060 364 amu
mass of ^{7}Li nucleus	7.018 222 amu
mass defect	0.042 142 amu
binding energy (1 amu = 931.14 MeV)	39.240 MeV

binoculars An optical instrument providing a telescope for each eye, thus giving distance perception as well as magnification.

Prism binoculars use a pair of prisms inside each telescope. These reflect rays by internal reflection. Their effect is to bring the inverted image upright, reduce the telescope length, and allow the object lenses to be farther apart than the eyes (thus improving stereoscopic vision). Binoculars are often described thus: 15 × 40. The first figure is the magnification; the second is the aperture of each object lens in mm.

Opera glasses are a simpler low-power device, consisting of two Galilean telescopes side by side. The telescopes produce upright images without the need for extra inverting lenses or prisms.

binocular vision Vision using two eyes. The brain forms a single three-dimensional view from the two separate images. This type of vision (*stereoscopic vision*) gives more information about distance and shape than monocular vision could.

bioelectricity /bÿ-oh-i-lek-**tris**-ă-tee/ Electricity generated in muscles, nerves, and other biological structures.

bioluminescence /bÿ-oh-loo-mă-**ness**-ĕns/ *See* luminescence.

biophysics /bÿ-oh-**fiz**-iks/ The use of physics in the study of biological phenomena. A common example is the physical explanation of the working of the eye.

Biot–Savart law /bee-**oh** sah-**var**/ The elemental field strength dB at a point distant r from a current element Idl in free space is given by:

$$dB = \mu_0 I dl \sin\theta / 4\pi r^2$$

The Biot–Savart law is named for the French physicists Jean Baptiste Biot (1774–1862) and Félix Savart (1791–1841).

bipolar transistor /bÿ-**poh**-ler/ *See* field-effect transistor.

biprism, Fresnel's /**bÿ**-priz-ăm/ A glass prism with a large angle, used to produce two coherent (virtual) sources for light interference experiments. As with Young's double slit arrangement, the wavelength λ of the incident monochromatic radiation is given by:

$$\lambda = yd/D$$

where y is the fringe separation, d is the source separation, and D is the source-screen distance. The fringes obtained with this arrangement are brighter than those in Young's experiment. The biprism is named for the French physicist Augustin Jean Fresnel (1788–1827) and Young's experiment for the British physicist, physician, and Egyptologist Thomas Young (1773–1829).

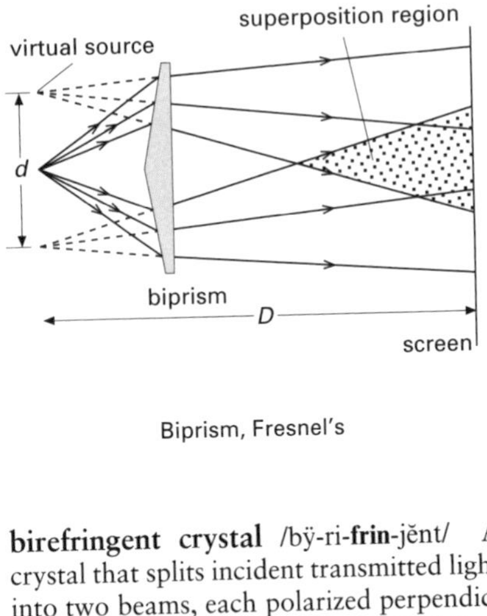

Biprism, Fresnel's

birefringent crystal /bÿ-ri-**frin**-jĕnt/ A crystal that splits incident transmitted light into two beams, each polarized perpendicularly to the other. The effect (called *birefringence* or *double refraction*) is particularly well-known in calcite (Iceland spar). It depends on the angle of incidence relative to the crystal axes, along which the speed of the light differs. The ordinary ray obeys the laws of refraction: it is polarized perpendicularly to the crystal axis. The extraordinary ray does not obey the laws of refraction (in the usual sense); hence its name. The study of the polarization properties of crystals is of great significance in geology, where it is used for the identification of minerals.

bismuth /**biz**-mŭth/ A brittle pinkish metallic element. Bismuth is widely used in alloys, especially low-melting alloys. The element has the property of expanding when it solidifies.

Symbol: Bi; m.p. 271.35°C; b.p. 1560±5°C; r.d. 9.747 (20°C); p.n. 83; r.a.m. 208.98037.

bistable circuit /bÿ-**stay**-băl/ (**flip-flop**) An electronic circuit, usually a MULTIVIBRATOR, that has two stable states and is switched from one to the other by a trigger pulse. Bistable circuits are used in computer logic for counting and storing binary digits (0 and 1). They form the basis of several different LOGIC GATES.

In a bistable multivibrator the input pulse is fed to the base terminal of one transistor (*TR1*) through a resistor *R1* and directly to the collector of the other transistor (*TR2*). The base of *TR2* and the collector of *TR1* are connected through a resistor *R2*. In logic circuits, bistable circuits may have two or perhaps more inputs. These are connected so that the output level (high or low) depends on whether one or both inputs are high. Sometimes a square-wave input is used as a clock or counter.

Bitter pattern A microscopic pattern of domain boundaries on a ferromagnetic surface, made visible by painting the surface with a colloidal suspension of very small iron particles. The production of Bitter patterns is similar to showing the shape of a magnet under a sheet of paper by using iron dust. Bitter patterns are named for the American physicist Francis Bitter (1902–67).

black body A body that absorbs all the radiation falling on it. The absorptance and emissivity of a black body are both equal to 1. In practice, a small hole in a uniform-temperature enclosure acts as a black body.

The radiation from a black body covers the whole wavelength range (sometimes the alternative term *full radiator* is used). The distribution of power with wavelength of this *black-body radiation* has a characteristic form. As the temperature increases the amount of radiation increases and the maximum in the curve moves to longer wavelengths. A black body radiates mainly infrared radiation below about 800 K. Visible radiation does not predominate until the temperature is above about 6000 K.

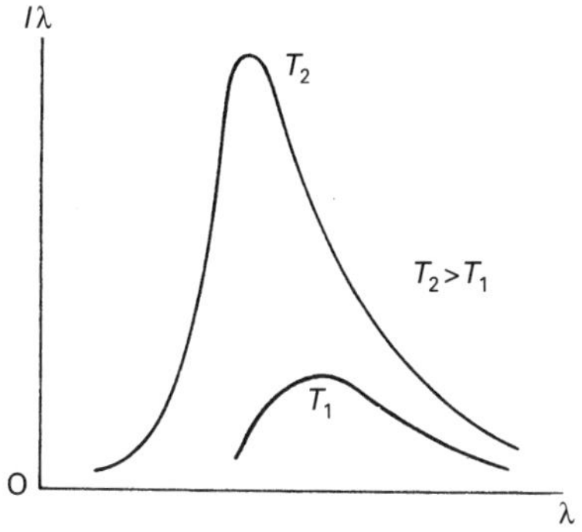

Black-body radiation

black hole A region of space–time in which the gravitational field is so strong that even light cannot escape from it. There is a considerable amount of observational evidence for the existence of black holes in the Universe. They are thought to arise from the gravitational collapse of very large stars that have exhausted their nuclear fuel. It has been suggested that supermassive black holes power QUASARS.

blanket A layer of fertile material surrounding the core in a breeder reactor.

blind spot The area of the retina where the optic nerve leaves the EYE. It is not sensitive to light as in this region there is no layer of rods and cones.

blink microscope A type of microscope used to compare two very similar photographs (such as particle tracks from a bubble chamber). The photographs are seen side by side, one with each eye, and are rapidly exposed and concealed. The brain tries to superimpose the two images, thus revealing any slight difference between them. *See also* microscope (compound).

Bloch wall /blok/ The boundary of a magnetic domain, over the width of which the atomic magnetic moment directions change. In iron, Bloch walls are around 100 nm thick. The wall is named for the Swiss-born American physicist Felix Bloch (1905–83).

block and tackle *See* pulley.

blooming A method of coating lenses to reduce back-reflection from their surfaces. It involves destructive interference in the thin layer. Each such layer can completely prevent reflection at only one wavelength (λ). λ is four times the layer thickness (t). For best effects the coating medium used should have a refractive constant $n = \sqrt{(n_1 n_2)}$, where n_1 and n_2 are the refractive constants of the media on each side. Single-layer blooming is normally used to prevent reflection of yellow light; bloomed surfaces thus reflect reds and blues, and appear purple. Multilayer blooming is sometimes employed, but is very costly.

blue shift *See* red shift.

Board of Trade unit (BTU) A unit of energy equivalent to the kilowatt-hour (3.6×10^6 joules). It was formerly used in the UK for the sale of electricity.

bohrium /**bor**-ee-ŭm, **boh**-ree-/ A synthetic radioactive element first detected by bombarding a bismuth target with chromium nuclei. Only a small number of atoms have ever been produced.

Symbol: Bh; p.n. 107; most stable isotope ^{262}Bh (half life 0.1s).

Bohr magneton Symbol: μ_B The unit of atomic magnetic moment, the moment of a single electron spin. It equals $eh/4\pi m_e c$, where e is the charge on an electron, h is the Planck constant, m_e is the electron rest mass, and c is the speed of light. Its value is $9.274\,009\,49 \times 10^{-24}$ joules per tesla (JT^{-1}). The unit is named for the Danish physicist Niels Bohr (1885–1962).

Bohr theory A pioneering attempt to apply quantum theory to the study of atomic structure, by Niels Bohr (1913). He assumed the newly introduced theory that an atom consisted of a massive positively charged nucleus with orbital electrons, and first considered in detail the simplest atom, that of hydrogen, with only one electron. The electron was supposed to move in circular orbits of radius r at speed v. By simple mechanics the total energy of such an orbit (kinetic plus potential) is shown to be $-e^2/8\pi\varepsilon_0 r$, where e is the electron charge. Bohr assumed that only certain orbits were possible, and tried to find a quantum rule to determine which ones by making assumptions (soon seen to be false) concerning the frequencies of radiation emitted and absorbed by the atoms. His calculations led however to the correct formula for the energies of the allowed states of the atom, and showed that the angular momenta were quantized in units of $h/2\pi$ where h is the Planck constant. For the *nth* quantum state the angular momentum is $nh/2\pi$ and the energy E_n is given by

$$E_n = -me^4/8\varepsilon^2{}_0h^2n^2$$

Assuming that the frequency of emitted radiation was given by $h\nu = E_{n1} - E_{n2}$, Bohr found that transitions to the orbit with $n = 2$ from higher values corresponded to the lines in the visible spectrum of hydrogen, while transitions to $n = 3$ gave lines in the near infrared.

The theory was soon successfully extended to the spectrum of singly-ionized helium and to the characteristic lines in the x-ray spectrum (*see* Moseley's law). Over the next twelve years the theory was developed far enough to suggest that very many facts in physics and chemistry might be explained in such terms, but innumerable difficulties were encountered with the detailed calculations. Various attempts to overcome these problems led to the modern theory of quantum mechanics pioneered by Heisenberg (1925), Schrödinger (1926), and Dirac (1928). Although Bohr's theory has been superseded it is of outstanding historical and philosophical interest.

boiling A change from liquid to gas occurring at a characteristic temperature (the *boiling point*). Boiling occurs when the saturated vapor pressure of the liquid equals the external pressure. Bubbles of vapor can then form in the liquid. The temperature at which this happens depends on external pressure; boiling points are therefore usually quoted at standard pressure. *See also* change of state; latent heat.

boiling-water reactor (**BWR**) A nuclear reactor in which water (in contact with the fuel elements) is used as both coolant and moderator. The water boils inside the reactor core. *Compare* pressurized-water reactor.

bolometer /boh-**lom**-ĕ-ter/ An instrument for measuring small amounts of radiant heat or microwaves. It depends on the change in resistance of a piece of metal foil or a superconductor when it absorbs radiant energy.

Boltzmann constant /**bohlts**-măn, -mahn/ Symbol: k The fundamental constant $1.380\ 650\ 5 \times 10^{-23}$ J K^{-1}, equal to the gas constant (R) divided by the Avogadro constant (N_A). The constant is named for the Austrian theoretical physicist Ludwig Edward Boltzmann (1844–1906). *See also* degrees of freedom.

bomb calorimeter A sealed insulated container, used for measuring energy released during combustion of substances (e.g. foods and fuels). A known amount of the substance is ignited inside the calorimeter in an atmosphere of pure oxygen, and undergoes complete combustion at constant volume. The resultant rise in temperature is related to the energy released by the reaction. Such energy values (*calorific values*) are often quoted in joules per kilogram (J kg^{-1}).

boron /**bor**-on, **boh**-ron/ A hard rather brittle metalloid element. It has the electronic structure $1s^22s^22p^1$. Only small quantities of elemental boron are needed commercially; the vast majority of boron supplied by the industry is in the form of borax or boric acid. Natural boron consists of two isotopes, ^{10}B (18.83%) and ^{11}B (81.17%). These percentages are sufficiently high for their detection by splitting of infrared absorption or by n.m.r. spectroscopy.

Symbol: B; m.p. 2300°C; b.p. 2658°C; r.d. 2.34 (20°C); p.n. 5; r.a.m. 10.811.

boron chamber A device for detecting low-energy neutrons, in which a compound of boron (usually the gas BF_3) fills an ionization chamber. The ^{10}B nuclei, which constitute 18% of natural boron, absorb neutrons and emit alpha particles, which are detected by the ionization they cause.

Bose–Einstein condensation /**bohz** **ÿn**-shtÿn/ A phenomenon in which several thousand atoms of certain elements are able to combine to form a single entity (a *superatom*) at very low temperatures. The phenomenon is important in the theory of superfluids.

Bose–Einstein condensation was first observed for atoms which are bosons in the late twentieth century. In the early twenty-first century it was observed for bosons formed by the pairing of atoms, which are fermions. Bose–Einstein condensation is named for the Indian physicist Satyendra Nath Bose (1894–1974) and the German-born physicist Albert Einstein (1879–1955), who discovered BOSE–EINSTEIN STATISTICS in 1924. The phenomenon of Bose–Einstein condensation was predicted by Einstein in 1924–5.

Bose–Einstein statistics The statistical rules for studying systems of identical bosons. It is assumed that (a) all identical particles are to be regarded as absolutely indistinguishable; and (b) any number of identical bosons can have the same set of quantum numbers in a given system. These rules were first introduced by Bose (1924) in his proof of Planck's radiation law, treating photons as quasi-particles. *See* Fermi–Dirac statistics; Maxwell–Boltzmann statistics.

boson /**boh**-zon/ A particle that obeys BOSE–EINSTEIN STATISTICS and has zero or integral spin. Unlike fermions, bosons are not conserved in number. That is bosons can be generated or destroyed singly, not in particle–antiparticle pairs, subject only to the basic conservation laws of mass, energy, momentum, charge, etc. *See* elementary particles; fermion; meson.

boundary layer A thin layer of fluid, such as the one next to a solid surface past which the fluid is moving. Friction with the surface slows flow within the boundary layer so that next to the surface the fluid is stationary. At the other edge of the boundary layer, the velocity approaches that of the main flow. Within it the effects of viscosity are significant, whereas in the main stream they can often be neglected.

Bourdon gauge /**boor**-dŏn/ (**Bourdon tube**) A type of fluid pressure gauge consisting of a coiled flattened tube. Pressure applied by a fluid inside the tube makes it tend to straighten, and this movement works a pointer that moves around a dial.

Boyle's law At a constant temperature, the pressure of a fixed mass of a gas is inversely proportional to its volume: i.e. $pV = K$, where K is a constant. (A graph of p against $1/V$ is a straight line.) The value of K depends on the temperature and on the gas. The law holds strictly only for ideal gases. Real gases follow Boyle's law at low pressures and high temperatures. Boyle's law is named for the Irish physicist Robert Boyle (1627–91). *See* gas laws.

Boyle temperature *See* gas laws.

Boys' experiment (1895) A method of determining the gravitational constant, G. A short beam (about 25 mm) with a gold sphere hanging from each end was suspended horizontally by a quartz torsion fiber. Measurements were made of the period of the torsional oscillations of the beam and of its angular deflection when large dense masses were placed near each sphere. In this way, G could be calculated. The method was more accurate than

Cavendish's similar experiment for *G*. The experiment is named for the British inventor Sir Charles Vernon Boys (1855–1944).

Bragg's law If a beam of x-rays of wavelength λ is directed at a crystal with parallel crystal planes that are distance *d* apart, then the reflected x-rays from each plane undergo interference. Constructive interference occurs at angles θ where $n\lambda = 2d\sin\theta$, *n* being an integer (1, 2, 3, etc.). θ is the angle between the crystal plane and the incident beam (called the *Bragg angle*). The equation is used in determining crystal structure from interference patterns produced by monochromatic x-rays. The law is named for the British physicist Sir Lawrence Bragg (1890–1971).

breeder reactor A NUCLEAR REACTOR that produces additional nuclei at a rate greater than that at which fuel is consumed. The core fuel consists of a fissile element, for example, uranium enriched to about 25% in the ^{235}U isotope. The core is surrounded by a blanket of fertile material, mostly ^{238}U in the form of natural or depleted uranium. Some of the surplus neutrons from the fission of ^{235}U convert ^{238}U into ^{239}Pu, which is fissile. A primary circuit of liquid sodium can be used through the core to carry heat away. The heat is transferred to a secondary circuit of sodium that boils water, the steam then operating turbines and generators as in a conventional power station. Such a reactor is also termed a *fast breeder reactor* or *fast reactor* because the neutrons moving through the core and blanket are fast moving, being of high energy (several MeV) as compared to those in thermal reactors (about 0.025 eV).

bremsstrahlung /**brem**-shtrah-lûng/ X-rays emitted when fast electrons are slowed down violently, as when electrons strike the target in an x-ray tube. The German word translates as 'braking radiation'. Bremsstrahlung is caused when an electron passes through the electric field of a nucleus and constitutes the continuous x-ray spectrum.

Brewster angle /**broo**-ster/ Symbol: i_B The angle of incidence, on a partially reflecting surface, at which the reflected radiation is fully plane-polarized. It is also the angle of incidence at which the reflected and refracted beams are perpendicular. Polarization by reflection is a refractive property of the surface.

$$_1n_2 = \tan i_B$$

The plane of polarization is parallel to the surface. The refracted radiation is partly polarized parallel to the normal. Formerly, the Brewster angle was called the *angle of polarization* or the *polarizing angle*. The Brewster angle is named for the Scottish physicist Sir David Brewster (1781–1868).

bridge circuit An arrangement of four electrical components in a square with an input across two opposite corners and an output across the other opposite corners. *See* Wheatstone bridge.

brightness A vague term describing the intensity of light. It can be applied to a source of light, to light itself, or to an illuminated surface. The brightness or intensity of light, in any of these three cases, relates to the rate of supply of energy (i.e. the power). The relation is complicated as it must take account of the sensitivity of the eye (or other detector) at different frequencies. *See also* photometry.

Brinell test /bri-**nel**/ A way of measuring the hardness of a material. A standard steel ball of known hardness is pressed into the material's surface with a known force. The size of the indentation indicates the hardness. The test is named for the Swedish metallurgist Johan August Brinell (1849–1925).

British thermal unit (**Btu**) A unit of energy. It was formerly defined by the heat needed to raise the temperature of one pound of air-free water by one degree Fahrenheit at standard pressure. Slightly different versions of the unit were in use depending on the temperatures between which the degree rise was measured. At 60.5°F it equals 1.054 615 kilojoules.

broken symmetry A situation in which the state of a system has a lower degree of symmetry than the symmetry of the equation that describes the system. There are many examples of broken symmetry in physics, particularly in the theory of condensed matter. Ferromagnetism and superconductivity are examples of broken-symmetry phenomena. Broken symmetry is also important in theories that attempt to unify different interactions and particle physics.

bromine /**broh**-meen, -min/ A deep red, moderately reactive element belonging to the halogens. Bromine is a liquid at room temperature (mercury is the only other element with this property).

Symbol: Br; m.p. –7.25°C; b.p. 58.78°C; r.d. 3.12 (20°C); p.n. 35; r.a.m. 79.904.

Brownian movement (**Brownian motion**) The random motion of small particles in a fluid – for example, smoke particles in air. The particles, which may be large enough to be visible with a microscope, are continuously bombarded by the invisibly small molecules of the fluid. Brownian movement is named for the Scottish botanist Robert Brown (1773–1858). *See also* kinetic theory.

brush An electrical contact to a moving part of an electric motor or generator.

brush discharge A form of bright gas discharge occurring near sharp points of high potential. The potential difference causing such a discharge is lower than that necessary for a spark or arc. The discharge is characterized by luminous streamers, which take on a treelike form.

bubble chamber A container of a liquid kept slightly above its boiling temperature by increased pressure and used to show tracks of ionizing radiation. The liquid is often liquid hydrogen. Just before the passage of a particle the pressure is momentarily reduced, and a photograph taken. Ions formed along the paths of charged particles or gamma-ray photons act as nuclei on which bubbles form. Magnetic fields can be applied causing curvature of the paths of charged particles. Bubble chambers are more useful than CLOUD CHAMBERS. The greater density of the liquid increases the chance that a nuclear reaction will occur.

bulk modulus *See* elastic modulus.

bumping When a gas-free liquid is heated in a smooth container the temperature may rise well above the boiling temperature at the applied pressure without boiling; the liquid is superheated. Then if a bubble forms the liquid will boil very violently and is said to bump. As this can be dangerous it is usual to place rough objects such as broken porcelain in flasks used for boiling certain liquids. Bubbles form readily on the rough surfaces and insure smooth steady boiling.

Bunsen burner /**bun**-sĕn/ A gas burner consisting of a vertical metal tube with an adjustable air-inlet hole at the bottom. Gas is allowed into the bottom of the tube and the gas-air mixture is burnt at the top. With too little air the flame is yellow and sooty. Correctly adjusted, the burner gives a flame with a pale blue inner cone of incompletely burnt gas, and an almost invisible outer flame where the gas is fully oxidized and reaches a temperature of about 1500°C. The Bunsen burner is named for the German chemist Robert Wilhelm Bunsen (1811–99). He was not the actual inventor of the Bunsen burner, but used it to great effect in pioneering work on spectroscopy.

Bunsen cell A type of primary cell in which the positive electrode is formed by carbon plates in nitric acid solution and the negative electrode consists of zinc plates in sulfuric acid solution.

buoyancy The tendency of an object to float. The term is sometimes used for the upward force (upthrust) on a body. *See* center of buoyancy.

butterfly effect Any effect in which a small change to a system results in a disproportionately large disturbance.

The term originates in the idea that the Earth's atmosphere is so sensitive to initial conditions that a butterfly flapping its wings in one part of the world may be the cause of a tornado happening in another part of the world. *See* chaos theory.

buzzer A electrical device in which a vibrating component, operated by an alternating electric current, produces a continuous buzzing sound.

BWR *See* boiling-water reactor.

bypass capacitor A capacitor that allows alternating currents to pass through it rather than through another component connected in parallel. The capacitance of the capacitor determines the frequency range of alternating current that can pass most easily.

C

cable, coaxial An electrical cable consisting of a central wire in an insulating sheath surrounded by a woven mesh of conductor, all inside an outer insulating sheath. Coaxial cables are not affected by external electric or magnetic disturbances or signals. They are used in transmitting high-frequency signals; for example, in television antenna connections. Normally the outer conductor is earthed.

cadmium /**kad**-mee-ŭm/ A transition metal obtained as a by-product during the extraction of zinc. It is used to protect other metals from corrosion, as a neutron absorber in nuclear reactors, in alkali batteries, and in certain pigments. It is highly toxic.

Symbol: Cd; m.p. 320.95°C; b.p. 765°C; r.d. 8.65 (20°C); p.n. 48; r.a.m. 112.411.

cadmium cell *See* Weston cadmium cell.

calcite /**kal**-sÿt/ (**Iceland spar**) A naturally occurring form of calcium carbonate. It is the best-known example of a mineral that shows double refraction (birefringence).

calcium /**kal**-see-ŭm/ A moderately soft, low-melting reactive metal. The electronic configuration is that of argon with an additional pair of 4s electrons.

Calcium is widely distributed in the Earth's crust and is the third most abundant element. At ordinary temperatures calcium has the face-centred cubic structure with a transition at 450°C to the close-packed hexagonal structure.

Symbol: Ca; m.p. 839°C; b.p. 1484°C; r.d. 1.55; p.n. 20; r.a.m. 40.0878.

calculus /**kal**-kyŭ-lŭs/ The branch of mathematics that deals with the differentiation and integration of functions. By treating continuous changes as if they consisted of infinitely small step changes, *differential calculus* can, for example, be used to find the rate at which the velocity of a body is changing with time (acceleration) at a particular instant.

Integral calculus is the reverse process, that is finding the end result of known continuous change. For example, if a car's acceleration a varies with time in a known way between times t_1 and t_2, then the total change in velocity is calculated by the integration of a over the time interval t_1 to t_2. *See* differentiation; integration.

calendar year *See* year.

californium /kal-ă-**for**-nee-ŭm/ A silvery radioactive transuranic element of the actinoid series of metals, not found naturally on Earth. Several radioisotopes have been synthesized, including californium-252, which is used as an intense source of neutrons in certain types of portable detector and in the treatment of cancer.

Symbol: Cf; m.p. 900°C; p.n. 98; most stable isotope ^{251}Cf (half-life 900 years).

Callendar and Barnes' experiment /**kal**-ĕn-der **barnz**/ The method was first used (1900) to determine Joule's equivalent. It was extended to the measurement of the specific heat capacities of liquids and gases over a range of temperatures. The values are determined very nearly under constant pressure.

The method introduced high-precision constant-flow calorimetry, pioneering the use of platinum-resistance thermometers and the measurement of electric current and potential difference with potentiome-

and potential difference with potentiometers for calorimetric measurements.

The fluid passes at a constant rate through a thermally insulated tube containing a spiral electrical heating wire along the axis. When steady state is reached the input and output temperatures are recorded. Then

$$VI = mc_p(\theta_2 - \theta_1) + w$$

where V and I are the p.d. across the heater coil and the current in it respectively, m is the mass of fluid passed per unit time, c_p is the specific heat capacity, θ_1 and θ_2 are the input and output temperatures and w is the (small) rate of heat loss. The rate of fluid flow is then changed and the values of V and I are adjusted until steady state is reached with the same values of θ_2 and θ_1. Assuming the rate of heat loss is the same this can be eliminated; hence c_p can be found. The experiment is named for the English physicist and engineer Hugh Callendar (1863–1930) and the Canadian physicist Howard Barnes (1873–1950).

calomel electrode A half cell having a mercury electrode coated with mercury(I) chloride, in an electrolyte consisting of potassium chloride and (saturated) mercury chloride solution. Its standard electrode potential against the hydrogen electrode is accurately known and it is a convenient secondary standard.

caloric theory /kă-**lor**-ik, -**loh**-rik/ An obsolete theory that heat was a weightless fluid. The theory was abandoned in the 19th century.

calorie /**kal**-ŏ-ree/ **1.** Symbol: cal Any of various c.g.s. units of energy approximately equal in SI derived units to 4.185 joules (J). Originally the calorie was defined as the heat needed to raise the temperature of one gram of water by one Celsius degree. Because the specific heat capacity of water changes slightly with temperature, the value determined for this calorie varied. Consequently new calories were defined whose values were fixed with reference to a specific temperature or to a mean value. For example, the mean or thermochemical calorie (cal_{TH}) was defined as one hundredth of the heat needed to raise one gram of water from zero degrees Celsius (0°C) to 100°C, and the 15°C calorie as the heat needed to raise one gram of water from 14.5°C to 15.5°C. Today, the thermochemical calorie is defined as 4.184 joules (J), the 15°C calorie used in North America as 4.1858J, and the international table calorie (cal_{IT}) as 4.186 8 joules.
2. Symbol: Cal A f.p.s. unit of energy equal to 1000 15° calories or the energy needed to raise the temperature of one kilogram of water at 14.5°C one Celsius degree. It is also called the *Calorie*, *kilocalorie*, or *dietary calorie*, the latter because it has been used to describe the equivalent amount of energy derived during digestion from a specified amount of food.

Calorie *See* calorie.

calorific value /kal-ŏ-**rif**-ik/ A measurement of the energy content of fuels. It is the amount of heat released when unit mass or volume of a fuel burns completely. *See* bomb calorimiter.

calorimeter /kal-ŏ-**rim**-ĕ-ter/ A device for measuring thermal energy; for example in determining thermal capacity, specific latent thermal capacity, energies of combustion, etc. *See also* bomb calorimeter.

camera An optical instrument able to record an image formed by visible light (or by other electromagnetic radiation). The record may be in the form of chemical changes in a photographic emulsion, or electrical signals as in a T.V. camera. A lens or system of lenses is used to focus the radiation onto the sensitive surface. Adjustment can be made for different object distances. The shutter allows light in for a set time (except in a video camera); the diaphragm varies the aperture.

camera, pin-hole *See* pin-hole camera.

camera obscura /**kam**-ĕ-ră ŏb-**skyoor**-ă/ A chamber or box in which an upside-down image of an external scene is formed by a small aperture or lens in the opposite side. It was the origin of the modern photographic camera. *See also* camera; pinhole camera.

Canada balsam /**bawl**-săm/ A transparent adhesive with a refractive constant of 1.53. As this constant is similar to that of many optical media, Canada balsam has various uses in optical instrumentation. *See*; for example; Nicol prism.

canal rays Streams of positive ions obtained from a discharge tube by boring small holes in the cathode. The ions being attracted to the cathode can thus pass through it forming positive rays.

candela Symbol: cd The SI base unit of luminous intensity; the luminous intensity in a given direction of a source that emits monochromatic radiation of frequency 540×10^{12} hertz and that has radiant intensity in that direction of 1/683 watt per steradian. Formerly, the unit was defined as the intensity (in the perpendicular direction) of the black-body radiation from a surface of 1/600 000 square meter at the temperature of freezing platinum and at a pressure of 101 325 pascals.

candle The former unit of luminous intensity, now replaced by the SI unit the candela. 1 candle = 1.02 cd. *See* candela.

candle-power An outdated measure of luminous intensity. The unit until 1948 was the international standard candle. This was first defined, in the UK, in the 1860 Gas Act; it was the brightness of a candle made in a certain way and burning at a certain rate.

capacitance /kă-**pass**-ă-tăns/ Symbol: *C* The ability of a system of conductors and insulators to store electric charge. If, in a given case, a charge *Q* is maintained by a potential difference *V*, then the capacitance is defined as the quotient *Q*/*V*. The unit of capacitance is the farad (F). *See also* capacitor.

capacitor /kă-**pass**-ă-ter/ (*formerly*, condenser) A device for storing electric charge. It usually consists of two parallel conductors separated by some insulating material (the dielectric). The capacitance of a capacitor increases:

1. the greater the common area of the conductors;
2. the smaller the distance between them;
3. the higher the relative permittivity of the dielectric.

The charge in a capacitor is stored partly by polarization of the particles of the dielectric. The capacitance (*C*) of a parallel-plate capacitor is given by

$$C = \varepsilon A/d$$

where ε is the permittivity of the material between the plates, *A* their common area, and *d* their separation.

An isolated sphere has capacitance:

$$C = 4\pi\varepsilon_1\varepsilon_0 r$$

where *r* is the radius of the sphere, and $\varepsilon_1\varepsilon_0$ the permittivity of the medium.

capillary action The effect that occurs when a fine tube, or capillary, is placed vertically with one end in a liquid. The height of the liquid inside the tube will be above or below the level of the liquid by an amount depending on the cohesion (surface tension) of the liquid and the adhesion between liquid and tube. The narrower the capillary is, the greater the difference in height. Because water 'sticks' to glass – its adhesion tends to be greater than its cohesion – it has a concave meniscus in a glass tube, and will be drawn up the tube by capillary action. Mercury, on the other hand, will be lowered inside a glass capillary, because here the cohesion is greater than the adhesion.

The distance *h* that a liquid of density ρ and surface tension γ will rise up a tube of radius *r* is given by the equation

$$h = 2\gamma \cos\alpha / rg\rho$$

where α is the contact angle between the meniscus and the tube. α is measured below the meniscus, so that for a convex meniscus, α is greater than 90° and *h* is negative.

capillary tube A tube with a narrow bore (internal diameter). Capillary tubing usually has relatively thick walls. Glass capillaries are used in mercury thermometers and experiments on surface tension. *See also* capillary action.

capture The absorption of one particle by another. For instance, a positive ion may capture an electron to form a neutral atom. In some nuclear reactions, an atomic nucleus may capture a neutron with emission of one or more gamma-ray photons.

carbon /kar-bŏn/ A nonmetallic element; carbon is a universal constituent of living matter and the principal deposits of carbon compounds are derived from living sources; i.e., carbonates (chalk and limestone) and fossil fuels (coal, oil, and gas). It also occurs in the mineral dolomite. Minute quantities of elemental carbon also occur as the allotropes graphite and diamond. A third allotrope, buckminsterfullerene (C_{60}), also exists. Naturally occuring carbon consists of three isotopes: ^{12}C (98.89%), ^{13}C (1.11%) and ^{14}C (minute traces in the upper atmosphere produced by slow neutron capture by ^{14}N atoms). ^{14}C is used for radiocarbon dating because of its long half-life of 5730 years.

Symbol: C; m.p. 3550°C; b.p. 4830°C (sublimes); C_{60} sublimes at 530°C; r.d. 3.51 (diamond), 2.26 (graphite), 1.65 (C_{60}) (all at 20°C); p.n. 6; r.a.m. 12.011.

carbon black (**soot**) A form of amorphous carbon produced by incomplete combustion of gas (or other organic matter). It is used in experiments as a coating for surfaces that need to be good absorbers of radiation, for example in detectors of thermal radiation, such as the thermopile. It is also used to increase the amount of thermal radiation emitted by a surface.

carbon cycle (**carbon-nitrogen cycle**) A series of nuclear reactions postulated to account for energy production in stars. In this series $^{1}{}_{6}{}^{2}C$ is an intermediary in the process by which hydrogen nuclei fuze to form helium with release of energy. The first step is the fusion of carbon and hydrogen nuclei:

$$^{12}_{6}C + ^{1}_{1}H \rightarrow ^{13}_{7}N + \text{gamma radiation}$$
$$^{13}_{7}N \rightarrow ^{13}_{6}C + \text{positron}$$
$$^{13}_{6}C + ^{1}_{1}H \rightarrow ^{14}_{7}N + \text{gamma radiation}$$
$$^{14}_{7}N + ^{1}_{1}H \rightarrow ^{15}_{8}O + \text{gamma radiation}$$
$$^{15}_{8}O \rightarrow ^{15}_{7}N + \text{positron}$$
$$^{15}_{7}N + ^{1}_{1}H \rightarrow ^{12}_{6}C + ^{4}_{2}He$$

The net result is:

$$4^{1}_{1}H \rightarrow ^{4}_{2}He$$

with a release of about 4.4 pJ of energy. *See also* proton-proton chain reaction.

carbon dating (**radiocarbon dating**) A method of dating – measuring the age of (usually archaeological) materials that contain matter of living origin. It is based on the fact that ^{14}C, a beta emitter of half-life approximately 5730 years, is being formed continuously in the atmosphere as a result of cosmic-ray action.

The ^{14}C becomes incorporated into living organisms. After death of the organism the amount of radioactive carbon decreases exponentially by radioactive decay. The ratio of ^{12}C to ^{14}C is thus a measure of the time elapsed since the death of the organic material.

Uncertainties arise because of uncertainty as to the past rate of production of ^{14}C, the possibility of exchange of carbon with carbon of a different age during the elapsed time, the possibility of contamination of the sample, and the effect of burning of fossil fuels on the composition of atmospheric carbon.

The method is most valuable for specimens of up to 20 000 years old, though it has been modified to measure ages up to 70 000 years. For ages of up to about 8000 years the carbon time scale has been calibrated by dendrochronology; i.e. by measuring the ^{12}C:^{14}C ratio in tree rings of known antiquity.

carbon microphone *See* microphone.

carbon–nitrogen cycle *See* carbon cycle.

Carnot cycle /kar-noh/ The idealized reversible cycle of four operations occurring in a perfect heat engine. These are the successive adiabatic compression, isothermal expansion, adiabatic expansion, and isothermal compression of the working substance. Heat Q_1 enters the working substance from a furnace during the expansion at the higher temperature T_1 and a smaller quantity Q_2 is discharged at the lower temperature T_2. The net work done by the engine in one cycle is W, which by conservation of energy is equal to $Q_1 - Q_2$. By definition, the efficiency η is equal to W/Q_1. Hence, $\eta = 1 - Q_1/Q_2$. Kelvin defined temperature T on the thermodynamic scale such that $T_2/T_1 = Q_2/Q_1$ in this ideal engine. Hence

$$\eta = 1 - (T_1/T_2)$$

This is the highest efficiency possible in principle. In practice, all real heat engines have lower efficiencies. The cycle is named for the French physicist Nicolas Leonard Sadi Carnot (1796–1832). *Compare* Otto cycle. *See also* Carnot's principle.

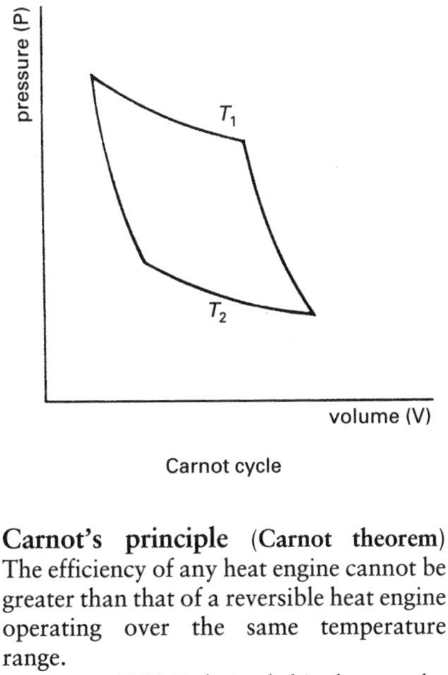

Carnot cycle

Carnot's principle (**Carnot theorem**) The efficiency of any heat engine cannot be greater than that of a reversible heat engine operating over the same temperature range.

Carnot (1824) derived this theorem by a fallacious argument before the formulation of the concept of energy. When the concept of energy and its conservation were established about 1849 Clausius and Kelvin showed that Carnot's principle was consistent with the new ideas provided one added a new principle, the second law of thermodynamics. *See* reversible change.

carrier 1. *See* carrier wave.
2. A charge carrier – either an electron or a positive hole.
3. The agent that carries the electric charge in a current, e.g. an electron, positive hole, or ion.

carrier wave An electromagnetic wave, usually in the long-wave radio to radar frequency range, that is used to carry information in a radio transmission. The information is superimposed on the carrier wave by MODULATION.

Cartesian coordinates /kar-**tee**-zhăn/ A method of defining the position of a point by its distance from a fixed point (origin) in the direction of two or more straight lines. On a flat surface, two straight lines, called the x-axis and the y-axis, form the basis of a two-dimensional Cartesian coordinate system. The point at which they cross is the origin (O). An imaginary grid is formed by lines parallel to the axes and one unit length apart. The point (2,3), for example, is the point at which the line parallel to the y-axis two units in the direction of the x-axis, crosses the line parallel to the x-axis three units in the direction of the y-axis. Usually the x-axis is horizontal and the y-axis is perpendicular to it. These are known as *rectangular coordinates*. If the axs are not at right angles, they are *oblique coordinates*

In three dimensions, a third axis, the z-axis, is added to define the height or depth of a point. The coordinates of a point are then three numbers (x,y,z). A right-handed system is one for which if the thumb of the right hand points along the x-axis, the fingers of the hand fold in the direction in which the y-axis would have to rotate to point in the same direction as the z-axis. A left-handed system is the mirror image of this. In a rectangular system, all three axs are mutually at right-angles. *See also* polar coordinates.

cascade connection A method of electrical connection in which the output terminals of the first device or network are connected to the input terminals of the second, those of the second to the third, and so on.

cascade liquefier A device for liquefying a series of gases in stages. The critical temperature of each gas is different. A gas with a high critical temperature is liquefied by compression and evaporated under a lower pressure, cooling a second gas below its critical temperature. The second gas can also be liquefied and evaporated to produce a still lower temperature, and so on. *See also* Linde process.

cascade process Any process that takes place in a number of stages.

An example is the diffusion process used to enrich uranium. The gas uranium hexafluoride diffuses through a long series of porous barriers. A very slight enrichment takes place at each, because of the very slightly higher speed of diffusion of $^{235}UF_6$ compared with that of $^{238}UF_6$. The gas-centrifuge enrichment process is also a cascade process.

Cassegrainian telescope /kass-ĕ-**grayn**-ee-ăn/ A type of reflecting telescope in which the converging mirror has a hole in its center. Light is reflected back onto a diverging mirror, then forward through the hole into the eyepiece. The telescope is named for the French sculptor Sieur Guillaume Cassegrain (*c.* 1650–1700). *See also* reflector.

catadioptric lens /kat-ă-dÿ-**op**-trik/ A compound lens in which both a mirror(s) and lenses are used to form an image.

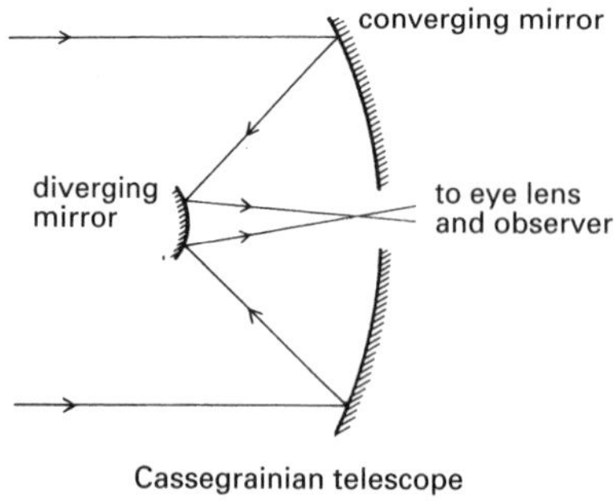

Cassegrainian telescope

catapault field The magnetic field that results when a conductor of electricity through which an electric current is flowing is put in a uniform external magnetic field.

cathetometer /kath-ĕ-**tom**-ĕ-ter/ An optical instrument for accurately measuring height or lengths; a small telescope or microscope mounted so that it can be slid along a graduated scale.

cathode /**kath**-ohd/ In electrolysis, the electrode that is at a negative potential with respect to the anode. In any electrical system, such as a discharge tube or a solid-state electronic device, the cathode is the terminal at which electrons enter the system.

cathode-ray oscilloscope (CRO) An instrument for investigating electrical signals. An electron beam is directed at the face of a cathode-ray tube, forming a small spot. The beam can be deflected in the horizontal and vertical directions by an electrostatic field from two sets of plates. The horizontal deflection is often a constant repeating sweep of the spot produced by an internal time base. The signal to be investigated is amplified and applied to the vertical deflecting plates. Thus, a graph of the signal is 'drawn' on the screen.

cathode rays Streams of electrons given off by the cathode of a gas-discharge tube at low pressure. If there is sufficient residual gas, the discharge occurs with a cold cathode. In this case, electrons are ejected from the cathode by positive-ion bombardment. In a high vacuum, the cathode is an electrically heated wire, which emits by the thermionic effect. *See also* secondary emission; thermionic emission.

cathode-ray tube (CRT) An electron tube that converts electrical signals into a pattern on a screen. The cathode-ray tube forms the basis of the CATHODE-RAY OSCILLOSCOPE and the television receiver. It consists of an electron gun, which produces an electron beam. The electrons are focused onto and moved across a luminescent screen by magnetic and/or electric fields, to give a small moving spot of light.

cation /**kat**-ÿ-ŏn, -on/ A positively charged ion, formed by removal of electrons from atoms or molecules. In electrolysis, cations are attracted to the negatively charged electrode (the cathode). *Compare* anion.

caustic The curve (surface in three dimenstions) on which rays meeting a curved reflector or lens meet after reflection or refraction. *See* focal point.

Cavendish's experiment /**kav**-ĕn-dish/ (1798) This was the first experiment to measure the gravitational constant *G* in the laboratory (rough measurements had already been made by observing the attraction of a pendulum by mountains, e.g. Bouger, 1749). Since this experiment by the English physicist and chemist Henry Cavendish (1731–1810), many investigators have performed similar laboratory experiments with improved accuracy (*see* Boys' experiment). These have shown that Newton's law applies with high precision on the laboratory scale, but, in recent years, small departures from this law have been established. *See also* Newton's law of universal gravitation.

In the original apparatus, a beam with small lead spheres at each end was suspended horizontally by a torsion wire. Large spheres of lead were then placed close to the smaller spheres on opposite sides of the beam in order to turn the beam through an angle. The lead spheres were then moved to the opposite sides of the beam, turning the beam in the opposite direction. *G* could be calculated from the total angular deflection. The method, which was the first attempt to measure *G*, was less accurate than Boys' experiment.

cavitation /kav-ă-**tay**-shŏn/ The formation of small cavities in a liquid, caused by a reduction in fluid pressure. The cavities may collapse, generating a large impulsive pressure which may erode and damage nearby solid surfaces (such as pump impellers, turbine blades, and ships' propellers). The phenomenon is exploited for cutting and drilling metals using ultrasonic waves.

cavity resonator *See* klystron; magnetron.

CD *See* compact disk.

celestial mechanics The study of the motion of planets and their satellites. Celestial mechanics is largely dealt with using Newtonian mechanics, with small corrections resulting from general relativity theory.

cell **1.** A system having two plates (electrodes) in a conducting liquid (electrolyte). An *electrolytic cell* is used for producing a chemical reaction by passing a current through the electrolyte (i.e. by ELECTROLYSIS). A *voltaic* (or *galvanic*) cell produces an e.m.f. by chemical reactions at each electrode. Electrons are transferred to or from the electrodes, giving each a net charge. *See also* accumulator; Clark cell; Daniell cell; Leclanché cell.
2. *See* photocell; photoelectric cell.
3. *See* Kerr effect.

Celsius scale A TEMPERATURE SCALE in which the temperature of melting pure ice is taken as 0° and the temperature of boiling water 100° (both at standard pressure). The *degree Celsius* (°C) is equal in magnitude to the kelvin provided that the scale has been calibrated so as to conform to the International Practical Temperature Scale at all temperatures. Before 1948 the Celsius scale was known as the *centigrade scale*. Celsius' original scale of 1742 was inverted (i.e. had 0° as the steam temperature and 100° as the ice temperature). The scale is named for the Swedish astronomer Anders Celsius (1701–44).

center of buoyancy (for an object in a fluid) The center of mass of the displaced fluid volume. For a floating object to be stable, the center of mass of the object must lie below the center of buoyancy; when the object is in equilibrium, the two lie on a vertical line. *See also* Archimedes' principle; flotation, law of.

center of curvature Each surface of a simple curved LENS or MIRROR is part of a sphere. The center of curvature c of such a surface is the center of the sphere of which the surface is part.

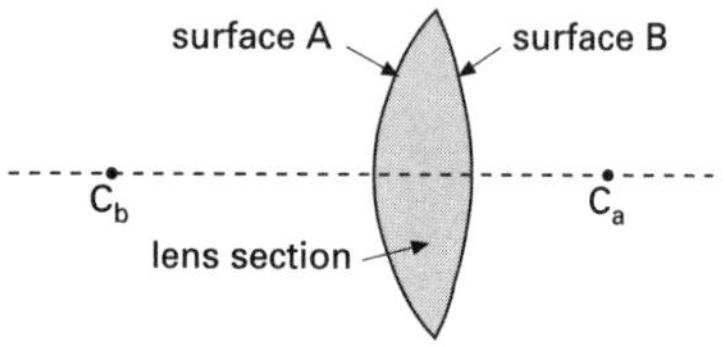

Center of curvature

center of gravity *See* center of mass.

center of mass A point in a body (or system) at which the whole mass of the body may be considered to act. Often the term *center of gravity* is used. This is, strictly, not the same, except if the body is in a constant gravitational field. The center of gravity is the point at which the weight may be considered to act.

The center of mass coincides with the center of symmetry if the body has a uniform density throughout. In other cases the principle of moments may be used to locate the point. The *centroid* of a plane or solid shape is the point at which the center of mass would be if the shape or solid were of uniform-density material.

center of oscillation The point on a stationary compound PENDULUM located vertically below the pivot at a distance that equals the length of the equivalent simple pendulum.

center of pressure If a surface lies horizontally in a fluid, the pressure at all points will be the same. The resultant force will then act through the centroid of the surface. If the surface is not horizontal, the pressure on it will vary with depth. The resultant force will now act through a different point; the center of pressure is not at the centroid. The center of pressure of a submerged surface is the point through which the resultant of the pressure forces acts.

centi- Symbol: c A prefix denoting 10^{-2}. For example, 1 centimeter (cm) = 10^{-2} meter (m).

centigrade scale *See* Celsius scale.

centimetric waves /sen-tă-**met**-rik/ Electromagnetic waves with wavelengths in the range 1–10 cm (i.e. part of the microwave region).

central bright image The image that occurs at the center of a diffraction pattern produced when light is diffracted at an opening. This image is brighter than all the other images produced by the diffraction.

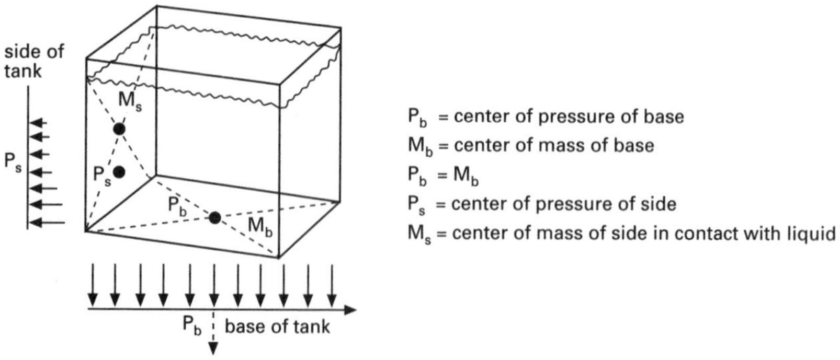

Center of pressure

central force A force that acts on any affected object along a line to an origin. Electric forces between charged particles are central; frictional forces are not.

centrifugal force A force supposed to act radially outward on a body moving in a curve. In fact there is no real force acting; centrifugal force is said to be a 'fictitious' force, and the use of the term is best avoided. The idea arises from the effect of inertia on an object moving in a curve. If a car is moving around a bend, for instance, it is forced in a curved path by friction between the wheels and the road. Without this friction (directed toward the center of the curve) the car would continue in a straight line. The driver also moves in the curve, constrained by friction with the seat, restraint from a seat belt, or a 'push' from the door. To the driver it appears that there is a force radially outward pushing his or her body out – the centrifugal force. In fact this is not the case; if the driver fell out of the car he or she would move straight forward at a tangent to the curve. It is sometimes said that the centrifugal force is 'reaction' to the centripetal force – this is *not* true. (The 'reaction' to the centripetal force is an outward push on the road surface.) *See also* centripetal force.

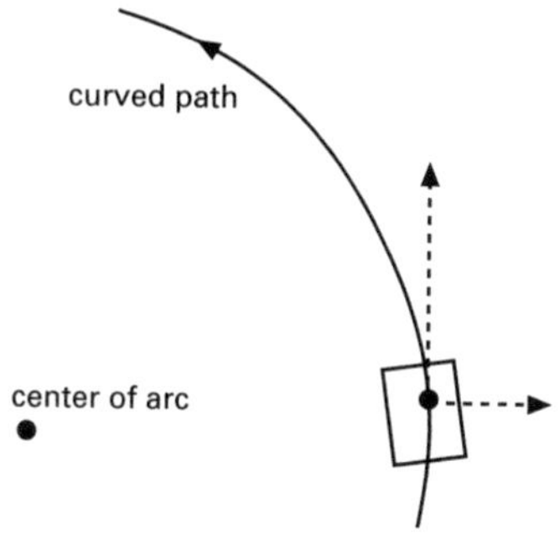

Centrifugal force

centrifuge /**sen**-tră-fyooj/ A device for separating the components of a suspension by spinning. In simple types, the liquid containing suspended particles is placed in a tube, which is whirled round in a horizontal circle. Particles less dense than the liquid move toward the axis of rotation; denser particles accumulate at the other end.

The device thus accelerates the process of separation that would result from gravitational forces. The gravitational pressure gradient is replaced by a much stronger pressure gradient related to centripetal forces. An *ultracentrifuge* is a high-speed centrifuge suitable for separating very small particles or even large molecules. Gas centrifuges are used in isotope separation.

centripetal force A force that causes an object to move in a curved path rather than continuing in a straight line. The force is provided by, for instance:

– the tension of the string, for an object whirled on the end of a string;
– gravity, for an object in orbit round a planet;
– electric force, for an electron in the shell of an atom.

The centripetal force for an object of mass m with velocity v, and path radius r is mv^2/r, or $m\omega^2r$, where ω is angular velocity. A body moving in a curved path with a velocity v has an acceleration because the direction of the velocity changes, even though the magnitude of the velocity may remain constant. This acceleration, which is directed toward the center of the curve, is the *centripetal acceleration*. It is given by v^2/r or ω^2r.

centroid /**sen**-troid/ *See* center of mass.

Cerenkov radiation /chĕ-**renk**-off/ A bluish light that is emitted when charged particles travel through a transparent medium at a speed that exceeds the speed of light in the medium. The radiation is named for the Russian physicist Pavel Alekseyevich Cerenkov (1904–90).

cerium /**seer**-ee-ŭm/ A ductile malleable gray element of the lanthanoid series of metals. It occurs in association with other lanthanoids in many minerals. It is used in several alloys (especially for lighter flints), as a catalyst, and in compound form in carbon-arc searchlights, etc., and in the glass industry.

Symbol: Ce; m.p. 799°C; b.p. 3426°C; r.d. 6.7 (hexagonal structure, 25°C); p.n. 58; r.a.m. 140.15.

cesium /**see**-zee-ŭm, **see**-see-/ (**caesium**) A soft golden highly reactive low-melting element. Cesium is used in photocells, as a catalyst, and in the cesium atomic clock. The radioactive isotopes ^{134}Cs (half life 2.065 years) and ^{137}Cs (half life 30.3 years) are produced in nuclear reactors and potentially dangerous atmospheric pollutants.

Symbol: Cs; m.p. 28.4°C; b.p. 678.4°C; r.d. 1.873 (20°C); p.n. 55; r.a.m. 132.91.

cesium clock An apparatus used to produce the steady frequency used in defining the second. It depends on the fact that, in a magnetic field, cesium-133 atoms can have two different energy levels between which transitions occur by absorption of radiofrequency radiation of a frequency of 9 192 631 770 hertz. In a cesium clock, the number of atoms in the higher state is detected, and the signal used to stabilize the oscillator producing the radiation.

c.g.s. system A system of units that uses the centimeter, the gram, and the second as the fundamental mechanical units. Much early scientific work was performed using this system, but it has now almost been abandoned.

In the c.g.s. system there are two sets of electrical units: *electrostatic units* in which the permittivity of free space (ε_0) is unity; and *electromagnetic units* in which the permeability of free space (μ_0) is unity.

chain reaction The progressive disintegration of fissile material (e.g. ^{235}U) by bombardment with neutrons, which in turn results in the production of more neutrons. These may, under suitable conditions, produce further fissions.

Fission of ^{235}U yields a varying number of neutrons depending on the energy of the incident neutrons. Neutrons may escape from the mass of uranium or be absorbed by nonfissile nuclei and so be ineffective in promoting the chain reaction. If one fission produces neutrons which, in turn, cause more than one fission, there is a branching chain reaction, which may be explosive. On the other hand, if insufficient neutrons are captured the chain reaction will die out. To sustain a controlled chain reaction in natural uranium, it is necessary to slow down the neutrons in a moderator. Slow neutrons are more likely to cause fission in ^{235}U than to be captured by the more abundant nonfissile isotope ^{238}U. A chain reaction can only be sustained in uranium without using a MODERATOR if the proportion of ^{235}U is considerably increased (*enrichment*). The artificial fissile nuclides ^{239}Pu, ^{241}Pu, and ^{233}U can also sustain chain reactions. *See* nuclear reactor.

change of state A change from one state of matter to another. *See* latent heat.

CHANGES OF STATE

solid–liquid	melting (fusion)
liquid–solid	freezing
liquid–gas	evaporation (boiling at boiling temperature)
gas–liquid	condensation
solid–gas	sublimation
gas–solid	condensation

chaos theory The theory of systems that exhibit apparently random unpredictable behavior. The theory originated in studies of the Earth's atmosphere and the weather. In such a system there are a number of variables involved and the equations describing them are nonlinear. As a result, the state of the system as it changes with time is extremely sensitive to the original conditions. A small difference in starting conditions may be magnified and produce a large variation in possible future states of the system. As a result, the system appears to behave in an unpredictable way and may exhibit seemingly random fluctuations (chaotic behavior). The study of such nonlinear systems has been applied in a number of fields, including studies of fluid dynamics and turbulence, random electrical oscillations, and certain types of chemical reaction. *See also* attractor; butterfly effect.

characteristic In an electronic device, such as a diode or a transistor, a graph that shows how a voltage or current between two of the terminals varies with respect to voltage or current at other terminals of the device.

charcoal An amorphous form of carbon made by heating wood or other organic material in the absence of air. *Activated charcoal* is charcoal heated to drive off absorbed gas. It is used for absorbing gases and for removing impurities from liquids.

charge (**electric charge**) Symbol: Q A basic property of some elementary particles of matter. Charge has no definition; rather it is taken as a basic experimental quantity. There are two types of charge conventionally called *positive* and *negative*. Like charges repel each other and unlike charges attract each other. The unit of charge is the coulomb (C). The charge on a body arises from an excess or deficit of negative electrons with respect to positive protons.

charge density The electric charge per unit volume of a material or per unit area of a surface or per unit length of a line. *Volume charge density* (symbol: ρ) is measured in coulombs per cubic meter (C m^{-3}). *Surface charge density* (symbol: σ) is measured in coulombs per square meter (C m^{-2}). The unit of *linear charge density*, λ, is the coulomb per meter (C m^{-1}). For a conductor, excess charge is distributed over the surface; the surface charge density increases with curvature.

Charles' law /sharlz, **charl**-ziz/ (**law of Charles and Gay-Lussac**) This was originally an experimental law relating the behavior of real gases to an arbitrary temperature scale (mercury-in-glass). For a given mass of gas at constant pressure, the volume increases by a constant fraction of the volume at 0°C for each Celsius degree rise in temperature. The constant fraction (α) has almost the same value for all gases – about 1/273 – and Charles' law can be written in the form

$$V = V_0(1 + \alpha_v\theta)$$

where V is the volume at temperature θ and V_0 the volume at 0°C. The constant α_v is the thermal expansivity of the gas.

A similar relationship exists for the pressure of a gas heated at constant volume:

$$p = p_0(1 + \alpha_p\theta)$$

Here, α_p is the pressure coefficient. Very nearly, $\alpha_p = \alpha_v$ for all gases. From the ideal gas equation $pV_m = RT$, we deduce that

$V_m \propto T$ at constant p

$p \propto T$ at constant Vm.

These equations are sometimes said to express Charles' law, although they are just consequences of the definition of T and cannot be derived from his experimental results. The law is named for the French physicist and physical chemist Jacques Alexandre César Charles (1746–1823), and for the French physicist and chemist Joseph-Louis Gay-Lussac (1778–1850). *See also* gas laws.

charm *See* quark.

chemical hygrometer *See* hygrometer.

chemiluminescence /kem-ă-loo-mă-**nes**-ĕns/ *See* luminescence.

chemisorption /kem-ă-**sorp**-shŏn, -**zorp**-/ *See* adsorption.

chirality /kÿr-**al**-ă-tee/ *See* optical activity.

Chladni figures /**klahd**-nee/ Patterns that reveal the modes of vibration of a glass or metal plate. They are made by sprinkling a fine powder on the plate. The powder accumulates along the nodes of the vibration, where the plate does not vibrate. The figures are named for the German physicist Ernst Florens Friedrich Chladni (1756–1827).

chlorine /**klor**-een, -in, **kloh**-reen, -rin/ A green reactive gaseous element belonging

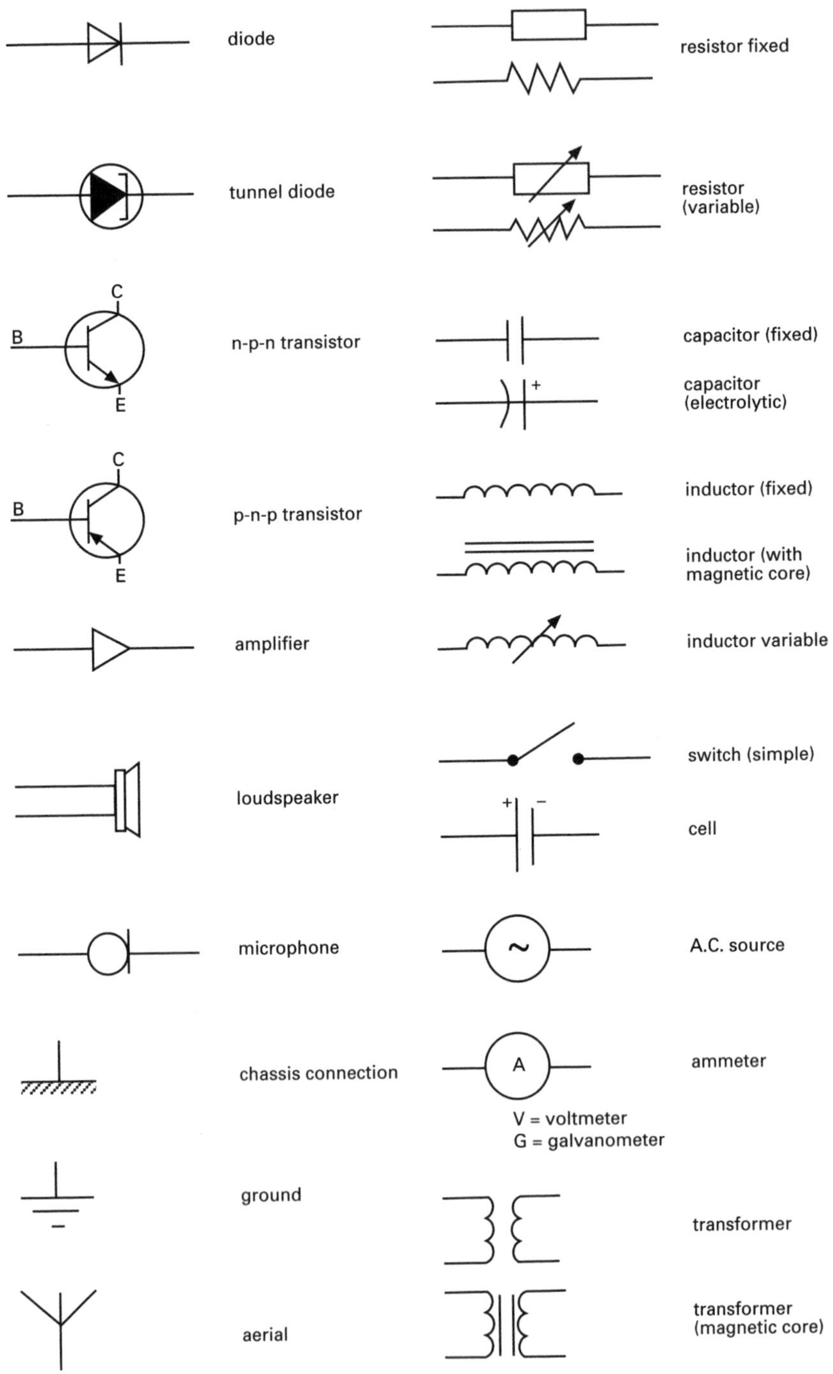

Circuit elements

to the halogens. It accounts for about 0.055% of the Earth's crust.

Symbol: Cl; m.p. –100.38°C; b.p. –33.97°C; d. 3.214 kg m^{-3} (0°C); p.n. 17; r.a.m. 35.4527.

choke An inductor used to reduce the high-frequency components of an alternating signal by presenting a higher impedance for them. Chokes are also used for smoothing fluctuations in the output current of a rectifier circuit.

choroid /**kor**-oid, **koh**-roid/ The middle of the three layers of the EYE.

chromatic aberration /krŏ-**mat**-ik/ *See* aberration.

chromium /**kroh**-mee-ŭm/ A transition metal; chromium is used in strong alloy steels and stainless steel and for plating articles. It is a hard silvery metal that resists corrosion at normal temperatures.

Symbol: Cr; m.p. 1860±20°C; b.p. 2672°C; r.d. 7.19 (20°C); p.n. 24; r.a.m. 51.9961.

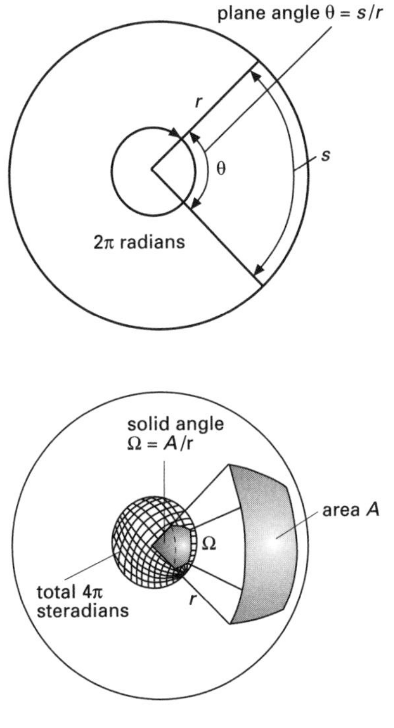

Circular measure

chromosphere /**kroh**-mŏ-sfeer/ The part of the atmosphere of the Sun that is directly above the photosphere and below the corona. It is roughly 10 000 kilometers thick.

circuit, electrical A combination of electrical components that form a conducting path. When the path is broken, for example by a switch, the circuit is said to be *open*. When a complete loop is formed allowing current to flow as a result of a potential difference in the circuit, it is said to be *closed*. A *short circuit* is one in which there is a low resistance path through which the current can flow. *See also* integrated circuit.

circuit breaker A safety device in an electric circuit that interrupts the flow of current in the event of a short circuit or other fault. *See also* fuse.

circuit element A resistor, capacitor, inductor, transistor, or other device used in making up electric circuits.

circular measure Measurement of angles in radians.

circular motion A form of periodic (or cyclic) motion, that of an object moving at constant speed v in a circle of constant radius r. For this to be possible, a positive central force must act. The terms used to describe circular motion are shown in the table overleaf. *See also* centripetal force; rotational motion.

circular polarization A type of polarization of electromagnetic radiation in which the plane of polarization rotates uniformly round the axis as the ray progresses. Circularly polarized light is equivalent to (and is produced by) the combination of two equal plane-polarized rays moving to-

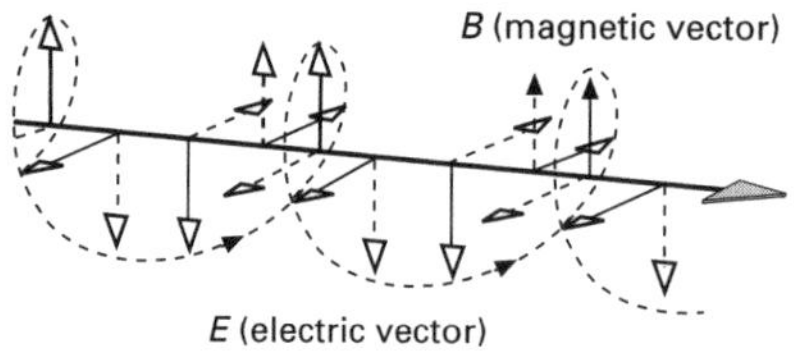

Circular polarization

gether but out of phase by 90°. *See also* plane polarization; polarization.

Clark cell A type of cell formerly used as a standard source of e.m.f. It consists of a mercury cathode coated with mercury sulfate, and a zinc anode. The electrolyte is zinc sulfate solution. The e.m.f. produced is 1.434 5 volts at 15°C. The Clark cell has been superseded as a standard by the Weston cadmium cell. The cell is named for the English engineer Josiah Clark (1822–98).

classical physics The part of physics that was developed before (and so does not include) quantum theory. For instance, Newton's laws of motion belong to classical mechanics; they are a special case of quantum mechanics in the case of a body having a small quantum mechanical wavelength.

cleavage The splitting of a crystal along planes of atoms, to form smooth surfaces.

clinical thermometer A thermometer designed for taking body ('blood') temperature; a mercury thermometer with a kink in the thread just above the bulb, and calibrated in the range 35–46°C (95–115°F). The mercury reaches a maximum value but does not then return to the bulb until shaken down past the kink.

closed circuit *See* circuit; electrical.

closed pipe A pipe with one end closed and the other end open. If air is blown into the pipe a stationary sound wave may be formed, with a node at the closed end and an antinode at the open end.

closed shell An ELECTRON SHELL in an atom or molecule that is completely filled with electrons. Closed shells of nucleons in nuclei also exist.

closed system (**isolated system**) A set of one or more objects that may interact with each other, but do not interact with the world outside the system. This means that there is no net force from outside or energy transfer. Because of this the system's angular momentum, energy, mass, and linear momentum remain constant.

close packing A type of packing of atoms in crystals, in which the atoms are regarded as spheres that have exactly the same size and fill space in the most efficient way. In a single layer each sphere is surrounded by six other spheres, arranged at the corners of a hexagon. A second layer fills as many of the hollows as possible. There are two ways in which a third layer can be placed. If it is directly above the first layer then there is *hexagonal close packing*, with the layer sequence being denoted ABABAB .... If the third layer is above the hollows of the first layer that have not been filled by the second layer, then there is *cubic close pack-*

QUANTITIES USED IN DESCRIBING CIRCULAR MOTION

Quantity	*Symbol*	*Formula*
angular displacement	θ	–
angular velocity	ω	$\omega = v/r$
centripetal acceleration	a	$a = v^2/r = \omega^2 r$
centripetal force	F	$F = mv^2/r = m\omega^2 r$
frequency	f	$f = \omega/2\pi$
period	T	$T = 1/f$

ing, with the layer sequence being denoted ABCABC....

cloud chamber A chamber used to show the tracks of ionizing radiation, especially alpha and beta particles.

A *diffusion cloud chamber* has felt strips near the top soaked in water and ethanol. The bottom of the chamber is held at a low temperature, and there is continuous diffusion of vapor down the chamber. At one particular level, water droplets condense only along the tracks of ionizing radiation. The *expansion cloud chamber* contains moist air, sometimes with ethanol vapor, which is cooled by a sudden adiabatic expansion, causing the air to become supersaturated with water vapor. Water droplets condense out preferentially on ions formed along the tracks.

It can be arranged that passage of an ionizing particle through the chamber is detected by a counter and the resulting pulse can be used to operate the pump, so that the expansion takes place just after the ion pairs have been formed. A camera may also be triggered to take a photograph of the tracks just as they become visible. Magnetic fields can be applied and the resulting curvature of the tracks provides information about the charge and energy of the particles.

See also bubble chamber.

coaxial /koh-**aks**-ee-ăl/ Denoting bodies or shapes that have a common axis of rotation or radial symmetry. *See also* cable; coaxial.

cobalt /**koh**-bawlt/ A lustrous silvery-blue hard ferromagnetic transition metal. It is used in alloys for magnets, cutting tools, and electrical heating elements and in catalysts and some paints.

Symbol: Co; m.p. 1495°C; b.p. 2870°C; r.d. 8.9 (20°C); p.n. 27; r.a.m. 58.93320.

Cockcroft–Walton accelerator /**kok**-roft **wawl**-tŏn/ The first particle accelerator (1932), used to achieve the first artificial nuclear disintegration by bombarding a lithium target with protons (to produce alpha particles). The device was a simple form of linear accelerator; potential differences up to 800 kV were achieved by a voltage multiplier device. The Cockcroft–Walton accelerator is named for the British physicist Sir John Cockcroft (1897–1967) and the Irish physicist Ernest Thomas Sinton Walton (1903–95).

coefficient of expansion *See* expansivity.

coefficient of friction *See* friction.

coercive force The magnetizing field intensity required to reduce the flux density in a material to zero. If the material is magnetically saturated the coercive force is a maximum; it is then referred to as the *coercivity*. *See also* hysteresis cycle.

coercivity /koh-er-**siv**-ă-tee/ *See* hysteresis cycle.

coherent Two sources of waves are said to be *coherent* if there is a constant relationship between the phases of the waves emitted by them. For most sources of light or similar electromagnetic radiations (other than radio waves) it is usually necessary to derive the radiations of the two sources from one original source, as in Young's interference experiment.

Radiation is said to be coherent if there is a regular phase relationship between widely different parts of a wavefront, or if such regularity is maintained for a considerable time. Radiation from a laser is exceptionally coherent.

coherent units A system or sub-set of UNITS (e.g. SI units) in which the derived units are obtained by multiplying or dividing together base units, with no numerical factor involved other than one.

cohesion *See* surface tension.

coincidence circuit A device that detects the occurrence of two events that either coincide or occur within a specified time interval of each other. Signals recording the two events are fed into the device, which produces an output only if the two impulses occur within the specified time. Such a device is the basis of a coincidence counter.

cold emission Emission of electrons from a solid by a process other than thermionic emission. The term is usually used to describe either field emission or secondary emission.

cold rolling The reduction in the thickness of a metal by applying pressure to it. This process usually strengthens the metal by introducing dislocations.

collector *See* transistor.

colligative properties /kŏ-**lig**-ă-tiv/ The properties of a solution that depend only on the number of solute particles (such as ions or molecules) present in the solution but not on the type of particles present. Examples of colligative properties include depression of the freezing point, elevation of the boiling point, reduction in vapor density, and osmotic pressure.

collimator /**kol**-ă-may-ter/ An arrangement for producing a parallel beam of radiation for use in a spectrometer or other instrument. A system of lenses and slits is utilized.

collision An interaction between objects in which they approach each other near enough to exert a mutual influence, although they do not necessarily come into actual contact. The objects may be subatomic particles, groups of particles, or rigid bodies and the interaction may involve an exchange of charge, energy, or momentum, which is usually conserved after the collision.

colloid /**kol**-oid/ A substance containing very small particles (sizes in the range 10^{-9}–10^{-5} m). Sols, gels, and emulsions are examples of colloids.

color A physiological sensation produced in the eye-brain system relating to the wavelength of visible radiation. Traditionally the visible spectrum is divided into seven color regions, although many people cannot distinguish the color 'indigo' between violet and blue. Individual eyes differ in their sensitivities to different wavelengths and the spectral colors also merge gradually into each other. In the retina of the eye, certain cells (cones) are responsible for color vision; they cannot operate at low light levels, which is why color vision is not so effective at night.

Monochromatic radiations of different wavelengths are called *hues*; the normal human eye can distinguish yellow hues only 1 nm apart. If a hue is mixed with white light, it is an 'unsaturated' hue, now called a *tint*. Different saturated hues mix together, either to produce the effect of other hues, or of nonspectral colors. The study of color mixing by addition of colored lights (hues) shows that any color sensation can be produced by mixing three primary hues. It is normal to choose a red, a green, and a blue for this purpose. Colored television pictures are produced on this basis.

While 'white' light is defined as the normal mixture of all visible radiation, the same sensation can be obtained by mixing complementary pairs of lights (hues).

The apparent colors of surfaces in white light depends on the wavelengths that are reflected. Thus yellow paint (pigment) reflects mainly yellow wavelengths (with perhaps some orange and green); the rest are absorbed. When two paints of different colors are mixed, the color reflected will be the sum of the wavelengths absorbed by neither pigment. This is 'color mixing by subtraction'. The three usual primary colors are the greenish-blue 'cyan' (white light with red absorbed); the crimson magenta (green absorbed), and yellow (blue absorbed).

color (in particle physics) A quality analogous to electric charge that characterizes the strong interactions of QUARKS. The word color in this context has nothing to do with the physiological sensation of color. However, some of the properties of color for strongly interacting particles are analogous to ordinary color. There are three 'primary' colors for quarks, denoted red, blue, and green.

color blindness *See* color vision.

colorimeter /kul-ŏ-**rim**-ĕ-ter/ An instrument able to give readings for any color in terms of the intensity of each of the three primary colors.

color vision The ability of the eye to distinguish different COLORS. The mechanism is not fully understood. It is thought that there may be three types of cone cell in the retina, able to react to red, green, and blue light. Nerve signals corresponding to the intensities of each of these give color impressions in the brain.

About 5% of males and 0.5% of females have inherited defects of color vision. In many cases, there is a reduced ability to distinguish colors, particularly red and green and especially in dim light. In more extreme cases there is very little discrimination in the whole range from red to green. The popular term *color blindness* is to be avoided as the lack of any color vision is extremely rare. A few individuals have exceptionally good color vision, especially in the blue and violet. These people can see a distinct color 'indigo'.

coma *See* aberration.

comet A small body that moves around the Sun with a highly eccentric orbit. The periods of the orbits of comets vary between about 150 years (*short-period comets*) and hundreds of thousands of years (*long-period comets*). A comet consists of a *nucleus* made of ice and dust surrounded by a *coma*, which consists of gas and dust. If the comet is near the Sun radiation pressure drives gas and dust away from the coma to form a 'tail', which always points away from the Sun.

common emitter *See* transistor.

communication satellite A man-made satellite that is put into a PARKING ORBIT and designed to allow television and telephone communication between different places.

commutator /**kom**-yŭ-tay-ter/ The part of a direct-current ELECTRIC MOTOR or generator that connects the coil(s) to the outside circuit, changing the connections round as the coil rotates.

compact disk (CD) A storage system in which data is recorded in the form of tiny pits or patches in a spiral track on the disk. The data is read by directing a low-power laser onto the track and detecting the reflected radiation. CDs are commonly used to store music, and also as computer storage devices. Some types are re-writable.

comparator /**kom**-pă-ray-ter, kŏm-**pa**-ră-/ An apparatus used in measurements of the expansivity of solids. A horizontal bar of the material is used with a fine scratch near each end. The bar is surrounded by a water bath and two vertical microscopes are used, focused on the scratches on each end. Changes of length, produced by changes in temperature, are measured by the microscopes; a standard bar at constant temperature is used for comparison.

compass A device used to show magnetic force field direction. It consists of a small magnet pivoted so that it is free to move in a horizontal plane. In the Earth's magnetic field a compass points along the magnetic meridian. A scale, graduated in degrees, may be placed underneath the magnet to assist in navigation. *See also* magnetometer.

complementary colors Two colored lights (hues) that mix to produce the sensation of white. There is an infinite number of pairs of complementary colors. The spectral hues can be plotted around a roughly triangular 'chromaticity curve' that has white at the center. The members of each pair of complementary colors lie at opposite sides of white on this curve. *See also* color.

complexity theory The study of highly complex physical systems. Key ingredients in complexity theory include chaos theory and the concepts of broken symmetry and self-organization. Computers are used extensively in analyzing such systems.

complex number A number z that can be written in the form $z = x + iy$, where x and y are real numbers and i is the square root of –1 (i.e. $i^2 = -1$), with x and iy being called the *real part* and the *imaginary part* respectively of the complex number. Complex numbers are used extensively in the physical sciences and in engineering.

component forces *See* component vectors.

component vectors The components of a given vector (such as a force or velocity) are two or more vectors with the same effect as the given vector. In other words the given vector is the resultant of the components. Any vector has an infinite number of sets of components. Some sets are more use than others in a given case, especially pairs at 90°.

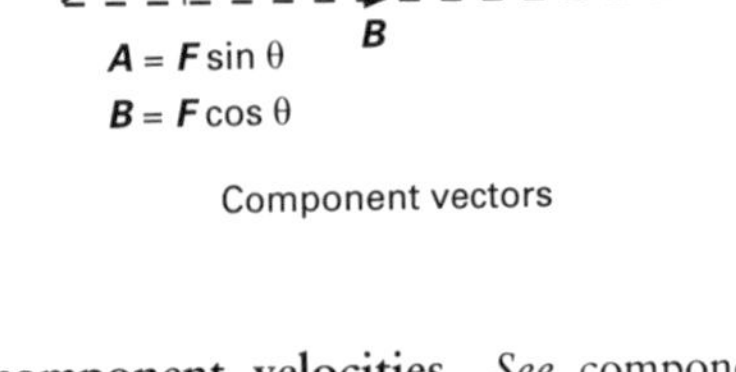

Component vectors

component velocities *See* component vectors.

composite material A material that is composed of two or more materials. For example, reinforced concrete is concrete that is set round steel wires. Other examples are various plastics reinforced with glass or carbon fiber. Composite materials are frequently produced so as to have better mechanical properties than single materials. For example, reinforced concrete improves the tensile properties of concrete.

compound lens Two or more lenses used together as a unit. For instance, the eyepiece of a telescope and the lens of a camera are both normally compound lenses; each has a number of elements (the single lenses) along the same optical axis. Insect compound eyes are not compound lenses in this sense – the elements are not on a single axis.

The elements of a compound lens may or may not touch. The function of compound (rather than single) lenses is usually to minimize aberrations.

See also doublet.

compound microscope *See* microscope, compound.

compound pendulum *See* pendulum.

compressibility /kŏm-pres-ă-**bil**-ă-tee/ The reciprocal of the bulk modulus of a material.

Compton effect /**komp**-tŏn/ An increase in the wavelength of x-rays or gamma rays when scattered by loosely bound electrons in substances, the electrons being ejected. Theoretically the phenomenon is treated as a collision between a photon and an electron, the latter being regarded as a free particle initially at rest. The photon transfers energy and momentum to the electron, hence the wavelength increases. The Compton effect is named for the American physicist Arthur Holly Compton (1892–1962). *See also* duality.

Compton wavelength The length-scale associated with a particle at which the description of the particle is dominated by relativistic quantum mechanics. It is defined by h/mc, where h is the Planck constant, m is the rest mass of the particle, and c is the speed of light in a vacuum. The Compton wavelength of an electron is 2.4263×10^{-12} m. It is named for Arthur Compton.

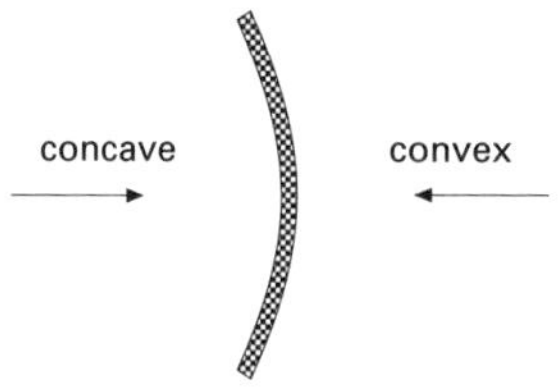

Concave and convex surfaces

concave /kong-**kayv**, **kong**-kayv/ Curving inward, away from the viewpoint. A *concave mirror* is one with a concave reflecting surface. A *concave lens* is either a biconcave or a plano-concave lens. *See also* convex; lens; mirror.

concavo-convex /kong-kay-voh-kon-**veks**/ Describes a LENS with one concave surface and one convex surface. Most spectacle lenses have this shape. Concavo-convex lenses can be converging or diverging. Lenses of this type are sometimes called *meniscus lenses*.

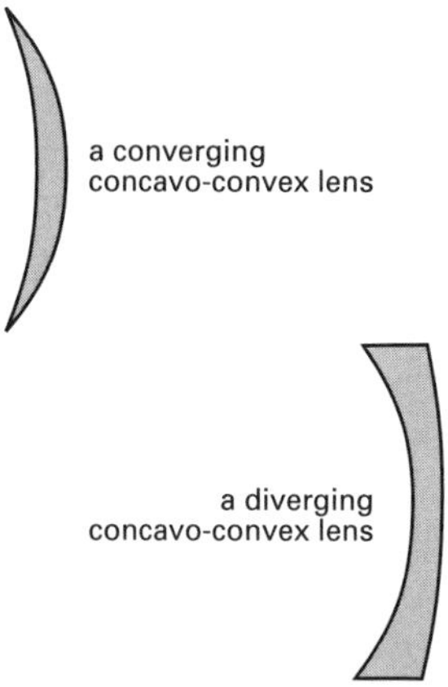

Types of concavo-convex lens

condensation /kon-den-**say**-shŏn/ A change from vapor to liquid – the reverse of evaporation. *See also* dew; relative humidity.

condensation pump *See* diffusion pump.

condenser 1. An early term for capacitor. 2. A lens, set of lenses, or mirror in an optical instrument used to concentrate light onto the object to be viewed. *See* microscope; compound; projectors.

condenser microphone A type of microphone in which the diaphragm is one plate of a charged capacitor ('condenser'). Sound waves move the diaphragm, thus altering the capacitance. The corresponding changes in potential difference generate the signal.

conductance /kŏn-**duk**-tăns/ Symbol G. The ratio I/V in an electrical conductor, where I is the current and V is the potential difference. The conductance is the reciprocal of the resistance. The SI unit is the siemens (symbol: S).

conduction /kŏn-**duk**-shŏn/ 1. A process of heat transfer through a substance without movement of the substance itself. The rate of transfer depends on the sample length and cross-sectional area, the temperature difference, and the nature of the material.

Good conductors, such as copper and silver, are often metals that have free electrons. The energy is transmitted by movement of these. In poor solid conductors (insulators) there are no conduction electrons and the transfer is only by vibrations of atoms or molecules in the crystal structure. *See also* conductivity; thermal.

2. Passage of charge through a sample under the influence of an electric field. The charge may be carried by electrons (e.g. in metals), by electrons and positive holes (in electrolytes and semiconductors), or by ions (in electrolytes and gases). *See also* energy bands.

conduction band *See* energy bands.

conductivity, electrical /kon-duk-**tiv**-ă-tee/ Symbol: σ The ability of a material to conduct electric current; the reciprocal of the resistivity. Conductivity does not depend on the dimensions of the conductor.

It is also the ratio of current density to electric field strength. The unit is the siemen per meter (S m^{-1}).

conductivity, thermal Symbol: λ, k A measure of the ability of a material to conduct energy. It is defined as the energy transfer per unit time per unit cross-sectional area of the conductor, per unit temperature gradient along the conductor. The unit of thermal conductivity is the watt per metre kelvin (W m^{-1} K^{-1}). For a parallel-sided block of material with no loss through the sides:

$$E/t = \lambda A(\theta_2 - \theta_1)/l$$

where E is the amount of energy transferred in a time t, A the cross-sectional area, l the length, and θ_2 and θ_1 the temperatures of the faces. In general:

$$\mathrm{d}E/\mathrm{d}t = -\lambda A\mathrm{d}\theta/\mathrm{d}t$$

See also Lees' disk; Searle's bar.

conductor, electrical All metals are good conductors of electricity, as is graphite. These are called electrical conductors. Most nonmetals are very poor conductors, and good insulators may have electrical conductivities of the order of 10^{-20} of those of metals at ordinary temperatures. The electrical conductivity of nonmetals usually increases rapidly with temperature so some become fairly good conductors at high temperatures. *See also* energy bands; semiconductor.

conductor, thermal A substance that has a relatively high thermal conductivity. In general, many metals are good conductors of energy – copper and silver are notable examples. Many non-metallic solids are poor conductors (i.e. they are *thermal insulators*). Gases are particularly poor conductors (or good insulators).

cones The cells in the retina of the EYE that are important for vision in normal light. They are also responsible for color vision. Their mechanism of action is not fully understood. *See also* color.

configuration /kŏn-fig-yŭ-**ray**-shŏn/ The arrangement of electrons in shells around the nucleus of an atom. The electron configuration largely determines the chemical behavior of an element as well as some physical properties, such as electrical conductivity. *See* electron shell.

conic /**kon**-ik/ A type of plane curve defined so that for all points on the curve the distance from a fixed point (the *focus*) has a constant ratio to the perpendicular distance from a fixed straight line (the *directrix*). The ratio is the eccentricity of the conic, e; i.e. the eccentricity is the distance from curve to focus divided by distance from curve to directrix.

The type of conic depends on the value of e: when e is less than 1 it is an ellipse; when e equals 1 it is parabola; when e is greater than 1 it is hyperbola. A circle is a special case of an ellipse with eccentricity 0.

The original definition of conics was as plane sections of a conical surface – hence the name *conic section*. In a conical surface having an apex angle of 2θ, the cross-section on a plane that makes an angle θ with the axis of the cone, (i.e. a plane parallel to the slanting edge of the cone) is a parabola. A cross-section in a plane that makes an angle greater than θ with the axis is an ellipse. A cross-section in a plane making an angle less than θ with the axis is a hyperbola and because this plane cuts both halves (nappes) of the cone, the hyperbola has two arms.

There are various ways of writing the equation of a conic. In Cartesian coordinates:

$$(1 - e^2)x^2 + 2e^2qx + y^2 = e^2q$$

where the focus is at the origin and the directrix is the line $x = q$ (a line parallel to the y-axis a distance q from the origin). The general equation of a conic (i.e. the *general conic*) is:

$$ax^2 + bxy + cy^2 + dx + ey + f = 0$$

where a, b, c, d, e, and f are constants (here e is *not* the eccentricity). This includes degenerate cases (*degenerate conics*), such as a point, a straight line, and a pair of intersecting straight lines. A point, for example, is a section through the vertex of the conical surface. A pair of intersecting straight lines is a section down the axis of the surface.

See also ellipse; hyperbola; parabola.

conjugate points Any pair of points relative to a lens, reflector, or other optical system such that either is imaged at the other. Compare the symmetry in object distance u and image distance v in the equation:

$$1/v + 1/u = 1/f$$

conservation law A physical law that some property of a system remains constant throughout a series of changes. Examples are the laws of conservation of mass, energy, momentum, and charge.

conservation of angular momentum, law of *See* constant angular momentum; law of.

conservation of charge The principle that the total net charge of any closed system is constant.

conservation of energy, law of *See* constant energy; law of.

conservation of linear momentum, law of *See* constant linear momentum, law of.

conservation of mass, law of *See* constant mass; law of.

conservation of mass and energy The law that the total energy (rest mass energy + kinetic energy + potential energy) of a closed system is constant. In most chemical and physical interactions the mass change is undetectably small, so that the measurable rest-mass energy does not change (it is regarded as 'passive'). The law then becomes the classical *law of conservation of energy*; the inclusion of mass in the calculation is necessary only in the case of nuclear changes or systems involving very high velocities. *See also* mass-energy equation; rest mass.

conservative field A field of force such that the work done on or by a body that is displaced in the field is independent of the path. Hence if a body is moved in a closed path (back to the starting point) the net work done is zero. Gravity and an electrostatic field in a vacuum are conservative. Any form of friction prevents the field from being conservative, although the total energy of any isolated system is conserved in all cases.

constant A quantity or factor that remains at the same value when others are changing. *See also* fundamental constants.

constantan /**kon**-stăn-tan/ An alloy of copper (50–60%) and nickel used in thermocouples and precision resistors. Its electrical resistance varies very little with changes in temperature.

constant angular momentum, law of (**law of conservation of angular momentum**) The total angular momentum of a system cannot change unless a net outside torque acts. *See also* constant (linear) momentum, law of.

constant energy, law of (**law of conservation of energy**) The total energy of a system cannot change unless energy is taken from or given to the outside. The law is equivalent to the first law of thermodynamics. *See also* mass-energy equation.

constant linear momentum, law of (**law of conservation of linear momentum**) The total *linear momentum* of a system cannot change unless a net outside force acts.

constant mass, law of (**law of conservation of mass**) The total mass of a system cannot change unless mass is taken from or given to the outside. *See also* mass-energy equation.

constant momentum, law of *See* constant (linear) momentum, law of.

constant pressure gas thermometer *See* gas thermometer.

constant volume gas thermometer *See* gas thermometer.

constructive interference *See* interference.

contact angle An angle formed between solid and liquid surfaces. For example, when a drop of mercury rests (in air) on top of a horizontal glass surface, the contact angle measured inside the drop from the glass to the mercury-air interface, is greater than 90°. This is because the mercury does not wet the glass – its cohesion is greater than the adhesion to glass. Water does wet the surface of clean glass and the contact angle is less than 90°. *See also* capillary action; meniscus.

contact potential A potential difference across the space between the free surfaces of two solids that are in contact.

containment /kŏn-**tayn**-mĕnt/ **1.** The process of containing radioactive material within a shield for safety reasons.
2. The process of containing the plasma in a fusion reactor so that it does not come in contact with the walls of the vessel. Magnetic fields are used.

continuous spectrum A SPECTRUM composed of a continuous range of emitted or absorbed radiation. Continuous spectra are produced in the infrared and visible regions by hot solids. *See* black body.

continuous wave An electromagnetic wave that is transmitted continuously over a period of time, as in radio communication. It is distinguished from a pulsed or intermittent wave as used in radar. *See also* radar.

control grid *See* grid.

control rods Rods of neutron-absorbing material that can be moved into or out of the core of a nuclear reactor to control the chain reaction. Control rods are made of boron, cadmium, or other substance that absorbs neutrons.

convection /kŏn-**vek**-shŏn/ The transfer of energy by flow of a liquid or gas. In *natural convection* the fluid flow is caused by temperature differences between one part of the fluid and another. For example, in an electric kettle the heating element raises the temperature of the water next to it, which expands and rises. Colder water then flows in beneath it, setting up a convection current. In *forced convection*, energy is carried away from the source by flow produced by a pump or fan.

In natural convection the rate of loss of energy from a body to the surroundings is proportional to $\theta^{5/4}$ where θ is the excess temperature over the surroundings. In forced convection it is proportional to θ. *See also* Newton's law of cooling.

conventional current An electric current considered to flow from points at positive potential to points at negative potential. In typical conductors, charge is carried by electrons, which flow in the opposite direction. In semiconductors, the charge may be carried by positive holes, flowing in the same direction as the conventional current.

convergent Coming together. A convergent beam becomes narrower as it travels. Its narrowest point is called the *focus*; after passing through the focus, the beam will be DIVERGENT (moving apart).

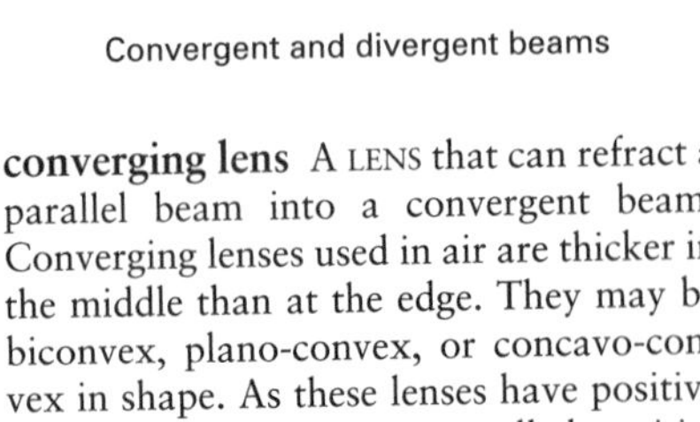

Convergent and divergent beams

converging lens A LENS that can refract a parallel beam into a convergent beam. Converging lenses used in air are thicker in the middle than at the edge. They may be biconvex, plano-convex, or concavo-convex in shape. As these lenses have positive power, they are sometimes called *positive lenses*. *Compare* diverging lens.

converging mirror (**converging reflector**)

A MIRROR that can reflect a parallel beam into a convergent beam. Converging reflectors always have concave surfaces. The section shape is the arc of a circle in simple cases; the arc of a parabola is needed for more precise work. As these mirrors have positive power, they are sometimes called *positive mirrors*. *Compare* diverging mirror.

converter A device for converting alternating current to direct current or vice versa.

converter reactor *See* nuclear reactor.

convex /kon-**veks**, **kon**-veks/ A *convex mirror* is one with a convex reflecting surface. A *convex lens* is either a biconvex LENS or a plano-convex lens. *See also* concave; mirror.

coolant /**koo**-lănt/ A fluid that removes energy from a source (e.g. a hot motor) by convection.

cooling correction A method of correcting for errors due to energy loss in experiments that involve temperature changes. In the method of mixtures, for instance, the final temperature reached by the mixture may be slightly less than the theoretical one as a result of cooling during the change. A graph of temperature against time throughout the course of the experiment (i.e. a *cooling curve*) allows a correction to be made based on Newton's law of cooling.

coordinates Numbers that define the position of a point, or set of points. *See also* Cartesian coordinates; polar coordinates.

coplanar forces Forces in a single plane. If only two forces act through a point, they must be coplanar. So too are two parallel forces. However, nonparallel forces that do not act through a point cannot be coplanar. Three or more nonparallel forces acting through a point may not be coplanar.

copper /**kopp**-er/ A transition metal; copper is used in electrical wires and in such alloys as brass and bronze. The metal itself is golden-red in color.

Symbol: Cu; m.p. 1083.5°C; b.p. 2567°C; r.d. 8.96 (20°C); p.n. 29; r.a.m. 63.546.

core **1.** A piece of iron or other magnetic material used to concentrate the magnetic lines of force in a transformer, electromagnet, or similar device.
2. The central part of a nuclear reactor in which the chain reaction occurs.

Coriolis force /kor-ee-**oh**-lis, ko-ree-/ (**Coriolis effect**) A 'fictitious' force used to describe the motion of an object in a rotating system. For instance, air moving from north to south over the surface of the Earth would, to an observer outside the Earth, be moving in a straight line. To an observer on the Earth the path would appear to be curved, as the Earth rotates. Such systems can be described by introducing a tangential Coriolis 'force'. The force is named for the French physicist Gustave-Gaspard Coriolis (1792–1843).

corkscrew rule *See* Maxwell corkscrew rule.

cornea /**kor**-nee-ă/ (*pl.* **corneas** or **corneae**) The transparent part of the EYE.

corona /kŏ-**roh**-nă/ The outer section of the atmosphere of the Sun. The inner corona (*K-corona*) has a temperature of about 2×10^{-6}K and extends to a height of about 75 000 km. The outer corona (*F-corona*) extends for millions of kilometers, and is much cooler.

corona discharge Small regions of glowing air around a charged conductor at a critical potential. It is often accompanied by hissing sounds and is caused by the ionization of the air around the conductor. Corona discharges are produced at sharp points, where the surface charge density and electric field, are high.

corpuscular theory /kor-**pus**-kyŭ-ler/ Until the early nineteenth century it was most usually believed that light consists of particles ('corpuscles'). Rectilinear propagation, reflection, and refraction could be explained by assuming Newton's laws of motion. Phenomena such as diffraction and Newton's rings, which are now believed to require wave theory, were too little understood to 'discredit' the corpuscular theory. The theory was abandoned following the thorough investigation of interference, diffraction and polarization, which demand a wave theory. Also Foucault (1850) showed that the speed of light in water is less than in air, in agreement with wave theory but not with corpuscular theory. The concept of the photon is sometimes regarded as a modern form of corpuscular theory. *See also* duality; wave theory of light.

cosmic radiation High-energy radiation reaching the Earth. The primary cosmic radiation mostly enters the solar system from remote regions in space, but part may be derived from the Sun. It consists mainly of atomic nuclei (particularly protons, some alpha particles, and small numbers of heavier nuclei) and there are some electrons and gamma rays. The nuclei that reach the atmosphere of the Earth have energies from 10^{10}eV up to 10^{19}eV. These particles collide with atomic nuclei in the upper atmosphere and give rise to the secondary radiation. This includes lower energy protons and neutrons, and many unstable particles, particularly pions.

Most of the secondary radiations either interact or decay in the atmosphere. The charged pions decay to give muons, which are very penetrating so that many reach sea-level and even penetrate deeply into the ground. At all levels the penetrating radiations are accompanied by a (relatively) less penetrating soft component. This consists of electrons, positrons, and gamma radiation.

cosmic string A hypothetical entity; a long one-dimensional defect in the fabric of space-time with an intense gravitational field. They have been postulated by cosmologists to explain various phenomena in cosmology but there is no evidence for their existence.

cosmological constant Symbol: λ. A constant that can be added to Einstein's field equations for general relativity theory, originally introduced by Einstein to account for a possible static Universe. In an expanding Universe the cosmological constant is not necessary, but it is believed that this constant may have a small nonzero value, with the physical consequence that the expansion of the Universe is accelerating. It is not known how to calculate the value of λ, although many attempts have been made.

cosmology /koz-**mol**-ŏ-jee/ The study of the Universe as a whole. Cosmology includes such questions as the evolution of the Universe and the formation of large-scale structures such as galaxies and clusters of galaxies. Cosmology is mostly dictated by general relativity theory but consideration of the EARLY UNIVERSE requires contributions from nuclear physics and the theory of elementary particles.

coudé telescope /koo-**day**/ A type of reflecting telescope. Light from the concave mirror is reflected back onto a convex mirror, then onto a plane mirror at an angle to the axis, and into the eyepiece. *See also* reflector.

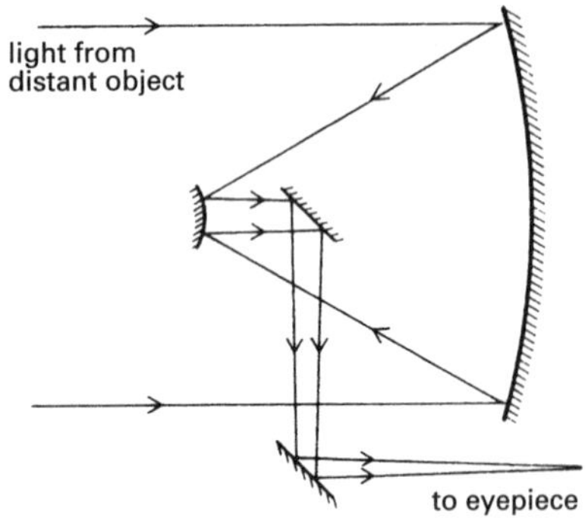

coudé telescope

coulomb /koo-**lom**/ Symbol: C The SI unit of electric charge, equal to the charge transported by an electric current of one ampere flowing for one second. 1 C = 1 A s. It is named for the French physicist Augustin de Coulomb (1736–1806).

coulombmeter /**koo**-lom-mee-ter/ *See* voltameter.

Coulomb's law The electric force between two charges is inversely proportional to the square of their distance apart. The full law giving the mutual force between two charges Q_1 and Q_2 is

$$F = Q_1Q_2/4\varepsilon_r\varepsilon_0 r^2$$

where r is the distance apart and $\varepsilon_r\varepsilon_0$ the permittivity of the medium between the charges.

counter A device for detecting and counting particles and photons. *See* Geiger counter; proportional counter; scaler; scintillation counter; semiconductor counter; spark.

couple A pair of equal parallel forces in opposite directions and not acting through a single point. Their linear resultant is zero, but there is a net turning-effect (moment). The resultant turning effect T (torque) is given by

$$T = d_1F_1 + d_2F_2$$
$$T = (d_1 - d_2)F$$
$$T = dF$$

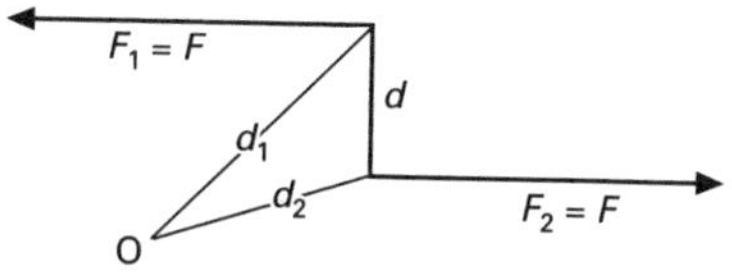

Couple

creep A slow deformation of a solid as a result of a continuous applied stress. *See also* slip.

critical angle *See* total internal reflection.

critical damping *See* damping.

critical density *See* critical state.

critical mass The minimum mass of material that can sustain a nuclear CHAIN REACTION. An explosive chain reaction (as in a nuclear fission weapon) results when subcritical masses are brought together to form a mass greater than the critical mass. In a nuclear reactor, the chain reaction is controlled by a moderator.

The critical mass depends on the purity of the material used, the geometry of the arrangement, and the presence or absence of reflectors of neutrons around the fissile material.

critical point *See* critical state.

critical pressure *See* critical state.

critical reaction A nuclear chain reaction initiated by a neutron in which, on average, one neutron is released from each nuclear reaction, thus enabling the chain reaction to be sustained. If less than one neutron is released on average then the reaction is said to be *subcritical* and the chain reaction fizzles out. If more than one neutron is released on average then the reaction is said to be *supercritical* and the chain reaction proliferates very rapidly, resulting in a nuclear explosion.

critical specific volume *See* critical state.

critical speed In fluid flow, the speed at which the behavior of the fluid switches from that of laminar flow to that of turbulent flow or vice versa. *See also* Reynolds number.

critical state (**critical point**) The state of a fluid under conditions in which the liquid and gas phases have the same density. The temperature, pressure, and density concerned are the *critical temperature*, *critical pressure*, and *critical density* respectively. The specific volume is the *critical specific volume*. Above its critical temperature a gas cannot be liquefied by increasing the pressure. The critical point corresponds to a point of inflection on an isothermal for the gas.

See also Andrews' experiments.

critical temperature *See* critical state.

critical volume *See* critical state.

CRO /see-ar-**oh**/ *See* cathode-ray oscilloscope.

cross produce *See* vector product.

cross-section In a collision process – for example, the bombardment of nuclei by neutrons – the apparent area that particles present to the bombarding particles. This is not its 'true' cross-sectional area, but depends on the probability of a reaction occurring. In particular, it varies with the energy of the incident particles. The measurement is used for other types of collision reactions besides neutron absorption, including reactions of atoms, ions, molecules, electrons, etc. The unit is the square meter (m^2).

crown glass *See* optical glass.

CRT /see-ar-**tee**/ *See* cathode-ray tube.

cryogenics /krÿ-ŏ-**jen**-iks/ The physics of very low temperatures, including techniques for producing low temperatures and the study of how materials behave when cooled.

cryometer /krÿ-**om**-ĕ-ter/ A thermometer designed to measure very low temperatures.

cryoscopy /krÿ-**os**-kŏ-pee/ The determination of freezing points.

cryostat /**krÿ**-ŏ-stat/ A container for keeping samples at a constant low temperature.

crystal A solid with a regular geometric shape. In crystals the particles (atoms, ions, or molecules) have a regular three-dimensional repeating arrangement in space. This is called the *crystal structure*. The crystal *lattice* is the arrangement of points in space at which the particles are positioned. The external appearance of the crystal is the *crystal habit*. The external shape of the crystal depends on the internal arrangement of the lattice. It also depends on the way in which the crystal is formed. *See also* crystalline.

crystal habit *See* crystal.

crystal lattice *See* crystal.

crystalline /**kris**-tă-lin, -lÿn/ Describing a solid that has a crystal structure; i.e. a regular internal arrangement of atoms, ions, or molecules. Note that a substance can be crystalline without necessarily having a geometrical external shape.

crystallography /kris-tă-**log**-ră-fee/ The study of the formation, structure, and properties of crystals. *See also* x-ray crystallography.

crystalloid /**kris**-tă-loid/ A substance that is not a COLLOID. It can pass through a semipermeable membrane. *See* colloid.

crystal microphone *See* microphone.

crystal oscillator A fixed-frequency OSCILLATOR in which the output is derived from a crystal of Rochelle salt or quartz specially cut so that its natural frequency of mechanical vibration is the desired value. This is usually in the range of about one kilohertz to several megahertz.

The action of crystal oscillators is based on the piezoelectric effect. A potential difference is applied to two electrodes attached to two parallel cut faces of the crystal. The electric field sets up a mechanical stress which causes the crystal to resonate at its natural frequency. This in turn generates an alternating potential difference across the crystal.

crystal rectifier A semiconductor diode. *See* diode; rectifier; semiconductor.

crystal structure *See* crystal.

crystal system A category of crystal classification based on the symmetry of its external form, which in turn reflects its internal structure. There are seven main crystal systems: cubic, hexagonal, monoclinic, orthorhombic, tetragonal, triclinic, and trigonal.

cubic expansivity *See* expansivity.

curie /**kyoo**-ree, kyoo-**ree**/ Symbol: Ci A unit of radioactivity, equivalent to the amount of disintegrations per second from the decay of one gram of radium. It is equal to 37 gigabecquerels. It is named for the French physicist Pierre Curie (1859–1906).

Curie point *See* Curie temperature.

Curie's law The susceptibility, χ, of some paramagnetic substances is inversely proportional to the thermodynamic temperature, T:

$$\chi = c/T$$

where c is a constant. More generally, the relationship is that of the *Curie–Weiss law*:

$$\chi = c/(T - \theta)$$

where θ is the *Weiss constant*, a fixed temperature characteristic of the material. The constant is named for the French physicist Pierre Weiss (1865–1940).

Curie temperature (**Curie point**) The temperature above which a ferromagnetic substance loses its ferromagnetism. It then shows paramagnetism only; the alignment of magnetic moments in domains is overcome by thermal vibrations. The Curie temperature for iron is 760°C; that for nickel is 356°C. Metals that are paramagnetic at room temperatures could become ferromagnetic if cooled. For example, the Curie temperature of dysprosium is –188°C.

Curie–Weiss law /vÿss/ *See* Curie's law.

curium /**kyoo**-ree-ŭm/ A highly toxic radioactive silvery element of the actinoid series of metals. A transuranic element, it is not found naturally on Earth but is synthesized from plutonium. Curium-244 and curium-242 have been used in thermoelectric power generators.

Symbol: Cm; m.p. 1340±40°C; b.p. ≅ 3550°C; r.d. 13.3 (20°C); p.n. 96; most stable isotope ^{247}Cm (half-life 1.56×10^7 years).

curl *See* vector calculus.

current Symbol: I A flow of electric charge. Current is the result of motion of electrons or ions under the influence of an e.m.f. It is measured in amperes (A).

current balance (**ampere balance**) A device for measuring electric current to a high degree of accuracy, by measuring the electromagnetic force produced between two conductors carrying the current.

current density Symbol: j The electric current per unit area of conductor cross-section. It is measured in amperes per square meter (A m^{-2}). In a conductor it is equal to nvQ, where n is the number of free charges per unit volume, v is their average drift velocity, and Q is their charge.

cybernetics /sÿ-ber-**net**-iks/ The science that deals with communication and control, particularly with regard to machines. It includes details about the input and output of information, FEEDBACK, and so on.

cycle A series of events that is regularly repeated (e.g. a single orbit, rotation, vibration, oscillation, or wave). A cycle is a complete single set of changes, starting from one point and returning to the same point in the same way.

cyclic accelerator /**sÿ**-klik, **sik**-lik/ *See* accelerator.

cyclotron /**sÿ**-klŏ-tron/ A device for accelerating charged particles to high energies. Protons can be given energies of over 10 MeV, deuterons over 20 MeV, and alpha particles about 40 MeV. Particles are injected near the center of an evacuated space between two D-shaped boxes placed between the poles of a strong permanent magnet. Within each 'dee' the particles describe a semicircular orbit. An alternating voltage is applied to the dees at such a frequency that the particles are accelerated by the potential difference each time they reach the gap, causing them to increase their path radii and speeds in steps. Eventually, after several thousand revolutions, they reach the perimeter of the dees at high speed, where a deflecting electric field directs them onto the target.

The energies that can be achieved are limited by the relativistic increase in mass of the particles. As their velocity approaches that of light, the increase in mass causes the period of rotation to increase so that they no longer reach the gaps in phase with the potential difference. *See also* synchrocyclotron.

cylindrical polar coordinates A method of defining the position of a point in space by its horizontal radius r from a fixed vertical axis, the angular direction θ of the radius from an axis, and the height z above a fixed horizontal reference plane. Starting at the origin O of the coordinate system, the point $P(r,\theta,z)$ is reached by moving out along a fixed horizontal axis to a distance r, following the circumference of the horizontal circle radius r centred at O through an angle θ, and then moving vertically upward by a distance z. For a point $P(r,\theta,z)$, the corresponding rectangular Cartesian coordinates (x,y,z) are:

$$x = r\cos\theta$$
$$y = r\sin\theta$$
$$z = z$$

Compare spherical polar coordinates.

D

Dalton's law (of partial pressures) The pressure of a mixture of gases is the sum of the partial pressures of each individual constituent. (The *partial pressure* of a gas in a mixture is the pressure that it would exert if it alone were present in the container.) Dalton's law is strictly true only for ideal gases. The law is named for the English scientist John Dalton (1766–1844).

damped oscillation An oscillation with an amplitude that progressively decreases with time. *See* damping.

damping The reduction in amplitude of a vibration with time by some form of resistance. A swinging pendulum will at last come to rest; a plucked string will not vibrate for long – in both cases internal and/or external resistive forces progressively reduce the amplitude and bring the system to equilibrium.

In most cases the damping force(s) will be proportional to the object's speed. In any event, energy must be transferred from the vibrating system to overcome the resistance.

Where damping is an asset (as in bringing a meter pointer to rest), the optimum situation occurs when the motion comes to zero in the shortest time possible, without vibration. This is *critical damping.* If the resistive force is such that the time taken is longer than this, *overdamping* occurs, without vibrations. Conversely, *underdamping* involves a longer time with several vibrations of decreasing amplitude.

Daniell cell /**dan**-yĕl/ A type of primary cell consisting of two electrodes in different electrolytes separated by a porous partition. The positive electrode is copper immersed in copper(II) sulfate solution. The negative electrode is zinc-mercury amalgam in either dilute sulfuric acid or zinc sulfate solution. With sulfuric acid the e.m.f. is about 1.08 volts; with zinc sulfate it is about 1.11 volts. The porous partition allows ions to move through while the reaction is taking place, and the cell must be dismantled when not in use.

At the copper electrode copper ions in solution gain electrons from the metal and are deposited as copper atoms:

$$Cu^{2+} + 2e^- \rightarrow Cu$$

The copper electrode thus gains a positive charge. At the zinc electrode, zinc atoms from the electrode lose electrons and dissolve into solution as zinc ions, leaving a net negative charge on the electrode:

$$Zn \rightarrow 2e^- + Zn^{2+}$$

The cell is named for the British chemist John Frederic Daniell (1790–1845).

dark energy The energy associated with the acceleration of the expansion of the Universe. About three-quarters of the energy of the Universe exists in the form of dark energy. The existence of dark energy can be taken into account using a nonzero value for the cosmological constant but the nature of this energy is not understood at the present time.

dark matter Invisible matter in the Universe that has been postulated in order to explain observations that cannot be accounted for in terms of detectable matter. The nature of dark matter is not known at the present time but it has been suggested that it consists of a mixture of neutrinos with nonzero masses (*hot dark matter*) and particles postulated to exist in SUPERSYMMETRY (*cold dark matter*).

dating Methods of determining the age of

archaeological or geological specimens, the Earth, meteorites, etc. The main methods of special interest in physics involve measurements of radioactivity. It is assumed that some radioactive nuclide is disappearing by decay and a product is formed. Measurements of the amount of parent nuclide, or product nuclide, or the ratio parent to product can give an estimate of the age in certain circumstances. *See* carbon dating; fission-track dating; potassium-argon dating; rubidium-strontium dating; thermoluminescent dating; uranium-lead dating.

daughter A given nuclide produced by radioactive decay from another nuclide (the *parent*).

day The time taken for the Earth to make one complete rotation on its axis. There are various methods of measuring this. The *solar day* is the time between two successive transits of the Sun across the meridian. Its value is not constant owing to the Earth's elliptical orbit. The *mean solar day* is the average of these values over one year. Its value is 86.4 kiloseconds. The *sidereal day* is measured with reference to a star. Its value is 86.164 09 kiloseconds.

d.c. /dee-**see**/ *See* direct current.

deadbeat Describing an instrument (e.g. a galvanometer) that is damped so that its oscillations die away very quickly. *See* damping.

dead room *See* anechoic chamber.

de Broglie wave /dĕ-**broh**-lyee/ *See* duality.

debye /dĕ-**bay**-ĕ/ Symbol: D A unit of electric dipole moment equal to $3.335\ 64 \times 10^{-30}$ coulomb meter. It is used in expressing the dipole moments of molecules. The unit is named for the Dutch-born American physicist and chemist Peter Debye (1884–1966).

Debye model A model for the specific heat of solids that was proposed by Peter Debye in 1912. Debye was able to give a good description of the specific heat of solids, particularly at low temperatures, in terms of the lattice vibrations of a solid having a distribution of frequencies up to a maximum cut-off frequency.

deca- Symbol: da A prefix denoting 10. For example, 1 decameter (dam) = 10 meters (m).

decay The spontaneous disintegration of unstable (radioactive) nuclei to give other nuclei, accompanied by the emission of particles and/or photons. The product of a given decay may be stable or may itself be radioactive. In a sample of radioactive material the radioactivity falls with time; the activity is proportional to the amount of substance present.

Decay follows an exponential law. At any time t the number (N) of original nuclei present is given by:

$$N = N_0 \exp -\gamma t$$

where N_0 is the original number and γ is a constant (the *decay constant* or *disintegration constant*) for a given nuclide. γ is related to the half-life by:

$$T = 0.693\ 15\ \gamma$$

decay constant *See* decay.

deci- Symbol: d A prefix denoting 10^{-1}. For example, 1 decimeter (dm) = 10^{-1} meter (m).

decibel /**des**-ă-bel/ Symbol: dB A unit of relative difference in power or intensity level, usually of a sound wave or electrical signal, measured on a logarithmic scale. The threshold of hearing is taken as 0 dB in sound measurement. Ten times this power level is 10 dB. The fundamental unit is the *bel*, but the decibel is almost exclusively used (1 db = 0.1 bel).

A power P has a power level in decibels given by:

$$10 \log_{10}(P/P_0)$$

where P_0 is the reference power.

declination Symbol: ε The angle between the geographic and magnetic meridians at a given point on the Earth's surface. Allowance must be made for this when using a compass to find geographical direction. At any point on the Earth's surface the declination changes slowly with time.

See also Earth's magnetism; magnetic variation.

defect An irregularity in the ordered arrangement of particles in a crystal lattice. There are two main types of defect in crystals. *Point defects* occur at single lattice points. A *vacancy* is a missing atom; i.e. a vacant lattice point. Vacancies are sometimes called *Schottky defects* (named for the Swiss–German physicist Walter Schottky (1886–1976)). An *interstitial* is an atom that is in a position that is not a normal lattice point. If an atom moves off its lattice point to an interstitial position the result (vacancy plus interstitial) is a *Frenkel defect* (named for the Russian mathematical physicist Jacov Ill'ich Frenkel (1894–1952)). All solids above absolute zero have a number of point defects, the concentration of defects depending on the temperature. Point defects can also be produced by strain or by irradiation. *Dislocations* (or *line defects*) are also produced by strain in solids. They are irregularities extending over a number of lattice points along a line of atoms.

definition The sharpness of focus of an image. No image can be perfectly sharp because of diffraction, even with a perfect optical system. *See also* resolving power.

deflection magnetometer *See* magnetometer.

degaussing /dee-**gow**-sing/ The neutralization of an object's magnetic field by the use of an equal and opposite field. It is used in color television sets to neutralize the Earth's magnetic field – current-carrying coils prevent the formation of color fringes on the image. A similar system is used to neutralize the magnetization of ships to prevent them from detonating magnetic mines.

degeneracy pressure The pressure that occurs in systems with degenerate fermions because of a combination of the Heisenberg uncertainty principle and the Pauli exclusion principle. It is degeneracy pressure that prevents the gravitational collapse of white dwarf stars or neutron stars to form black holes. The degenerate fermions are electrons in the case of a white dwarf and neutrons in the case of a neutron star.

degenerate Describing different quantum states that have the same energy. For instance, the three p orbitals in an atom all have the same energy but different values of the magnetic quantum number m. Differences in energy occur if a magnetic field is applied – the degeneracy is then said to be 'lifted'. The degeneracy of a system is closely related to the symmetry of that system.

degradation The irreversible loss of energy available to do work, with a consequent increase in entropy (in an isolated system). For example, consecutive collisions can lead to loss of energy of motion.

degree **1.** An interval on a scale of measurement, such as a temperature scale.
2. A unit of plane angle; 1/360 of a complete turn.

degrees of freedom The independent ways in which particles can take up energy. In a monatomic gas, such as helium or argon, the atoms have three translational degrees of freedom (corresponding to motion in three mutually perpendicular directions). The mean energy per atom for each degree of freedom is $kT/2$, where k is the Boltzmann constant; the mean energy per atom is thus $3kT/2$.

A diatomic gas also has two rotational degrees of freedom (about two axes perpendicular to the bond) and one vibrational degree. The rotations also each contribute $kT/2$ to the average energy. The vibration contributes kT ($kT/2$ for kinetic energy and $kT/2$ for potential energy). Thus, the average energy per molecule for a diatomic molecule is $3kT/2$ (translation) + $2kT/2$ (rotation) + kT (vibration) = $7kT/2$.

Linear triatomic molecules also have two significant rotational degrees of freedom; non-linear molecules have three. For non-linear polyatomic molecules, the number of vibrational degrees of freedom is $3N - 6$, where N is the number of atoms in the molecule.

The molar energy of a gas is the average energy per molecule multiplied by the Avogadro constant. For a monatomic gas it is $3RT/2$, etc. *See also* phase space; specific thermal capacity.

demagnetization /dee-mag-nĕ-ti-**zay**-shŏn/ The removal of the magnetic field of a magnet. In demagnetization, the domains are disordered so that they lie in a random pattern. This can occur as a result of rough treatment, or by raising the tempeature of the specimen and allowing it to cool while lying in an East-West direction. The most efficient method is to place the specimen in a coil through which is passed an alternating current. The current is then reduced to zero, removing the magnetism. A similar effect is obtained by taking the specimen out of a coil carrying an alternating current.

demodulation /dee-moj-ŭ-**lay**-shŏn/ (**detection**) The process in which the original modulating signal is reproduced in its original form from a modulated carrier wave by a radio circuit. *See also* detector; modulation.

dendrite /**den**-drÿt/ A branching crystal with a tree-like structure.

densitometer /den-să-**tom**-ĕ-ter/ An instrument for measuring the density of exposure of a photographic emulsion. Densitometers have a light source and photocell to measure the transmission (or reflection) of the sample. They are used in obtaining quantitative measurements from photographic records of spectra, diffraction experiments, etc.

density **1.** Symbol: ρ The mass of a substance per unit volume. This is also known as *mass density*. The units are kg m^{-3}.
2. Symbol: ρ_L The mass of a material per unit length. It is known as *linear density*.
3. Symbol: ρ_A The mass of a material per unit area. It is known as *area density*.
4. *See* relative density.
5. *See* charge density.

depletion layer A region near the p–n junction of a semiconductor diode in which the density of electric charge carriers is lower than in the bulk of the p and n parts of the semiconductor. A depletion layer is about 10^{-3} mm thick.

depolarizer A substance used in a voltaic cell to prevent polarization. Hydrogen bubbles forming on the electrode can be removed by an oxidizing agent, such as manganese(IV) oxide. *See also* polarization (electrolytic).

depression of freezing point The reduction in the freezing point of a liquid (solvent) when a substance (solute) is dissolved in it. The depression of freezing point is proportional to the concentration of the solute, for dilute solutions and non-volatile solvents at constant pressure.

depth of field (**depth of focus**) The object distance range of an optical system (e.g. a camera) that gives acceptably sharp images. The smaller the aperture, the greater the depth of field.

depth of focus *See* depth of field.

derivative *See* differentiation.

derived unit A unit defined in terms of BASE UNITS, and not directly from a standard value of the quantity it measures. For example, the newton is a unit of force defined as a kilogram meter second^{-2} (kg m s^{-2}). *See also* SI units

desorption /di-**sorp**-shŏn, **-zorp-**/ The reverse of adsorption. *See* adsorption.

destructive interference *See* interference.

detector **1.** (demodulator) An electronic circuit for extracting the original modulat-

ing signal from a carrier wave. *See also* modulation.
2. A device that responds to any physical effect, used to indicate the presence of a signal or to measure it. An example is an antenna inducing a voltage at the frequency of a received radio wave.

determinism The idea that knowing the exact state of a system and the law governing its change enables the state of that system to be calculated exactly for any future time. In principle, Newtonian mechanics is a deterministic system but in practice the sensitivity to initial conditions characteristic of chaotic systems undermines the predictability of most systems described by Newtonian mechanics. In addition, determinism is also undermined by the Heisenberg uncertainty principle of quantum mechanics.

deuterated /**dew**-ter-ay-tid/ *See* deuterium.

deuterium /dew-**teer**-ee-ŭm/ Symbol: D, ^2H A naturally occurring stable isotope of hydrogen in which the nucleus contains one proton and one neutron. The atomic mass is thus approximately twice that of ^1H; deuterium is known as 'heavy hydrogen'. Chemically it behaves almost identically to hydrogen, forming analogous compounds, although reactions of deuterium compounds are often slower than those of the corresponding ^1H compounds. Its physical properties differ slightly from the properties of ^1H. Deuterium oxide, D_2O, is called 'heavy water'; it is used as a moderator and coolant in some types of nuclear reactor. Chemical compounds in which deuterium atoms replace the usual ^1H atoms are said to be *deuterated.*

The natural abundance of deuterium is somewhat below 0.015% but this varies slightly with source.

deuteron /**dew**-ter-on/ A nucleus of a deuterium atom; i.e. a combination of a proton and a neutron.

deviation The turning of a ray during reflection or refraction. The *angle of deviation* is the angle through which the ray's direction changes – an undeviated ray has zero angle of deviation. In the case of a reflecting surface, the angle of deviation (d) can vary between almost zero and 180°. A refracting surface can produce values of d between zero and the complement of the critical angle (180° – c), where c is the critical angle. The angle of deviation produced by a prism varies with the angle of incidence and has a minimum value (D) when the ray passes symmetrically through the prism. D is related to the prism's refractive constant (n) and its angle (A):

$$n = \sin[(A + D)/2]/\sin(A/2)$$

This equation can be used for finding the refractive constant of the material of the prism using a spectrometer to measure D.

dew Water droplets that condense from air onto a cooler surface, such as may happen outdoors at night. After dark, a leaf, for example, will radiate and become cooler. If it is surrounded by calm warm moist air, condensation will occur, just as it does on a cold window pane in a relatively warm humid room. If the surface temperature is below the freezing temperature of water, the condensation will be as hoarfrost. *See also* relative humidity.

Dewar flask /**dew**-er/ (**vacuum flask**) A double-walled container of thin glass with the space between the walls evacuated and sealed to stop conduction and convection of energy through it. The glass is often silvered to reduce radiation. The flask is named for the Scottish scientist Sir James Dewar (1842–1923).

dew temperature (point) *See* relative humidity.

dextrorotatory /deks-troh-**roh**-tă-tor-ee, -toh-ree/ *See* optical activity.

dialysis /dȳ-**al**-ă-sis/ (*pl.* **dialyses**) A method of separating a COLLOID from a CRYSTALLOID using a semipermeable membrane. The membrane allows the passage of small crystalloid molecules but does not allow large colloid molecules to diffuse through. *See* electrodialysis.

diamagnetism /dÿ-ă-**mag**-nĕ-tiz-ăm/ The magnetic behavior of substances (e.g. bismuth and lead) with a negative susceptibility. The relative permeability is less than one. Usually the value of susceptibility is small for diamagnetic substances and relative permeability is slightly less than one. Diamagnetism results from the motion of electrons in atoms; under the influence of an applied field these change so as to oppose the field. The magnetization of a diamagnetic material is thus opposite to that of a paramagnetic or ferromagnetic material. If a bar of a diamagnetic substance is placed in a non-uniform field it settles with its axis perpendicular to the field.

Diamagnetism is a weaker effect than paramagnetism or ferromagnetism. It is shown by all substances and is independent of temperature, although in some materials it is masked by other types of magnetic behavior. *See also* magnetism.

diaphragm /**dÿ**-ă-fram/ **1.** A device used in optical instruments to reduce or control the aperture of the system. An *iris diaphragm* is a variable aperture made up of a number of movable overlapping plates. Diaphragms (or *stops*) are used to control the light admitted. They are also employed to reduce aberration in lens and mirror systems by restricting the light to the region close to the axis of the system. *See also* aberration.
2. A thin sheet of metal or plastic that vibrates to produce, or is vibrated by, sound waves, as in a loudspeaker or some types of microphone.

diascope /**dÿ**-ă-skohp/ An optical projector for translucent objects (e.g. slides). The name is rather old-fashioned.

diatomic /dÿ-ă-**tom**-ik/ Describing a molecule that consists of two atoms. Hydrogen (H_2), oxygen (O_2), nitrogen (N_2), and carbon monoxide (CO) are examples of diatomic gases.

diatonic scale /dÿ-ă-**tonn**-ik/ A musical scale in which notes are arranged in groups, with notes in a group lying between two limits that differ in pitch by an octave, where an octave is defined as being an interval between two corresponding notes in successive groups in which the frequencies of the notes have the ratio 2:1. The diatonic scale is used extensively in European music.

dichroic medium /dÿ-**kroh**-ik/ A birefringent material in which radiation polarized in one plane is freely transmitted, but radiation polarized perpendicular to this is absorbed. Tourmaline is a natural mineral with this property; Polaroid is a synthetic dichroic substance.

dielectric /dÿ-i-**lek**-trik/ A nonconductor of electricity. The name is usually used where electric fields are possible, as with the insulating material between the plates of a capacitor. *See also* relative permittivity.

dielectric constant An early term for relative permittivity.

dielectric heating The heating effect caused by the rapid alternation of the electric field across an insulator (dielectric) (as with the material between the plates of a capacitor in an a.c. circuit). The effect is used as a heating technique for certain materials. The loss of useful energy that results is *dielectric loss*.

dielectric loss *See* dielectric heating.

dielectric strength The maximum electric field strength that a dielectric can withstand without causing it to break down and pass a spark discharge. It is usually measured in kilovolts per meter.

differential equation An equation that contains derivatives. An example of a simple differential equation is:

$$dy/dx + 4x + 6 = 0$$

To solve such equations it is necessary to use integration. The equation above can be rearranged to give:

$$dy = -(4x + 6)dx$$

integrating both sides:

$$\int dy = \int -(4x + 6)dx$$

which gives:

$$y = -2x^2 - 6x + C$$

where C is a constant of integration. The value of C can be found if particular values of x and y are known: for example, if $y = 1$ when $x = 0$ then $C = 1$, and the full solution is

$$y = -2x^2 - 6x + 1$$

Note that the solution to a differential equation is itself an equation. Differentiating the solution gives the original equation. Equations like that above, which contain only first derivatives (dy/dx) are said to be *first order*, if they contain second derivatives they are *second order*; in general, the *order* of a differential equation is the highest derivative in the equation. The *degree* of a differential equation is the highest power of the highest order derivative.

The differential equation in the example given is a first order and first degree equation. Differential equations are used extensively in all branches of physics.

differentiation /dif-ĕ-ren-shee-**ay**-shŏn/ A process for finding the rate at which one variable quantity changes with respect to another. For example, a car might travel along a road from position x_1 to position x_2 in a time interval t_1 to t_2. Its average speed is $(x_2 - x_1)/(t_2 - t_1)$, which can be written $\Delta x/\Delta t$, where Δx represents the change in x in the time Δt. However, the car might accelerate or decelerate in this interval and it may be necessary to know the speed at a particular instant, say t_1. In this case the time interval Δt is made infinitely small, i.e. t_2 can be as close as necessary to t_1. The limit of $\Delta x/\Delta t$ as Δt approaches zero is the instantaneous velocity at t_1. The result of differentiation (i.e. the *derivative*) of a function $y = f(x)$ is written dy/dx or $f'(x)$. On a graph of $f(x)$, dy/dx at any point is the slope of the tangent to the curve $y = f(x)$ at that point. *See also* integration.

diffraction /di-**frak**-shŏn/ The effects of an obstacle on passing radiation, making the shadow edge less sharp than expected. Some radiation bends round the obstacle

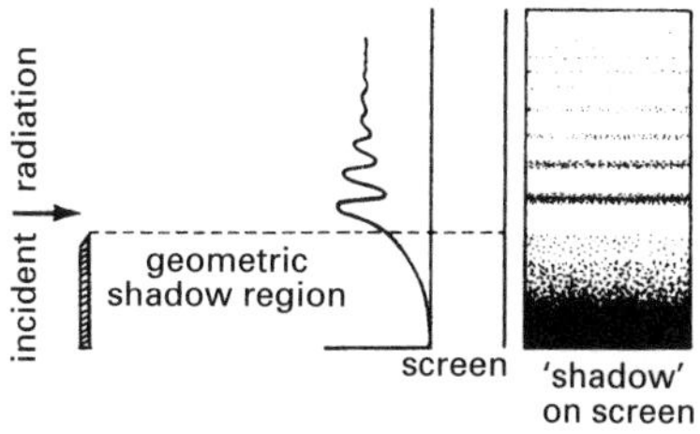

Diffraction at a straight edge

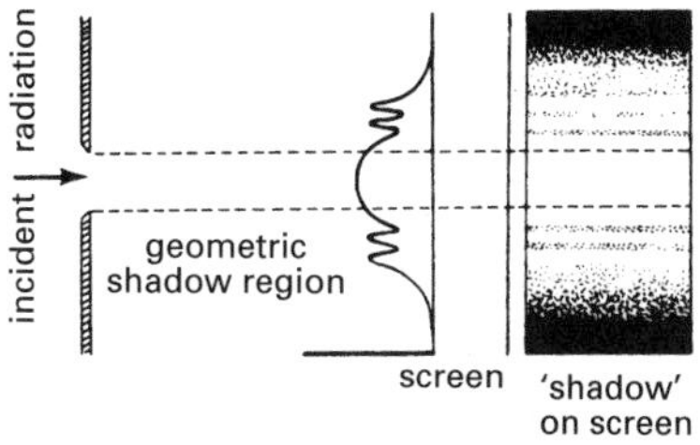

Diffraction at a wide slit (width 5λ).

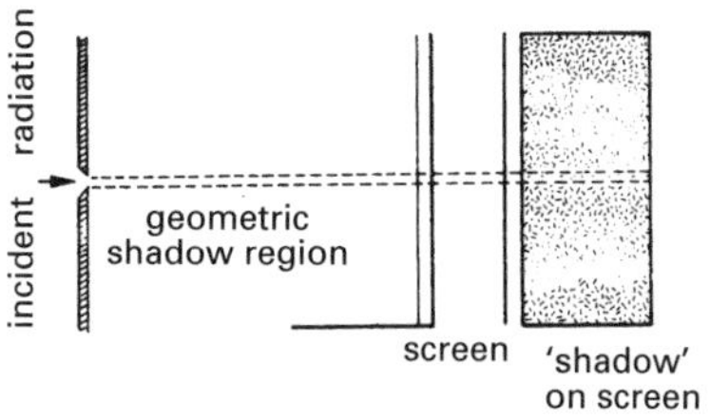

Diffraction at a narrow slit (width λ).

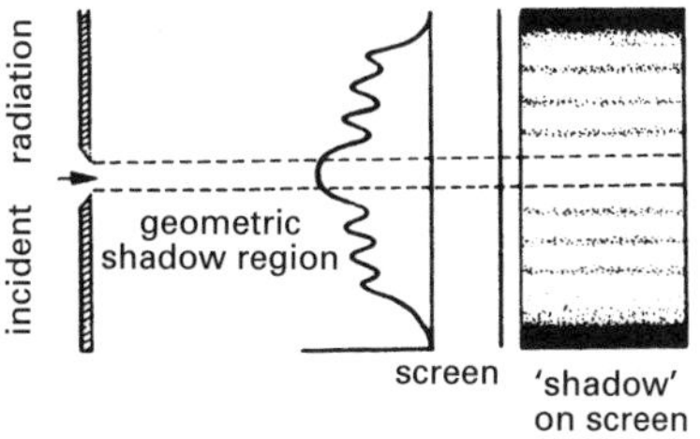

Diffraction at a slit (width 2½λ).

Diffraction at a slit (width 2½λ)

into the shadow region; in some parts of the non-shadow region the radiation is reduced in intensity. Diffraction is observed with all waves, particularly if the obstacle is comparable in size to the wavelength. Under these conditions, fringes (diffraction patterns) are formed. *See also* Fraunhofer diffraction; Fresnel diffraction.

diffraction grating The best gratings are made of glass (for transmitted radiation) or metal (for reflected radiation) with many thousand fine parallel grooves (lines) ruled on the surface with a diamond point. For some purposes cheap plastic casts of such gratings are suitable. Interference between the waves diffracted by the grooves causes maxima of intensity in directions determined by the wavelength, hence spectra are produced.

Transmission gratings are used mainly for visible light. Generally reflection gratings are necessary for ultraviolet and infrared. Using normal incidence on a transmission grating there is an undeviated transmitted beam (zero order) with spectra on each side. If d is the distance between the centers of adjacent lines and λ is the wavelength, spectra are produced at angles $\pm\theta$ where

$$d \sin\theta = m\lambda \ (m = 1, 2, 3, \text{etc.})$$

The value of m is called the *order* of the spectrum. As $\sin\theta$ cannot exceed 1 the number of orders of spectra is limited.

diffuse reflection *See* reflection.

diffuse refraction *See* refraction.

diffusion /di-**fyoo**-zhŏn/ **1.** Movement of a gas, liquid, or solid as a result of the random thermal motion of its particles (atoms or molecules). A drop of ink in water, for example, will slowly spread throughout the liquid. Diffusion in solids occurs very slowly at normal temperatures. *See also* Graham's law.
2. The random spreading out of a beam of radiation on REFLECTION or transmission. Diffusion occurs when a beam of light passes through a translucent medium (e.g. wax or frosted glass) or is reflected by a rough surface (e.g. paper or matt paint). *See also* refraction.

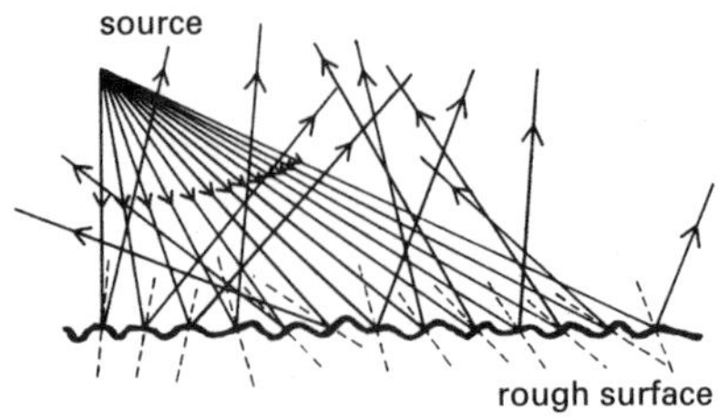

Diffusion of radiation

diffusion cloud chamber *See* cloud chamber.

diffusion pump (**condensation pump; vacuum pump**) A type of pump that can produce a high vacuum. Mercury or oil at low pressure diffuses through a small orifice and carries along in its flow molecules of gas (air) from region already at low pressure (produced by a backing pump). The diffused vapor and gas molecules condense out on the cooled walls of the pump.

digital recording A technique for recording sound that does not actually record sound directly but samples the pressure of the sound wave thousands of times per second and converts the values of the pressures found in this way into numbers. These numbers can be recorded, transmitted, and then turned back into sound.

dihedral /dȳ-**hee**-drăl/ Formed by or having two intersecting planes or sides. For example, the angle between two intersecting plane faces of a crystal is a *dihedral angle*.

dilatometer /dil-ă-**tom**-ĕ-ter/ An instrument for measuring thermal expansion. Depending on the type, it may measure linear or cubic expansivity of a gas, liquid, or solid. *See also* comparator; Regnault's apparatus.

dimensional analysis A technique for checking the validity of a solution to a problem, a unit, or an equation in physics.

DIMENSIONS AND UNITS OF SOME PHYSICAL QUANTITIES

Quantity	*Dimension*	*Unit*	*Normal unit*
mass	[M]	kg	kg
length	[L]	m	m
time	[T]	s	s
area	[L^2]	m^2	m^2
volume	[L^3]	m^3	m^3
density	[ML^{-3}]	kg m^{-3}	kg m^{-3}
acceleration	[LT^{-2}]	ms^{-2}	ms^{-2}
force	[MLT^{-2}]	kg ms^{-2}	N
pressure	[$ML^{-1}T^{-2}$]	kg $ms^{-2}m^{-2}$	Pa
momentum	[MLT^{-1}]	kg ms^{-1}	(N s)
pulsatance	[T^{-1}]	s^{-1}	Hz

It depends on the fact that, once a unit system has been defined fully (as in SI), each quantity can be expressed in terms of the base quantities in only one way. Here those base quantities are called *dimensions*. The table shows, first, the three dimensions used in mechanics, and, second, some other quantities in terms of these. For instance, the relationship between pressure, force, and area can be checked. Force divided by area has dimensions MLT^{-2}/L^2, i.e. dimensions $ML^{-1}T^{-2}$, as for pressure.

dimensions *See* dimensional analysis.

diode /dȳ-ohd/ An electronic device with two electrodes that exhibits rectifying action when a potential difference is applied. Current flows for one direction of the potential (the forward direction), but when the potential is reversed the current is normally zero or very small. There are two principal types of diode: **1.** The *semiconductor diode* consists of a single p-n junction. The current across the junction increases exponentially with voltage in the forward direction; in the reverse direction there is only a very small (leakage) current until breakdown occurs.
2. The *thermionic tube diode* consists of a heated negative cathode emitting electrons (by thermionic tube emission) in a vacuum with a positive anode, which accepts electrons when a positive potential is applied to it. The emitted electrons form a 'cloud' around the cathode, called the *space charge*. Since electrons repel each other owing to their like charges, the space charge limits the current as the voltage is increased. When the voltage is high enough the current reaches a maximum saturation value.

Some other types of SEMICONDUCTOR diode are the light-emitting diode (LED), the tunnel diode, and the Zener diode. *See also* rectifier; thermionic tube.

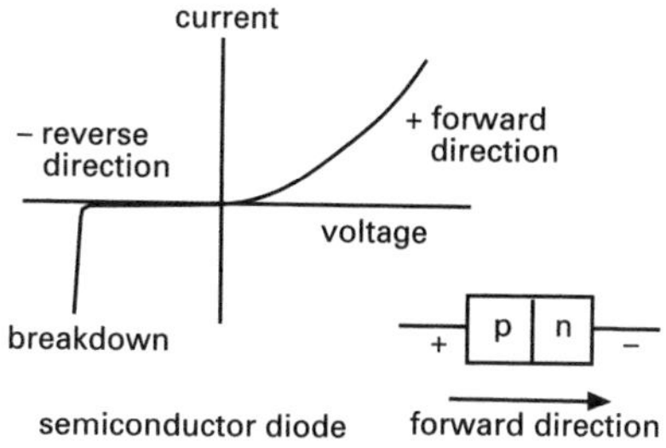

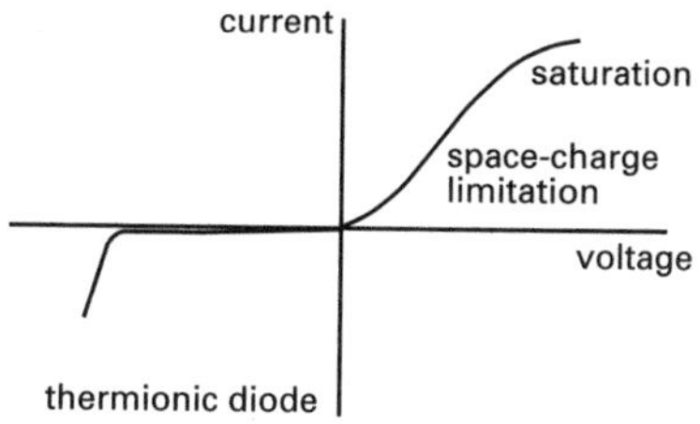

Current-voltage characteristics of diodes

diopter /dÿ-**op**-ter/ Symbol: D The unit of power of a lens or mirror, equal to the reciprocal of the focal distance in meters. A 5 D lens has a focal distance of 0.2 m; a diverging mirror of focal distance 1250 mm has a power of –0.8 D.

Power measures the ability of a reflector, refracting surface, or lens to converge a parallel beam.

dip, angle of *See* inclination.

dip circle *See* inclinometer.

dipole /**dÿ**-pohl/ A pair of opposite electric charges.

dipole antenna *See* half-wave antenna.

dipole moment Symbol: μ The product of the positive charge of a dipole and its distance from the negative charge (*see* dipole). It is measured in coulomb meters (SI units), though often expressed in debyes (1 debye = 3.336×10^{-30} coulomb meter).

direct current (**d.c.**) Electric current in one direction only. *Compare* alternating current.

direct current electric motor An electric motor that is run by direct current. The electric current flows round an armature coil which is connected to the source of the d.c. power, such as a battery, by a split-ring commutator. Since the armature is in an external magnetic field, the field exerts a couple on the armature when current flows. The commutator reverses the current direction when the motion of the sides of the coil changes from up to down. This means that the force in the armature is always in the same direction, thus ensuring constant rotation.

disappearing-filament pyrometer *See* optical pyrometer.

discharge, electrical The passage of an electric current through a gas at low pressure or a dielectric. It is usually accompanied by luminous effects. The gas molecules become ionized and the recombination of oppositely charged ions gives rise to the luminous effects.

See also canal rays; corona discharge; electric arc; electric spark; glow discharge; positive column.

discharge of a capacitor A process in which the electric charge in a capacitor is discharged by forming a circuit consisting of the capacitor connected to a resistor. The mathematical description of this process is analogous to that of radioactive decay, with the decay of the charge being exponential with time. *See also* time constant.

discharge tube A gas-filled or evacuated glass tube with a pair of sealed-in electrodes. A high voltage causes an electric discharge between the electrodes. An example is the familiar neon tube used for advertising signs. *See also* Geissler tube.

discriminator /dis-**krim**-ă-nay-ter/ A circuit that selects certain electrical signals and rejects others, according to a particular property of the signal, such as amplitude, frequency, or phase. For example, a pulse-height analyzer selects signals having a particular amplitude range, and a frequency discriminator picks out different frequencies.

disintegration The breaking up of unstable nuclei into two or more fragments, either spontaneously or as a result of bombardment by fast-moving particles. The disintegration of ^{7}Li by proton bombardment in Cockcroft and Walton's experiment is an example of the latter type; two ^{4}He nuclei (alpha particles) are produced.

disintegration constant *See* decay.

dislocation *See* defect.

disordered solid A solid that does not have the order of a perfect crystal (or a crystal with a very small concentration of impurities or defects). There are several types of disordered solids. In a *random alloy* made up of two metals A and B the pattern in which A and B occur is random. In a solid in which a crystal has a high concentration of impurities, the random distribution of impurities in the solid gives another type of disordered solid. In an *amorphous solid* (e.g. a glass) the atoms of the solid form a random network rather than a regular array.

dispersion During refraction, rays can be deviated (changed in direction) by angles related to the refractive index of the medium. However, the refractive constant of any medium varies with the wavelength of the transmitted radiation. Thus, if a narrow beam of radiation of mixed wavelengths is deviated by refraction, the angle of deviation will vary with wavelength. This is called dispersion; the radiation is spread out. It can be used to produce the color spectrum of white light using a prism. It can also be a problem, causing chromatic aberration in lenses.

Dispersion is actually an effect in which radiations having different wavelengths travel at different speeds in the medium. Normally the refractive index increases as the wavelength decreases, but there are regions, where the radiation is partially absorbed, in which refractive index decreases with decreasing wavelength. *See* anomalous dispersion.

dispersion forces *See* van der Waals' forces.

dispersive power Symbol: ω A measure of the variation for a medium of refractive constant n with wavelength λ. It is given by the ratio:

$$\omega = (n_b - n_r)/(n_y - 1)$$

Here, n_b, n_r, and n_y are absolute refractive constants for blue, red, and yellow light. More accurate values are obtained by using three monochromatic spectral lines of known wavelength: F (blue), C (red), and D (yellow).

displacement Symbol: s The vector form of distance, measured in meters (m) and involving direction as well as magnitude.

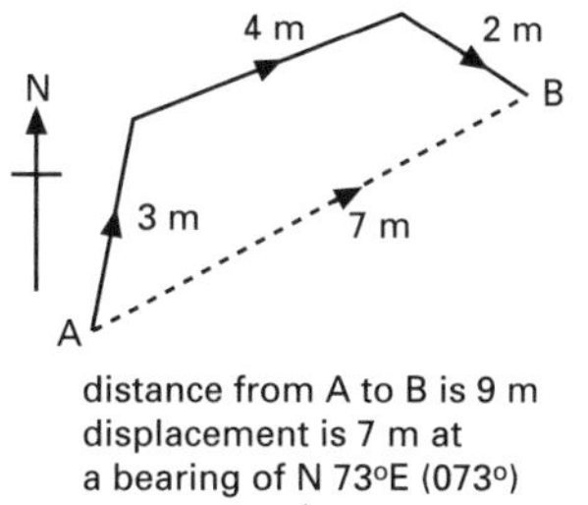

Displacement and distance

dissipation The removal of energy from a system to overcome some form of resistive force (mechanical or electrical). Without resistance (as in motion in a vacuum and the current in a superconductor) there can be no dissipation. Dissipated energy normally appears as thermal energy.

distance Symbol: d The length of the path between two points. The SI unit is the meter (m). Distance may or may not be measured in a straight line. It is a scalar; the vector form is displacement.

distance ratio (velocity ratio) For a MACHINE, the ratio of the distance moved by the effort in a given time to the distance moved by the load in the same time.

distillation A method of separating mixtures of liquids or purifying liquids by heating them to the boiling point, condensing the vapor produced, and collecting the liquid or liquids that form.

distortion *See* aberration.

div *See* vector calculus.

divergent Moving apart. *See* convergent.

diverging lens A LENS that can refract a parallel beam into a divergent beam. Diverging lenses used in air are thinner in the middle than at the edge. They may be biconcave, plano-concave, or concavo-convex in shape. As these lenses have negative power, they are sometimes called *negative lenses*. *Compare* converging lens.

diverging mirror (**diverging reflector**) A MIRROR that can reflect a parallel beam into a divergent beam. Diverging reflectors always have convex surfaces. As these reflectors have negative power, they are sometimes called *negative mirrors*. *Compare* converging mirror.

diverging reflector *See* diverging mirror.

D-layer *See* ionosphere.

domain A region inside a ferromagnetic material in which all the atomic magnetic fields point the same way. When the material is unmagnetized, the directions of the separate domains are random; when a strong external magnetic field is applied, the domains line up in the same direction. The domains with magnetic axes nearly parallel to that of the magnetizing field grow at the expense of those perpendicular to the field. The substance now has a resultant magnetic moment. The domains can grow until the magnetic axes are in line with the external field if this is strong enough. The material is then magnetically saturated. The presence of magnetic domains can be shown by Bitter patterns, the Barkhausen effect, and by various other techniques.

donor *See* semiconductor.

dopant /**doh**-pănt/ *See* semiconductor.

doping *See* semiconductor.

Doppler broadening /**dop**-ler/ The increase in the width of a spectral line as a result of the motion of the particles emitting or absorbing the radiation. It is named for the Austrian physicist Christian Johann Doppler (1803–53). *See* monochromatic radiation.

Doppler effect **1.** An apparent change in the frequency of a wave motion, caused by relative motion between the source and the observer. It is found with sound waves: as the source and observer move together the apparent frequency of the sound is higher than that produced; as they move apart it is lower. An example is that of an approaching motorbike – the pitch of the engine sound appears to drop suddenly as it passes.

The apparent frequency (f') of sound as a result of relative motion between source and observer in a direct line is given by:

$$f' = (V + U_o)f/(V - U_s)$$

where V is the speed of sound, U_o the observer's velocity, U_s the source velocity, and f the source frequency.

2. A similar effect for electromagnetic radiation. The radiation emitted by a source with a motion component towards the observer will have increased frequency: if the source is moving away from the observer, the observed frequency is reduced. This has great significance in astronomy. Thus the Doppler effect on radar pulses reflected by the planets gives information on their rotation. The displacement of lines in the spectra of stars and galaxies (i.e. the red shift or the blue shift) can tell the astronomer how these bodies are moving.

For electromagnetic radiation the theory of the effect is not the same as for sound. This is because there is no fixed medium to give a frame of reference. Relativity theory gives the following result:

$$\nu = \nu_0\sqrt{[(1 - v/c)/(1 + v/c)]}$$

Here ν is the observed frequency, ν_0 is the emitted frequency; v is the speed of separation of source and observer, related to the speed of light c. If v is small compared to c the expression simplifies to:

$$\nu = \nu_0(1 - v/c)$$

Doppler shift *See* red shift.

dose The amount of energy from ionizing radiation absorbed by unit mass of material. The SI unit of absorbed dose is the gray. When considering effects on living organisms, the dose in grays is multiplied by a factor expressing the relative effectiveness of the radiation to obtain the dose equivalent which in SI units is expressed in sieverts. For example, alpha rays cause about twenty times as much biological damage as do beta rays of the same energy.

All persons are exposed to ionizing radiation derived mostly from natural causes, such as cosmic rays, radioactivity in the environment, and radioactivity in the body itself. Additional radioactivity is caused by nuclear explosions and accidents and by the use of luminous paints. Some individuals receive additional doses in medical examination or treatment, or from working with x-radiation or radioactive substances. Persons employed in the nuclear industry or as radiologists or radiographers are protected by a statutory code that defines working conditions and sets legal limits to the dose equivalent allowable. National codes are based on the rulings of the International Commission on Radiological Protection. Exposure to ionizing radiation can cause harm (particularly cancer) to the exposed persons and genetic changes, which may affect their descendants.

dosimeter /doh-**sim**-ĕ-ter/ An instrument that measures the amount of ionizing radiation received by a person or a region. A simple type is the film badge, which is worn by personnel and developed daily (radiation fogs the film) to check on absorbed dose.

dot product *See* scalar product.

double refraction *See* birefringent crystal.

double slit experiment *See* Young's double slit.

doublet A compound lens with two elements, often in contact. The purpose is usually to reduce aberrations. If the lenses are in contact, the power of the doublet is the sum of the powers of the elements.
Thus, $P = P_1 + P_2$
or, $1/f = 1/f_1 + 1/f_2$ (f is focal distance).
A 5 diopter (D) lens combined with a –2.5 D lens forms a doublet of power 2.5 D, or focal distance 400 mm.

drag A force exerted by a fluid (gas or liquid) on an object in its flow. It acts in the direction of the flow and therefore tends to resist the motion of the object through the fluid.

drain *See* field-effect transistor.

drift velocity The average velocity of an electron through a conductor when a current is flowing. If there are n electrons per unit volume, the cross-sectional area of the wire is A, V is the average speed of an electron passing through a conductor, and e is the charge of an electron, then the current I is given by $I = nAVe$.

dry cell A voltaic cell in which the electrolyte is in the form of a jelly or paste. Dry cells are used for flashlights and other portable appliances. *See also* Leclanché cell.

duality Wave-particle duality: the concept that some entities show both wave and particle properties. There are two cases. **1.** Electromagnetic radiation. Einstein (1905) argued that, whereas the propagation of these radiations is fully described by the theory of electromagnetic waves, interaction with matter occurs as if the radiation travels in packets (photons) each with a quantum of energy $h\nu$, where h is the Planck constant and ν is the frequency of the wave. His examples were the Planck radiation law, photoelectric emission, and photoionization of gases. The principle has been verified for radiations of all wavelengths from microwaves to gamma rays, in many different phenomena involving emission, absorption or scattering. Whereas such wave properties as frequency, phase speed, and electric and magnetic fields are measurable quantities, photons are only observable when being created or destroyed.
2. Particle waves. In 1923 de Broglie suggested that atomic particles might have wave-like properties and that the wavelength (λ) would have the same relationship to the momentum (p) as applies to electromagnetic radiation: $\lambda = h/p$. This concept has been used very successfully in the form of quantum theory called *wave mechanics*. Beams of particles including electrons, neutrons, and gas molecules incident on crystals form diffraction patterns analogous to those given by x-rays. The wavelength formula is fully verified.

See also wave theory of light.

dubnium /**dub**-nee-ŭm/ A radioactive synthetic element made by bombarding ^{249}Cf nuclei with ^{15}N atoms. It was first reported by workers at Dubna, a town near Moscow.

Symbol: Db; p.n. 105; most stable isotope ^{262}Db (half-life 34 s).

ductility /duk-**til**-ă-tee/ The property of a metal that allows it to be drawn out into wire. Compare malleability.

Dulong and Petit's law /doo-**long** ănd pĕ-**teez**/ The molar thermal capacity of a solid element is approximately equal to $3R$, where R is the gas constant (25 J k^{-1} mol^{-1}). The law applies only to elements with simple crystal structures at normal temperatures. At lower temperatures the molar heat capacity falls with decreasing temperature (it is proportional to T^3). Molar thermal capacity was formerly called *atomic heat* – the product of the atomic weight (relative atomic mass) and specific thermal capacity. The law is named for the French physicists Pierre-Louis Dulong (1785–1838) and Alexis-Thérèse Petit (1791–1820).

dynamic equilibrium *See* equilibrium.

dynamics The study of how objects move under the action of forces. This includes the relation between force and motion and the description of motion itself. *See* equations of motion; Newton's laws of motion.

dynamo A small device for converting mechanical energy into electrical energy. The motion of a coil rotating in a magnetic field induces an alternating current in the windings. This current itself causes a turning force opposing the initial motion. To keep the dynamo turning, an energy input is required from an external source, such as a gasoline or diesel motor. Many bicycles have a small dynamo built into the hub of the wheel (or driven by the tire) to power the cycle lamps.

See also generator.

dynamometer /dȳ-nă-**mom**-ĕ-ter/ An apparatus for measuring force, usually the output power (brake horsepower) of an engine or motor.

dyne /dȳn/ Symbol: dyn The former unit of force used in the c.g.s system. It is equal to 10^{-5} N.

dysprosium /dis-**proh**-see-ŭm, -shee-ŭm/ A soft malleable silvery element of the lanthanoid series of metals. It occurs in association with other lanthanoids. One of its few uses is as a neutron absorber in nuclear reactors; it is also a constituent of certain magnetic alloys.

Symbol: Dy; m.p. 1412°C; b.p. 22562°C; r.d. 8.55 (20°C); p.n. 66; r.a.m. 162.50.

E

ear The organ used to hear sounds. It is divided into three parts: the *outer ear*, the *middle ear*, and the *inner ear*. Sound waves cause the ear-drum, in the outer ear, to vibrate. The middle ear transfers the energy of sound waves into the inner ear. The receptors for hearing are in the inner ear.

early Universe The era in the evolution of the Universe soon after the BIG BANG when it was very hot and dense. As the Universe expanded it cooled, giving rise to a sequence of phase transitions associated with broken symmetries. The light elements such as helium were mostly formed in the early Universe. It has been suggested that in the early Universe *inflation*, i.e. very rapid expansion, occurred. The very early Universe, occurring immediately after the big bang, requires a fundamentally new physics involving the combination of quantum mechanics and general relativity theory before it will be understood.

Earnshaw's theorem /**ern**-shaw/ The principle that in a system of bodies in which the interactions between the bodies are described by an inverse square law, such as Newton's law of gravitation or Coulomb's law of electrostatics, the bodies cannot be in a stable static equilibrium state. This result means that the planets in a planetary system such as the Solar System and the electrons in an atom cannot be stationary. The law was found by the Reverend Samuel Earnshaw (1805–88) in 1842.

Earth inductor A type of fluxmeter for measuring large uniform fields (normally that of the Earth). A large coil of known area and number of turns is flipped through 180° and the induced e.m.f. is measured using a ballistic galvanometer.

earthing *See* grounding.

Earth's magnetism (**geomagnetism**) A magnetic field can be observed at any point on or near the Earth's surface. This field is rather like that which would be obtained if a giant bar magnet were present inside the planet; it is closer to the field of a magnetized steel sphere. A magnetic compass can be used to show the direction of the field – of great importance in navigation. The Earth's north magnetic pole is currently off the northern coast of Canada; the south magnetic pole is on the edge of Antarctica. The strength and direction of the Earth's field at any point are defined in terms of three magnetic elements – the inclination, the declination, and the horizontal component of the field strength. These elements are not constant. The source of the Earth's field is thought to be due to electric currents circulating in the molten nickel-iron core. *See* magnetic variation.

echelon /**esh**-ĕ-lon/ A type of diffraction grating consisting of a number of equal thin glass sheets stacked on a slant. In use the light is reflected from the stepped side of the stack; *d* in the grating equation is very large, so that very high spectral orders are possible.

echo The result of reflection of sound waves (or other waves), usually by a hard surface. This causes an observer to detect two signals; the first a direct signal and the second a fainter reflected signal. The delay between the two indicates the distance of the reflector.

echolocation /ek-oh-loh-**kay**-shŏn/ A method of finding the range and bearing of something (the 'target') by transmitting high-frequency sounds and detecting their echoes on their return. *See* sonar.

echo sounder *See* sonar.

eclipse The 'hiding' of one heavenly body behind another. The eclipsed object, the eclipsing object, and the observer are in a straight line.

In a *solar eclipse*, the Moon appears to pass across the Sun, so that the Moon's shadow sweeps over the Earth's surface. If the Moon appears to cover the Sun completely, there is a *total eclipse* of the Sun – the observer is in the umbra of the Moon's shadow. If the Moon only partly covers the Sun, there is a *partial eclipse* – the observer is in the penumbra of the shadow. Solar eclipses can happen only at New Moon; the Moon's shadow is narrow, so total eclipses of the Sun are rare at any one place on the Earth. It is a coincidence that Sun and Moon appear to be of similar size from the Earth. However as neither the Earth–Moon distance nor the Sun–Earth distance is constant, the sizes of Sun and Moon seen from the Earth do vary. Sometimes there is an *annular eclipse*, when the Moon is not 'big' enough to cover the Sun's disk. Then the vertex of the lunar umbra does not reach the Earth's surface.

As the Earth's shadow is larger than the Moon's, *lunar eclipses* are more commonly seeen than solar eclipses. Lunar eclipses can occur only at Full Moon, when the Earth's shadow sweeps the Moon's surface. They also last longer than solar eclipses, as Earth and Moon are now moving in the same direction through space. Another big difference is that lunar eclipses do not clearly show umbra and penumbra. This is because sunlight is refracted into the umbra cone by the Earth's atmosphere, acting as a ring-shaped lens. This refraction also explains the reddish color of the Moon during eclipse.

eddy-current heating *See* induction heating.

eddy currents Induced currents set up in a conductor by a changing magnetic field. They occur in transformers and other electrical devices. The currents produce a heating effect corresponding to a loss of useful energy (*eddy-current loss*). Metal cores in electrical machines are usually laminated (built of thin sheets) to reduce such losses; the surface layers between the laminations have high electrical resistance.

One useful application of eddy currents is in damping moving-coil instruments – the coil is wound on a piece of soft iron. When the coil moves in the instrument's magnetic field, eddy currents are induced in the iron so as to oppose the motion, causing the coil to settle quickly. As Lenz's law would predict, eddy currents always flow so as to oppose the effect producing them.

Edison cell *See* nickel-iron accumulator.

effective resistance The resistance of part of an electric circuit in which there is an alternating current. It is the power dissipated divided by the square of the current. If power is in watts and current in amperes, the units are then ohms. Effective resistance expresses a combination of two effects: the resistance from scattering of electrons by lattice atoms (as for d.c.) and the resistance from the magnetic effects of the changing current.

effective value *See* root-mean-square value.

efficiency Symbol: η The ratio of the work done by a system or device to the energy input. Similarly, the efficiency is equal to the useful power output divided by the power input. There is no unit; however efficiency is often quoted as a percentage. There are several branches of physics in which efficiency is used.

1. For MACHINES the efficiency is the force ratio divided by the distance ratio.

2. For an electric cell delivering current through a load resistance (R), the efficiency is given by $\eta = R/(R + r)$, where r is the internal resistance of the cell. Note that the

higher the value of *R*, the higher the efficiency.
3. For an electric motor, the efficiency is the mechanical power developed divided by the electrical power input.
4. For a heat engine, the efficiency is the work done divided by the heat input. *See* Carnot cycle.

effort The force that moves a load in a simple machine such as a lever or pulley. The ratio of the load to the effort is the force ratio.

effusion The flow of gas molecules through relatively large holes.

EHF *See* extremely high frequency.

einsteinium /ÿn-**stÿ**-nee-ŭm/ A radioactive transuranic element of the actinoid series, not found naturally on Earth. It can be produced in milligram quantities by bombarding ^{239}Pu with neutrons to give ^{253}Es (half-life 20.47 days). Several other short-lived isotopes have been synthesized.

Symbol: Es; m.p. 860 ± 30°C; p.n. 99; most stable isotope ^{254}Es (half-life 276 days). The element is named for the German–Swiss–American theoretical physicist Albert Einstein (1879–1955).

Einstein model A model for the specific heat of solids that was proposed by Albert Einstein in 1907, in which it was assumed that the specific heat is due to the vibrations of the atoms of the solid, with these vibrations all having the same frequency. This model gives rough qualitative agreement with experiment, but not quantitative agreement at low temperatures. An improvement that gives much better agreement with experiment is the DEBYE MODEL.

Einstein–Podolsky–Rosen experiment /**ÿn**-stÿn pŏ-**dol**-skee **roh**-zĕn/ *See* Bell's paradox.

Einstein's equation *See* photoelectric effect.

Einstein's field equation An equation put forward by Albert Einstein (1879–1955) in 1915 as the fundamental equation of general RELATIVITY theory. It generalizes Newton's theory of gravity by relating mass (energy) to the curvature of space–time.

Einstein shift *See* red shift.

elastance /i-**las**-tăns/ The reciprocal of capacitance. It is measured in reciprocal farads (F^{-1}).

elastic collision A collision for which the restitution coefficient is equal to one. Kinetic energy is conserved during an elastic collision. *See also* restitution; coefficient of.

elasticity /i-las-**tis**-ă-tee/ The tendency of a material to return to its original dimensions after a deforming stress has been removed. Up to a certain point (the proportional limit) the material obeys Hooke's law. The strain produced is proportional to the stress and the sample returns to its original dimensions if the stress is removed. Above the proportional limit the material no longer obeys Hooke's law, i.e. stress is not directly proportional to strain. At slightly higher stresses the *elastic limit* is reached. Above this the sample no longer returns to its original dimensions after the stress is removed; i.e. a permanent deformation occurs. The *yield point* is the point at which the material begins to 'flow' – i.e. the strain increases with time (up to the breaking point) without further increase in the stress.

elastic limit *See* elasticity.

elastic modulus The ratio of the stress on a body to the STRAIN produced. There are various moduli of elasticity depending on the type of STRESS applied. The *Young modulus* (symbol: *E*) refers to tensile or compressive stress when the cross-section is free to change. It is the force per unit cross-sectional area divided by the fractional elongation of the sample. The *axial modulus* refers to tensile or compressive stress when constraints prevent change in the cross-section. The bulk modulus (symbol: K) is used for a bulk stress – i.e. pressure changes on the body. It is minus the pressure change divided by the fractional change in volume. The *shear modulus* (or *rigidity modulus*) (symbol: G) refers to a shearing (or twisting) stress. It is the tangential force per unit area divided by the angular deformation. Elastic moduli have the same units as stress (newtons per square meter). *See also* elasticity; Poisson ratio.

elastomer /i-las-**tom**-ĕ-ter/ A substance that is elastic – that is, it returns to its original shape when a deforming stress is removed. *See* elasticity.

E-layer *See* Heaviside layer.

electret /i-**lek**-trit/ A piece of permanently electrified material, having a positive charge at one end and a negative charge at the other.

electrical energy Energy resulting from the position of an electric charge in an electric field.

electrical storm A storm associated with the separation of electric charge in clouds. When the resulting electric field is strong enough a flash of lightning occurs, followed by thunder.

electric arc A luminous continuous electrical discharge between two electrodes in a gas, having high current density and low potential difference. The electrodes are raised to high temperatures causing thermionic emission, which maintains the current. Often there is sufficient evaporation of electrode material for its emission spectrum to be prominent in the light. This *arc spectrum* does not include some of the lines produced by high-voltage discharges. Arcs can be used as sources in spectrographic studies. The heating effect of arcs is used in arc welding. Arcs, particularly between carbon electrodes, are also used as high-intensity light sources called *arc lamps*, which are used for example in searchlights.

electric bell A device in which a bell is struck by a hammer whose movement is produced by an electromagnet. In a direct-current (battery-operated) bell, the electromagnet consists of two coils wound on two cores connected by an iron yoke. When the button is pressed to close the circuit, the coils are energized and attract an iron bar (armature). The armature movement breaks the circuit, the coil de-energizes again, and the armature moves back (by a spring) and closes ('makes') the circuit again, and so on, as long as the switch is closed. An alternating-current bell does not have this type of *make-and-break contact.* The armature is a permanent magnet, which is alternately attracted and repelled by the reversing poles of an electromagnet through which the alternating current passes.

electric charge *See* charge.

electric constant *See* permittivity.

electric current *See* current.

electric displacement Symbol: D The electric flux density in a dielectric material:

$$D = \varepsilon_r\varepsilon_0 E$$

where E is the electric field strength in free space and $\varepsilon_r\varepsilon_0$ the permittivity of the material.

electric field A region in which a force would be exerted on an electric charge. Like magnetic fields, electric fields can be represented by lines of force. The electric field is completely defined in magnitude and direction at any point by the force upon a unit positive charge situated at that point. Electric fields can be produced by electric charges or by changing magnetic fields.

electric field strength (**electric intensity**) Symbol: E A measure of the strength of an electric field at a point, defined as the force per unit charge on an electric charge placed at that point: i.e. $E = F/Q$. The electric field strength is the gradient of the electrical potential at a point ($-\mathrm{d}V/\mathrm{d}r$); the unit used is the volt per meter ($\mathrm{V\ m^{-1}}$).

electric flux Symbol: Ψ A measure of an electric field in a vacuum or in a dielectric. The flux in a field can be represented by the number of lines of electric force through a given perpendicular area. The *electric flux density* at a point is the number per unit area at that point. Electric flux density is related to electric displacement.

electric image A method of solving electrostatic problems. A charge +*Q* a certain distance from a conducting surface acts as if there were an 'image' of the charge of magnitude −*Q* positioned on the opposite side of the surface. For a plane conductor the image is as far behind the surface as the charge is in front.

electric intensity *See* electric field strength.

electricity The nature and effects of moving or stationary electric charges, and the branch of physics that studies them.

electric meter (**watt-hour meter**) An instrument that measures electrical energy consumption in watt-hours.

electric motor An electrical machine such that an electric current drives the motor enabling it to do work mechanically. Current passed through a coil forms an electromagnet, which causes the coil (or *rotor*) to rotate in the magnetic field of a permanent magnet or another electromagnet. The rotation causes a back e.m.f. in the coil, which restricts the amount of electrical work done. When the motor is doing external mechanical work the magnitude and phase of the back e.m.f. changes correspondingly, affecting the work done by the electric power source. Electric motors can be designed to work with direct-current supply, or, more commonly, with alternating current. *See also* induction motor; synchronous motor.

electric polarization *See* polarization.

electric potential Symbol: *V* The potential at a point in an electric field is the energy required to bring unit electric charge from infinity to the point. The unit of electric potential is the volt, *V*. *See also* potential difference.

electric power *See* power.

electric spark A discharge of electricity in a gas accompanied by light and sound. The discharge occurs between two points of opposite high electric potentials.

The spark can only occur when the potential difference exceeds a certain value, called the sparking potential. The sudden expansion and then contraction of the gas after it has been heated by the spark gives rise to the characteristic clicking sound.

electrochemical equivalent /i-lek-trŏ-**kem**-ă-kăl/ Symbol: *z* The mass of an element released from a solution of its ions when a current of one ampere flows for one second during electrolysis.

electrochemical series *See* electromotive series.

electrochemistry /i-lek-trŏ-**kem**-iss-tree/ The study of the formation and behavior of ions in solutions. It includes electrolysis and the generation of electricity by chemical reactions in cells.

electrode Any part of an electrical device or system that emits or collects electrons or other charge carriers. An electrode may also be used to deflect charged particles by the action of the electrostatic field that it produces.

electrodeless deposition A method of coating objects with a thin layer of metal (similar to electrodeposition) without using an electric current. An example is the deposition of a layer of silver on glass or plastic to make mirrors. The method relies on the choice of electrolyte and the chemistry involved.

electrodeposition /i-lek-troh-dep-ŏ-**zish**-ŏn/ (**electroplating**) The process of depositing a layer of solid (metal) on an electrode by electrolysis. Positive ions in solution gain electrons at the cathode and are deposited as atoms. Copper, for instance, can be deposited on a metal cathode from an acidified copper sulfate solution.

electrode potential A measure of the tendency of a metal to lose electrons to a surrounding solution. It is the potential difference, at equilibrium, between the metal and a solution of its ions.

The electrode potential cannot be measured directly and is usually compared against that of a hydrogen electrode. Metals that have a greater tendency than hydrogen to form positive ions have a positive electrode potential. Electrode potentials are quoted as measured under standard conditions. *See also* half cell.

electrodialysis /i-lek-troh-dÿ-**al**-ă-sis/ (*pl.* **electrodialyses**) A method of removing salts (crystalloids) from a colloidal solution using two semipermeable membranes. The solution is placed between the membranes, which have on their other sides pure solvent containing electrodes with a voltage applied between them. *See* colloid; dialysis.

electrodynamics /i-lek-troh-dÿ-**nam**-iks/ The study of the relationship between mechanical forces and magnetic and electric forces. Such effects are the basis of, for example, the electrical generator and the electric motor. In contrast to ELECTROSTATICS, electrodynamics deals with moving electric charges and the associated magnetic fields.

electroforming /i-**lek**-trŏ-for-ming/ A method of using ELECTRODEPOSITION to make metal objects. A non-conducting mold of the object to be made (called a mandrel) is made conducting by coating it with graphite or by silvering, and the mandrel is used as the cathode in an electroplating bath. Metal builds up on the mold and when it is thick enough, the mold is removed by melting it (if wax) or dissolving it (if plastic).

electroluminescence /i-lek-troh-loo-mă-**ness**-ĕns/ *See* luminescence.

electrolysis /i-lek-**trol**-ă-sis/ The production of chemical change by passing charge through certain conducting liquids (electrolytes). The current is conducted by migration of ions – positive ones (cations) to the cathode (negative electrode), and negative ones (anions) to the anode (positive electrode). Reactions take place at the electrodes by transfer of electrons to or from them.

In the electrolysis of water (containing a small amount of acid to make it conduct adequately) hydrogen gas is given off at the cathode and oxygen is evolved at the anode.

At the cathode the reaction is:

$$H^+ + e^- \rightarrow H$$
$$2H \rightarrow H_2$$

At the anode:

$$OH^- \rightarrow e^- + OH$$
$$2OH \rightarrow H_2O + O$$
$$2O \rightarrow O_2$$

In certain cases the electrode material may dissolve. For instance in the electrolysis of copper(II) sulfate solution with copper electrodes, copper atoms of the anode dissolve as copper ions

$$Cu \rightarrow 2e^- + Cu^{2+}$$

See also Faraday's laws (of electrolysis).

electrolyte /i-**lek**-trŏ-lÿt/ A liquid containing positive and negative ions, that conducts electricity by the flow of those charges. Electrolytes can be solutions of acids or metal salts ('ionic compounds'), usually in water. Alternatively they may be molten ionic compounds – again the ions can move freely through the substance. Liquid metals (in which conduction is by free electrons) are not classified as electrolytes. *See also* electrolysis.

electrolytic capacitor A type of capacitor consisting of two metal plates, one with a very thin oxide layer, immersed in a suitable electrolyte. Electrolytic capacitors provide a high capacitance for a small size. It is important that they are always charged up with the correct plates positive and negative, otherwise the oxide film may break down.

ELECTROMAGNETIC SPECTRUM
(note: the figures are only approximate)

Radiation	*Wavelength (m)*	*Frequency (Hz)*
gamma radiation	-10^{-10}	$10^{19}-$
x-rays	$10^{-12}-10^{-9}$	$10^{17}-10^{20}$
ultraviolet radiation	$10^{-9}-10^{-7}$	$10^{15}-10^{17}$
visible radiation	$10^{-7}-10^{-6}$	$10^{14}-10^{15}$
infrared radiation	$10^{-6}-10^{-4}$	$10^{12}-10^{14}$
microwaves	$10^{-4}-1$	$10^{9}-10^{13}$
radio waves	$1-$	-10^{9}

electrolytic cell *See* cell; electrolysis.

electrolytic separation The separation of isotopes by electrolysis. *See* isotope separation.

electromagnet /i-lek-troh-**mag**-nĕt/ A coil of insulated wire wrapped around a core of soft iron. When there is a current in the wire, a magnetic field results; the core becomes magnetized. The core loses its magnetism when the current is switched off. Electromagnets are used in telephones, electric bells, relays, and many other devices. *See also* solenoid.

electromagnetic induction /i-lek-troh-mag-**net**-ik/ The induction of an electromotive force (e.m.f.) between the ends of a conductor by change in relation to an external magnetic field. The 'laws of electromagnetic induction' follow:
1. (*Faraday*) An electromotive force is induced in a conductor when there is a change in the magnetic field around it.
2. (*Faraday*) The electromotive force induced is proportional to the rate of change of the field.
3. (*Lenz*) The direction of the induced electromotive force tends to oppose the change.

Electromagnetic induction is used in generators, transformers, microphones, and many other devices. The effect can also cause problems, as in the formation of eddy currents.

The magnitude of the induced e.m.f. is proportional to the rate of change of total flux linkage, Φ. This is called *Neumann's law*, named for the German physicist Franz Neumann (1798–1895). Thus,

$$E = -\mathrm{d}\Phi/\mathrm{d}t$$

where E is the e.m.f. in volts, Φ the flux linkage in webers, and t the time in seconds. The minus sign expresses Lenz's law.

See also mutual induction; self-induction.

electromagnetic interaction *See* interaction.

electromagnetic moment *See* magnetic moment.

electromagnetic pump A pump with no moving parts, used for circulating electrically conducting fluids. A direct current is passed between two electrodes inserted in the fluid circuit and a constant magnetic field is directed perpendicular to this current. There is a force on the fluid mutually perpendicular to these two directions, and therefore parallel to the direction of flow, thus propelling the fluid. One of its applications is in pumping liquid sodium coolant through nuclear reactors.

electromagnetic radiation Energy waves propagated by electric and magnetic fields that oscillate at right angles to one another as they travel through space. Radiations of all frequencies (ν) travel in free space with the same speed $c = 2.997\ 925 \times 10^8$ m s^{-1}. The wavelength (λ) in free space is given by $\lambda\nu = c$. The spectrum includes *radio waves*, *microwaves*, *infrared*, *light*, *ultraviolet*, *x-* and *gamma-radiation*. *See* duality; Poynting vector; wave theory of light.

electromagnetic spectrum As electromagnetic radiation has wave properties, it must be possible to have a whole series that varies in frequency. This is in fact the case. Visible light is only a small part of that spectrum. The other radiations behave in the same way as light – showing reflection, refraction, absorption, diffraction, interference, and polarization in the right circumstances. All travel at the same speed through a vacuum – $2.997\ 925 \times 10^8$ m s^{-1} (about 300 000 km s^{-1}).

The table outlines the spectrum, with indication of the frequencies and wavelengths of each sector. Note that this *is* a spectrum – there are no sharp boundaries between the types, but a gradual change in behavior as the frequency changes.

electromagnetic units *See* c.g.s. system.

electromagnetic waves *See* electromagnetic radiation.

electromagnetism /i-lek-troh-**mag**-nĕ-tiz-ăm/ **1.** The various phenomena resulting from the interplay of electric and magnetic fields.
2. The magnetism produced by an ELECTROMAGNET.

electrometer /i-lek-**trom**-ĕ-ter/ An electronic instrument for measuring low potential differences and currents, while drawing very little current from the circuit. Electrometers are d.c. amplifiers with very high input impedances. Early types (*valve voltmeters*) used special valve circuits for amplifying steady signals. In *vibrating-reed electrometers* a capacitor is used with one plate mechanically vibrated at constant frequency. A steady low potential difference can thus be converted into an alternating one, which can then be amplified. More modern types of electrometer use transistor circuits. Electrometers are often used for measuring very low currents (less than 10^{-9} A) by passing the current through a standard high resistance. Ionization currents are measured in this way.

electromotive force /i-lek-trŏ-**moh**-tiv/ (**e.m.f.**) Symbol: E When a component of an electric circuit (e.g. a cell) does work W in driving a charge Q round the circuit it is said to have an electromotive force E given by $E = W/Q = P/I$ where P is the power and I is the current. As for potential difference, the SI unit is the volt. In some cases the system may work in reverse. For example, work may be done in recharging an accumulator by driving the current backward through it. Then, ideally, the energy stored in the component is equal to EQ. Hence e.m.f. is to be distinguished from potential difference, which in current electricity always refers to energy dissipation. If the current through a resistance is reversed the potential difference is reversed, but energy is dissipated just as before.

Electromotive forces can be generated in a circuit also by electromagnetic induction and by the thermoelectric effects, as in the thermocouple.

If a cell with internal resistance r is connected across an external resistance R the current I is given by

$$I = E/(R + r)$$

The potential difference between the terminals V is IR, hence $V = E - Ir$. If R is increased to infinity, I becomes zero, hence the value of E is equal to the value of V on open circuit. As explained above, although their values may be equal, E and V are distinct in character.

electromotive series (**electrochemical series**) The arrangement of chemical elements in order of their standard electrode potentials. Elements that tend to lose electrons from their atoms and acquire a positive charge are electropositive. Those that gain electrons are below hydrogen in the table and are electronegative.

A standard hydrogen electrode, at which the reaction $H^+ + e \rightarrow \frac{1}{2}H_2$ takes place, is taken as having zero electrode potential. Elements higher in the series acquire positive charge more easily than those lower down, and tend to displace lower elements from solution. For example, if a strip of zinc is placed in a solution of copper sulfate, copper will be deposited and the zinc will dissolve as ions.

electron /i-**lek**-tron/ An elementary particle of negative charge ($-1.602\ 177\ 33 \times 10^{-19}$ C) and rest mass $9.109\ 389\ 7 \times 10^{-31}$ kg. Electrons are present in all atoms in shells around the nucleus. They are emitted from the nucleus in a type of beta decay. They are classified as leptons.

See also elementary particles.

electron affinity 1. The work needed to remove an electron from a negative ion and move it to infinity.
2. A measure of the electronegativity of an atom (i.e. its tendency to acquire electrons). An element's position in the ELECTROMOTIVE SERIES gives an indication of its electron affinity.

electron-beam welding A technique in which welding is carried out in a vacuum chamber using a narrow beam of high-energy electrons. These electrons are emitted from a cathode, then accelerated and focused using a magnetic field. Electron-beam welding has many specialized industrial applications.

electron capture 1. The transformation of a proton into a neutron in an atomic nucleus, accompanied by the emission of x-rays. It occurs when the nucleus 'captures' an orbital electron.
2. The capture of an electron by an atom or molecule to form a negative ion. *See* ionization; neutron capture.

electron charge The charge on an electron, a fundamental constant equal to $-1.602\ 177\ 33 \times 10^{-19}$ coulombs.

electron configuration The arrangement of electrons (in shells) around the nucleus of an atom. Configurations are represented by a number (the principal quantum number), a letter (representing the azimuthal quantum number) and a numerical superscript (indicating the number of electrons in that set). For example, $2p^6$ represents a full third shell (principal quantum number 2 and azimuthal quantum number 1), containing six electrons.

electron diffraction The diffraction of a beam of electrons by matter. Since diffraction is a characteristic property of waves the occurrence of electron diffraction shows the wave nature of electrons, in accord with the expectations of quantum mechanics. Electron diffraction is used to investigate molecules in the gas phase and can be used to establish molecular geometry and bond lengths. It is also an important technique for studying the structure of surfaces and reactions at surfaces. In this case the electrons used have a low energy so as not to penetrate the bulk of the material. The technique is called *low-energy electron diffraction* or *LEED*.

electron gun A device that generates a stream of electrons. In a cathode-ray tube, an electron gun produces the moving spot of light on the screen. An electron gun generally consists of a heated cathode in a vacuum tube. The electrons are emitted by thermionic emission and are accelerated towards a positive electrode (anode) by high voltage. They pass through a hole in the anode to form a narrow beam, which can be focused by electron lenses.

electronic energy levels *See* energy levels.

electronics /i-lek-**tron**-iks/ The study and use of circuitry involving such components as semiconductors, thermionic tubes and other vacuum devices, resistors, capacitors, and inductors.

electron lens A device for focusing a beam of electrons. Electron lenses use electric fields, produced by a system of metal electrodes, or magnetic fields produced by coils.

electron microscope An apparatus for producing a magnified image by using a beam of electrons focused by electron lenses.

In the *transmission electron microscope* the electron beam passes through the sample and is focused onto a fluorescent screen to produce a visual image. Magnifications above 200 000 can be obtained.

The *scanning electron microscope* can be used with thicker samples. A beam of primary electrons is scanned across the specimen and the secondary electrons emitted are focused onto a screen. The magnification and resolution are lower than in the transmission microscope, but three-dimensional images can be obtained.

In optical microscopes the ultimate resolution depends on the wavelength of the light. Better resolution is obtained using high-energy electrons because their wavelength can be shorter at sufficiently high energies.

See de Broglie wave.

electron optics The use of electric and magnetic fields to focus and direct beams of electrons. Similar methods are used with beams of positive or negative ions.

electron probe microanalysis (EPM) A method of quantitative analysis used for small samples in which a beam of electrons is focused onto the surface of the sample so that x-rays of characteristic intensities are emitted.

electron radius Symbol r_e The radius of an electron, a fundamental constant equal to $2.817\,77 \times 10^{-15}$ m. The concept of an electron radius occurred in theories of the electron put forward in the late nineteenth and early twentieth centuries. This concept is now thought to be invalid since it is forbidden both by special relativity theory and by quantum mechanics. *See also* Compton wavelength.

electron rest mass Symbol m_e The mass of an electron at rest, a fundamental constant equal to $9.109\,558 \times 10^{-31}$ kg.

electron shell A group of electrons, around an atomic nucleus, that have the same principal quantum number. Starting with the innermost shell, the quantum numbers are 1, 2, 3, etc. and are sometimes designated K, L, M, etc. Each shell may be sub-divided into sub-shells, designated s, p, d, f, corresponding to azimuthal (orbital) quantum numbers (l) of 0, 1, 2, 3.

electron spin resonance (ESR) A branch of microwave spectroscopy that uses microwave radiation (of known wavelength and frequency) to increase the energy of unpaired electrons in a strong magnetic field. It is employed to investigate chemical bonding, paramagnetism, and electron bands in conductors.

electron tube An electronic component that depends on flow of electrons in a gas or vacuum; for example, a thermionic tube or discharge tube.

electronvolt /i-lek-tron-**vohlt**/ Symbol: eV A unit of energy equal to $1.602\,191\,7 \times 10^{-19}$ joule. It is defined as the energy required to move an electron charge across a potential difference of one volt. It is normally used only to measure energies of elementary particles, ions, or states.

electrophoresis /i-lek-troh-fŏ-**ree**-sis/ The movement of charged colloid particles under the influence of an electric field. Its principal application is in the separation of proteins.

electrophorus /i-lek-**trof**-ŏ-rŭs/ A simple device for supplying electrical charge by electrostatic induction. It consists of a flat insulating plate and a separate metal plate with an insulating handle. The insulating plate is charged by friction; the metal plate is placed on it and momentarily earthed, leaving it with an induced charge.

electroplating /i-**lek**-trŏ-play-ting/ *See* electrodeposition.

electroscope /i-**lek**-trŏ-skohp/ An instrument for detecting electrical potential difference; it can also detect the presence and sign of an electric charge. A common type consists of a pair of light conducting leaves (often of gold foil) suspended side by side from an insulated rod. When a charge is applied to the plate on the top of the rod, the leaves swing apart because they receive electric charges of the same sign and repel each other.

electrostatic field /i-lek-trŏ-**stat**-ik/ An electric field produced by stationary (static) electric charge.

electrostatic generator A machine designed to produce a continuous supply of electric charge by electrostatic means. Typical examples are the Wimshurst machine and the Van de Graaff generator.

electrostatic induction The production of electric charge at some point by separation of positive and negative charges in an electric field. If a body with a positive charge (say) is brought close to a neutral conductor, the free electrons in the conductor flow such that they tend to accumulate in the end near the body. This end thus has a negative charge, the other is positive. There is a resulting force of attraction between the body and the conductor. Similarly, an insulator is charged by induction by polarization.

electrostatic precipitation A method of removing unwanted solid or liquid particles from gases (as in pollution control). The particles are given a charge by an electric field, and then attracted to charged plates.

electrostatics /i-lek-trŏ-**stat**-iks/ (**static electricity**) The study of the effects associated with electric charge at rest. The fundamental principle of electrostatics is that similarly charged bodies repel each other, whereas oppositely charged bodies attract each other. Electric charges at rest generate electric fields. Electrostatics is governed by COULOMB'S LAW.

electrostatic shield A hollow conducting container surrounding an apparatus, used for protection against electric fields. Faraday demonstrated that the electric field inside a closed conductor is zero. (Electrostatic shields are often called *Faraday cages*.)

electrostatic units *See* c.g.s. system.

electrostriction /i-lek-troh-**strik**-shŏn/ A change in the dimensions of a dielectric when it is placed in an electric field. It can be regarded as the converse of the piezoelectric effect.

electroweak theory /i-lek-trŏ-**week**/ A theory unifying the electromagnetic and the weak INTERACTIONS.

element /**el**-ĕ-mĕnt/ In optics, a single lens forming part of a doublet or other compound lens.

elementary particles (**fundamental particles; subatomic particles**) Indivisible particles from which all matter is formed. At one time atoms were believed to be the elementary particles. Later it was realized that they were divisible; that is, made of protons, neutrons, and electrons. More recently many new particles have been identified and attempts made to group them into families and explain their relationships with each other.

Elementary particles are characterized by their mass, charge, spin, and certain other properties ('strangeness', 'charm', etc.). One form of classification is into leptons and hadrons, distinguished by the way in which the particles interact. Hadrons are further divided into baryons and mesons. Finally, baryons are subdivided into nucleons and hyperons. Another classification is into *fermions* (having half-integral spin) and *bosons* (having integral spin).

See table overleaf. See also antiparticle; quark; resonance.

elements, magnetic *See* magnetic elements.

elevation of boiling point An increase in the boiling point of a liquid that generally occurs when a solid is dissolved in it. This increase is proportional to the concentration of particles dissolved. The elevation of the boiling point of a liquid is an example of a COLLIGATIVE PROPERTY.

ellipse A conic with an eccentricity between 0 and 1. An ellipse has two foci. A line through the foci cuts the ellipse at two *vertices*. The line segment between the vertices is the *major axis* of the ellipse. The

ELEMENTARY PARTICLES

Particle	Charge	Symbol	Mass	Spin
leptons				
electron	-1	e^-	0.511	½
neutrino	0	ν	0	½
	0	ν_μ	0	½
muon	-1	μ^-	105.66	½
baryons				
proton	+1	p	938.26	½
neuron	0	n	939.55	½
xi particle	0	Ξ^o	1314.9	½
	-1	Ξ^-	1321.3	½
sigma particle	+1	Σ^+	1189.5	½
	0	Σ^o	1192.5	½
	-1	Σ^-	1197.4	½
lambda particle	0	$\wedge$	1115.5	½
omega particle	-1	Ω^-	1672.5	¾
mesons				
kaon	-1	k^-	493.8	0
	+1	k^+	493.8	0
pion	+1	Π^+	139.6	0
	0	Π^o	135	0
	-1	Π^-	139.6	0
phi particle	0	ϕ	1020	1
psi particle	0	ψ	3095	1
eta particle	0	η^o	548.8	0

Note: it is common to measure the mass of particles in units of energy/c^2, where *c* is the speed of light. The values above are in units of MeV/c^2 = 178 x 10^{-30} kg.

point on the major axis mid-way between the vertices is the *center* of the ellipse. A line segment through the center perpendicular to the major axis is the *minor axis*. Either of the chords of the ellipse through a focus parallel to the minor axis is a *latus rectum*. The area of an ellipse is πab, where a is half the major axis and b is half the minor axis. (Note that for a circle, in which the eccentricity is zero, $a = b = r$ and the area is πr^2.)

The *sum property* of an ellipse is that for any point on the ellipse the sum of the distances from the point to each focus is a constant. The ellipse also has a reflection property; for a given point on the ellipse the two lines from each focus to the point make equal angles with a tangent at that point.

In Cartesian coordinates the equation:

$$x^2/a^2 + y^2/b^2 = 1$$

represents an ellipse with its center at the origin. The major axis is on the x-axis and the minor axis on the y-axis. The major axis is $2a$ and the minor axis is $2b$. The foci of the ellipse are at the points $(+ea,0)$ and $(-ea,0)$, where e is the eccentricity. The two directrices are the lines $x = a/e$ and $x = -a/e$. The length of the latus rectum is $2b^2/a$. *See also* conic.

ellipsoid /i-**lip**-soid/ A three-dimensional figure that can be formed by rotating an ellipse about one of its axes of symmetry. In the case of rotation about the major axis the ellipsoid is called a *prolate ellipsoid*, with the ellipsoid formed by rotation about the minor axis being called an *oblate ellipsoid*.

elliptical polarization A type of polarization of electromagnetic radiation in which the radiation can be considered to consist of two plane-polarized radiations traveling together, out of phase by 90° and of unequal amplitude.

emanation The name formerly given to certain isotopes of radon, which diffused out of the parent material containing radium, from which they are formed by alpha decay. ^{220}Radon is a member of the ^{232}thorium decay series and it was therefore known as *thoron* or thorium emanation. The radon isotope that is a member of the actinium series was known as actinium emanation or *actinon*. Summarizing:
radium emanation is now known as ^{222}Rn;
thoron emanation is now known as ^{220}Rn;
actinium emanation is now known as ^{219}Rn.

e.m.f. *See* electromotive force.

emission spectrum *See* spectrum.

emissive power *See* emissivity.

emissivity /em-ă-**siv**-ă-tee/ (**emissive power**) Symbol: ε The power radiated by a surface compared to that radiated by a black body, all other things being equal. There is no unit. Strictly this should be defined for a given wavelength λ (and called *spectral emissivity*). Kirchhoff's radiation law equates the emissivity of a surface to its absorptance. Thus a perfect reflector has zero absorptance and zero emissivity; a black body has an absorptance of 1 and an emissivity of 1.

emittance /i-**mit**-ăns/ *See* exitance.

emitter *See* transistor.

e.m.u. Electromagnetic unit. *See* c.g.s. system.

emulsion A dispersion of droplets of one liquid in another.

endoergic /en-doh-**er**-jik/ Describing a nuclear process that absorbs energy. *Compare* exoergic.

endothermic /en-doh-**th'er**-mik/ Describing a process in which heat is absorbed (i.e. heat flows from outside the system, or the temperature falls). The dissolving of a salt in water, for instance, is often an endothermic process. *Compare* exothermic.

energy Symbol: *W* A property of a system – its capacity to do work. Energy and work have the same unit: the joule (J). It is convenient to divide energy into KINETIC ENERGY (energy of motion) and POTENTIAL ENERGY ('stored' energy). Names are given to many different forms of energy (chemical, electrical, nuclear, etc.); the only real difference lies in the system under discussion. For example, chemical energy is the kinetic and potential energies of electrons in a chemical compound. *See also* internal energy; mass-energy.

energy bands Ranges of energy that are either allowed or forbidden for valence electrons in a solid. In a solid each quantum state that contains a valence electron in an atom, and each empty state, contributes to a state that extends over a volume containing many atoms. The energies for these states are distributed continuously over bands, which are typically a few electronvolts wide.

In a metal there is one (*valence-conduction*) band of energies within which there are far more quantum states than there are electrons. All the states of low energies are normally filled, and the high ones are empty. Over a small range above and below the Fermi level there are comparable numbers of occupied and empty states. When there is a potential gradient in the metal, electrons move freely from occupied states to adjacent empty states at the same energy (kinetic plus potential), so metals are good conductors.

In a nonmetal there is a lower allowed band (the *valence band*) and an upper allowed band (the *conduction band*). Between these is a range of energies (the *forbidden band*) in which there are few or no quantum states. In pure substances, the number of electrons is exactly equal to the number of states in the valence band. At low temperatures, the valence band is filled and the conduction band is empty. If there is a potential gradient, no conduction can result as there are no empty quantum states of the same energy adjacent to occupied ones, and the Pauli principle excludes more than one electron in a state. Hence non-metals at low temperatures are usually very good insulators. At higher temperatures a certain proportion of the electrons move from the valence band into the conduction band. Conduction is now possible in both bands as there are adjacent occupied and empty states of the same energy. The mechanism of conduction in the almost full valence band is complicated; the solid behaves as if positive charges (*holes*) were moving in the opposite direction to the electrons.

If the *energy gap* between the top of the valence band and the bottom of the conduction band is more than about 2e V the substance is regarded as an insulator; if it is less it is regarded as a semiconductor. Impurities can give rise to very narrow bands of allowed energies within the forbidden band. This can have very large effects upon the properties, and doping with suitable substances can cause either n-type or p-type conductivity.

energy level The energy of a particle, such as an electron, in a system. *See also* quantum state.

engine A device that turns thermal energy into work. Practical engines work by burning fuel and are also known as HEAT ENGINES.

enrich To increase the fraction of one isotope in a mixture of isotopes of the same element. It usually refers to the enrichment of uranium by the ^{235}U isotope (natural abundance about 0.7%) to obtain suitable fissile material for nuclear reactors or nuclear weapons. It can also be used to describe the process of obtaining heavy water from natural water. *See* isotope separation.

enrichment *See* enrich.

entanglement *See* Bell's paradox.

enthalpy /en-**thal**-pee, **en**-thal-/ Symbol: H The sum of the internal energy (U) and the product of pressure (p) and volume (V) of a system: $H = U + pV$. In a chemical reaction carried out at constant pressure, the change in enthalpy measured is the internal energy change plus the work done by the volume change:

$$\Delta H = \Delta U + p\Delta V$$

entropy /**en**-trŏ-pee/ Symbol: S In any system that undergoes a reversible change, the change of entropy is defined as the heat absorbed divided by the thermodynamic temperature: $\delta S = \delta Q/T$. A given system is said to have a certain entropy, although absolute entropies are seldom used: it is change in entropy that is important. The entropy of a system measures the availability of energy to do work.

In any real (irreversible) change in a closed system the entropy increases. Although the total energy of the system has not changed (first law of thermodynamics) the available energy is less – a consequence of the second law of thermodynamics.

The concept of entropy has been widened to take in the general idea of disorder – the higher the entropy, the more disordered the system. For instance, a chemical reaction involving polymerization may well have a decrease in entropy because there is a change to a more ordered system. The 'thermal' definition of entropy is a special case of this idea of disorder – here the entropy measures how the energy transferred is distributed among the particles of matter.

epidiascope /ep-ă-**dÿ**-ă-skohp/ A form of projector able to project images of both translucent objects (e.g. slides) and flat opaque objects (e.g. diagrams and pictures). It is a combined episcope and diascope. Epidiascopes are now rarely used.

episcope /**ep**-ă-skohp/ *See* opaque projector.

epitaxy /**ep**-ă-taks-ee/ (epitaxial growth) The growth of a crystal of one substance on the surface of a crystal of a base substance substance (called the *substrate*) in which the crystal structure of the substance that is being grown has the same crystal structure as the substrate. Epitaxy is used extensively in the semiconductor industry to produce thin layers of material with specific properties.

EPR experiment *See* Bell's paradox.

equation of state An equation that relates the pressure, volume, and temperature of a substance. The simplest equation of state is the ideal gas equation (*see* gas laws) in which it is assumed that the molecules of the gas do not occupy any volume or interact with each other. There are several equations of state for gases that are based on less idealized assumptions. An example is the VAN DER WAALS' EQUATION.

equations of motion Equations that describe the motion of an object with constant acceleration (a). They relate the velocity v_1 of the object at the start of timing to its velocity v_2 at some later time t and the object's displacement s.

The five variables are given in the table. Each of the five equations relates four of the variables, as in the second table, in which the omitted variable is shown in brackets.

QUANTITIES USED IN THE EQUATIONS OF MOTION

Variable	*Symbol*	*Unit symbol*
acceleration	a	m /s^2 (ms^{-2})
first velocity	v_1 (u)	m /s (ms^{-1})
second velocity	v_2 (v)	m /s (ms^{-1})
time	t	s
displacement	s	m

Equations of motion

Variables	*Equation*
$a, v_1, v_2, t, (s)$	$v_2 = v_1 + at^2$
$v_1, v_2, t, s, (a)$	$s = (v_1 + v_2)t/2$
$a, v_1, t, s, (v_2)$	$s = v_1t + at^2/2$
$a, v_2, t, s, (v_1)$	$s = v_2t - at^2/2$
$a, v_1, v_2, s, (t)$	$v_2^2 = v_1^2 + 2\ as$

equator, magnetic *See* isoclinic line.

equilibrant /i-**kwil**-ă-brănt/ A single force that is able to balance a given set of forces and thus cause equilibrium. It is equal and opposite to the resultant of the given forces.

equilibrium /ee-kwă-**lib**-ree-ŭm, ek-wă-/ (*pl.* **equilibriums** or **equilibria**) **1.** A state of

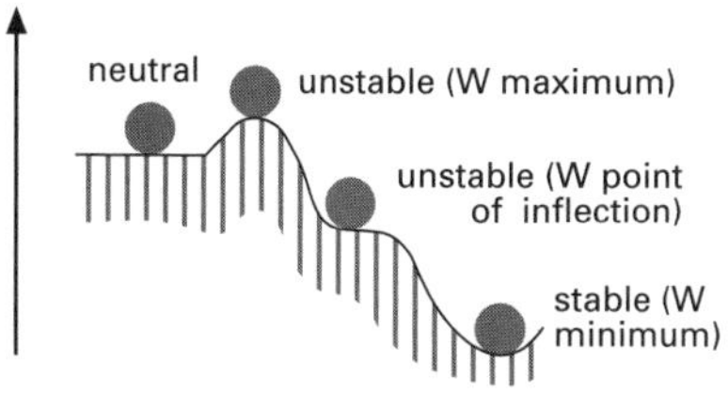

Positions of equilibrium

constant momentum. An object is in equilibrium if:
1. its linear momentum does not change (it moves in a straight line at constant speed and has constant mass, or is at rest);
2. its angular momentum does not change (its rotation is zero or constant).
For these conditions to be met:
1. the resultant of all outside forces acting on the object must be zero (or there are no outside forces);
2. there is no resultant turning-effect (moment).
An object is not in equilibrium if any of the following are true:
1. its mass is changing;
2. its speed is changing;
3. its direction is changing;
4. its rotational speed is changing.
2. A state of a system in which the properties do not change with time. For example a saturated vapor in contact with its liquid is at equilibrium. The pressure, density, temperature, etc., are unchanging. In this case molecules are still leaving the liquid to the gas phase, but are re-entering from the vapor at the same rate. This type of system is said to be in *dynamic equilibrium*. If two bodies at different temperatures are placed together, heat transfer occurs until they

have the same temperature. They are then at *thermal equilibrium*.

A system in which the free energy is at a minimum value is said to be in *thermodynamic equilibrium*. Both the above cases are examples of thermodynamic equilibrium.

equipartition of energy /ee-kwă-par-**tish**-ŏn, ek-wă-/ The principle that the total energy of a molecule is, on average, equally distributed among the available DEGREES OF FREEDOM. It is only approximately true in most cases.

equipotential /ee-kwă-pŏ-**ten**-shăl, ek-wă-/ A line or surface such that all points on it are at the same potential. Equipotentials in a field cut the lines of force at right angles.

equivalence principle A fundamental principle of mechanics, which can be stated in several ways. One statement of the principle is that the inertial mass and the gravitational mass of a body are equal.

erbium /**er**-bee-ŭm/ A soft malleable silvery element of the lanthanoid series of metals. It occurs in association with other lanthanoids. Erbium has uses in the metallurgical and nuclear industries and in making glass for absorbing infrared radiation.

Symbol: Er; m.p. 1529°C; b.p. 2863°C; r.d. 9.066 (25°C); p.n. 68; r.a.m. 167.26.

erect Describing an image that is the same way up as the object.

erecting prism (**inverting prism**) A prism used to invert an image in an optical system without change of size or shape. Erecting prisms operate by internal reflection.

erg A former unit of energy used in the c.g.s. system. It is equal to 10^{-7} joule.

error A difference between the true value of a quantity and the value of that quantity found in an experiment or observation. The effects of *random errors* can be overcome by taking several measurements and averaging out the results. *Systematic errors* occur if the scales or positions in the measuring apparatus are incorrect.

escape velocity The minimum velocity that an object must have in order to escape from the surface of a planet (or moon) against the gravitational attraction. The escape velocity is equal to $\sqrt{(2GM/r)}$, where G is the gravitational constant, M is the mass of the planet, and r is the radius of the planet.

e.s.u. Electrostatic unit. *See* c.g.s. system.

ether /**ee**-th'er/ (**aether**) A hypothetical fluid, formerly thought to permeate all space and to be the medium through which electromagnetic waves were propagated.

europium /yû-**roh**-pee-ŭm/ A silvery element of the lanthanoid series of metals. It occurs in association with other lanthanoids. Its main use is in a mixture of europium and yttrium oxides widely employed as the red phosphor in television screens. The metal is used in some superconductor alloys.

Symbol: Eu; m.p. 822°C; b.p. 1597°C; r.d. 5.23 (25°C); p.n. 63; r.a.m. 151.965.

eutectic /yoo-**tek**-tik/ Describing a mixture of substances with the lowest possible freezing temperature – i.e., no other mixture of the same substances freezes at a low temperature.

evaporation **1.** A change of state from solid or liquid to gas (or vapor). Evaporation can take place at any temperature; the rate increases with temperature. It occurs because some molecules have enough energy to escape into the gas phase (if they are near the surface and moving in the right direction). Because these are the molecules with higher kinetic energies, evaporation results in a cooling of the liquid. *See also* vapor pressure.

2. The process of vaporizing a metal at a high temperature in a vacuum. It is used to coat objects with thin metal films.

exa- Symbol: E A prefix denoting 10^{18}.

For example, 1 exaohm (EΩ) = 10^{18} ohms (Ω).

exchange force A type of force such as that postulated between nucleons in a nucleus. This strong NUCLEAR FORCE operates within a range of up to about 2×10^{-13} cm. The theory holds that protons and neutrons are constantly being transformed into each other, the mediating particle being a meson.

excitation /eks-ÿ-**tay**-shŏn/ **1.** The production of a magnetic field by an electric current in the winding of an electromagnet. **2.** The process of producing an excited state of an atom, molecule, etc.

excitation energy The energy that is required to change an atom, molecule, etc., from one quantum state to a state with a higher energy. The excitation energy (sometimes called *excitation potential*) is the difference between two energy levels of the system.

excitation potential *See* excitation energy.

excited state A state of an atom, molecule, or other system, with an energy greater than that of the ground state. *Compare* ground state.

exclusion principle (**Pauli principle**) The rule that no two identical fermions can have the same set of quantum numbers in a system. For example: all the conduction electrons in a piece of copper must have different sets of quantum numbers; all the protons in a nucleus must have different sets, etc. The rule was proposed by the Swiss physicist Wolfgang Pauli (1900–58) in 1925 for electrons in atoms in order to explain atomic structure. *See* Fermi–Dirac statistics.

exclusive OR gate *See* logic gate.

exitance /**egz**-ă-tăns, **eks**-ă-tăns/ Symbol: *M* The (radiant or luminous) flux emitted from a surface per unit area. *Radiant exitance* (M_e) is measured in watts per square meter (W m^{-2}); luminous exitance (M_v) is measured in lumens per square meter (lm m^{-2}). Exitance was formerly called *emittance*.

exoergic /eks-oh-**er**-jik/ Denoting a nuclear process that gives out energy. *Compare* endoergic.

exothermic /eks-oh-**th'er**-mik/ Denoting a chemical reaction in which heat is evolved (i.e. heat flows from the system or the temperature rises). Combustion is an example of an exothermic process. *Compare* endothermic.

exotic nuclei Types of nucleus that have the highest and lowest possible neutron numbers for a given atomic number. Such nuclei are very unstable but consideration of them is very important in the theory of possible nuclei

expansion, thermal The change of size of a sample (solid, liquid, or gas) when samples increase in size (expand) when warmed and decrease (contract) when cooled. The effect is due to increased motion (energy) of the atoms or molecules at higher temperatures. (Gases also expand when the pressure is reduced.) *See also* Boyle's law; Charles' law; expansivity.

expansion cloud chamber *See* cloud chamber.

expansivity /eks-pan-**siv**-ă-tee/ A measure of the tendency of a substance to undergo thermal expansion. A solid bar, for example, of length l_1 and temperature θ_1, increases to length l_2 when its temperature rises to θ_2:

$$l_2 = l_1(1 + \alpha(\theta_2 - \theta_1))$$

Here, α is the *linear expansivity* of the material of the bar. In fact, the expansivity itself varies with temperature, so the mean value of α is obtained between temperatures θ_1 and θ_2. A more accurate equation is:

$$l = l_0(1 + a\theta + b\theta^2 + c\theta^3 + \ldots)$$

l_0 is the length at 0°C, *a*, *b*, *c*, etc., are constants.

Solids also have a *superficial expansivity* (β) relating to the increase in area. It can be shown that, to a good approximation, $\beta = 2\alpha$, and

$$A_2 = A_1(1 + \beta(\theta_2 - \theta_1))$$

Similarly, the cubic expansivity (γ) relates to increase in volume; $\gamma = 3\alpha$:

$$V_2 = V_1(1 + \gamma(\theta_2 - \theta_1))$$

For liquids, only the cubic expansivity (γ) is useful. There are two values defined. If the expansivity is measured by observing the volume change in a container, the *apparent expansion* is found. This is because the size of the container also changes. The *absolute expansion* is the actual change of volume, allowing for the container. Expansivities were formerly called *coefficients of expansion*. *See also* comparator, Regnault's apparatus.

extensive variable A quantity that is proportional to the size of the system. Mass, energy, and volume are examples of extensive variables. For example, if two systems with the same mass are brought together to form a larger system, then the mass doubles. *Compare* intensive variable.

external combustion engine *See* internal combustion engine.

extraordinary ray *See* birefringent crystal.

extremely high frequency (EHF) A radio frequency in the range between 300 GHz and 30 GHz (wavelength 1 mm–1 cm).

extrinsic semiconductor /eks-**trin**-sik, -zik/ *See* semiconductor.

eye A sense organ able to produce optical images and to send corresponding nerve signals to the brain. Light enters through the transparent *cornea* and passes through the watery liquid (*aqueous humor*), lens, and gelatinous substance (*vitreous humor*) onto the retina at the back of the eye. The cornea, aqueous humor, and lens form a converging system, which allows an image to be focused on the retina. The focal distance of their system can be varied by the ciliary muscles, which control the curvature of the lens. The quantity of light admitted is varied by the circular iris; this contracts or relaxes, changing the aperture of the eye (the pupil). Light-sensitive cells in the retina produce electrical signals, which are carried along the optic nerve to the brain. The cells are most concentrated in the *yellow spot* (*macula lutea*), and especially in its central part (*fovea*). Images produced here are seen in greatest detail. Two types of cell are found in the retina: the cones, which are used for photopic ('daylight') vision and COLOR VISION; and rods, which are used for peripheral and scotopic ('night-time') vision. The cones are particularly concentrated in the yellow spot whereas the rods cover a large area around this. *See also* accommodation; hyperopia; myopia; presbyopia.

eye lens The lens nearest the eye in a compound EYEPIECE. *Compare* field lens.

eyepiece (**ocular**) The lens or combination of lenses nearest the eye in an optical instrument. It is used to produce a final virtual magnified image of the previous image in the system. Various compound lens systems may be used to reduce aberrations. Often crosswires or a graticule is included in the structure of the eyepiece to assist with location or measurement.

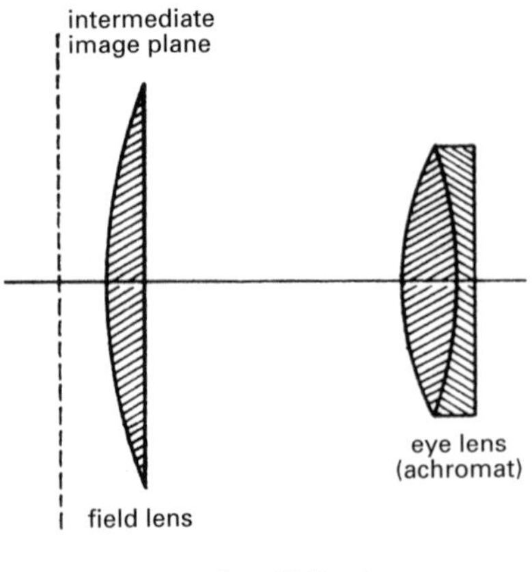

Eyepiece (Kellner)

F

Fabry–Perot interferometer /**fab**-ree **pair**-roh/ A type of interferometer first constructed by Charles Fabry and Alfred Perot in the early 19th century, in which monochromatic light passes through a pair of parallel half-silvered plates, with it being possible to change the distance between the plates by adjusting one of the plates. This produces an interference pattern consisting of circular fringes. The device has been used in spectroscopy.

Fahrenheit scale A TEMPERATURE SCALE in which the ice temperature is taken as 32° and the steam temperature is taken as 212° (both at standard pressure). It is not used for scientific purposes. To convert between degrees Fahrenheit (F) and degrees Celsius (C) the formula $C/5 = (F - 32)/9$ is used. The scale was devised by the German physicist Daniel Fahrenheit (1686–1736).

fallout Radioactivity precipitated from the atmosphere, usually originating from nuclear explosions, occasionally from releases of radioactivity from nuclear establishments. It consists of fission fragments, the most worrying of which are ^{131}I (half-life 8 days), which becomes concentrated in the thyroid gland if ingested, and ^{90}Sr (half-life 28 years), which accumulates in bone. Both may reach the human population in milk from cows grazing on land contaminated by fallout and by other less significant routes.

farad /**fa**-răd, -rad/ Symbol: F The SI unit of capacitance. When the plates of a capacitor are charged by one coulomb and there is a potential difference of one volt between them, then the capacitor has a capacitance of one farad. 1 farad = 1 coulomb volt^{-1} (C V^{-1}). It is named for the English physicist and chemist Michael Faraday (1791–1867).

faraday /**fa**-ră-day/ An old unit of electrical charge equal to the FARADAY CONSTANT.

Faraday cage *See* electrostatic shield.

Faraday constant Symbol: *F* The electric charge of one mole of electrons, equal to $9.648\ 670 \times 10^4$ coulombs per mole. It is the Avogadro constant multiplied by the electron charge. *See* Faraday's laws (of electrolysis).

Faraday disk *See* homopolar generator.

Faraday effect The rotation of the plane of polarization of radiation in a dielectric medium in a magnetic field. The angle of rotation (θ) is proportional to the field strength B, and to the length l of the path in the medium in the field: θ is proportional to Bl. *See also* Kerr effect.

Faraday's laws (of induction) Three laws describing the electromagnetic induction of an e.m.f. in a conductor by a changing magnetic flux:
(1) If the number of lines of force linked with a conducting circuit is changing, a current is induced in the circuit.
(2) The direction of the induced current is such that the magnetic field it produces tends to keep the flux linkage constant. (This is a statement of Lenz's law).
(3) The total quantity of electricity passing round the circuit is directly proportional to the total change in lines of force (and inversely proportional to the resistance of the circuit).

Faraday's laws (of electrolysis) Two laws

resulting from the work of Michael Faraday on electrolysis:
(1) The amount of chemical change produced is proportional to the charge passed.
(2) The amount of chemical change produced by a given charge depends on the ion concerned. More strictly it is proportional to the relative ionic mass of the ion, and the charge on it. Thus the charge Q needed to deposit m grams of an ion, relative ionic mass, M, carrying Z charges, is given by:

$$Q = FmZ/M$$

F, the Faraday constant, has a value of one faraday, i.e. $9.648\ 670 \times 10^4$ coulombs per mole.

Faraday's laws of electromagnetic induction *See* electromagnetic induction; Faraday's laws (of induction).

far infrared The longer-wavelength part of the infrared region of the electromagnetic spectrum; i.e. the part farthest in wavelength from the visible region and nearest to the radio-wave region.

far point The farthest point from the eye at which an image can be focused on the retina. The eye is completely relaxed (minimum power) in focusing an object at its far point. In the normal eye the far point is at infinity. In the myopic (short-sighted) eye, light from distant objects is focused in front of the retina. The far point is then only a few meters from the eye. *See also* myopia.

far sight *See* presbyopia.

fast neutron A high-energy neutron which, when produced by NUCLEAR FISSION, is too energetic to produce further fission and thereby sustain a nuclear chain reaction. *See also* moderator; slow neutron.

fast reactor *See* nuclear reactor.

fatigue Physical weakness in a material resulting from many small repeated stresses.

feedback The action of returning some of the output energy of a device (e.g. an amplifier) to the input. There are two types – *positive feedback* and *negative feedback* – according to the relative phases of the feedback voltage and the input signal. The feedback is negative if these voltages are out of phase, and positive if the voltages are in phase.

Although negative feedback reduces the input voltage and input energy, and hence also the gain of the amplifier, the advantage is that noise and distortion are also reduced, while stability is improved. Positive feedback increases the gain and energy output of the amplifier, with the result that the amplifier becomes an oscillator. Feedback is made through a loop circuit using a passive component.

Negative feedback is also used in other control systems. For example, the stabilizer fitted to an ocean-going ship utilizes negative feedback by generating a voltage proportional to the instability (e.g. the angle of tilt) and using this voltage to drive the control system. Thus the effect of the voltage is to reduce the cause which generated it. Another, non-electrical, example is the centrifugal governor.

femto- Symbol: f A prefix denoting 10^{-15}. For example, 1 femtometer (fm) = 10^{-15} meter (m).

Fermat's principle The basic law of geometrical optics. It states that a ray will take a path through a medium such that the time required to cover a given distance is minimum. The law leads to the concept of rectilinear propagation and the laws of reflection and refraction. It is named for the French mathematician Pierre de Fermat (1601–65).

fermi /**fer**-mee/ A unit of length equal to 10^{-15} meter. It was formerly used in atomic and nuclear physics.

Fermi–Dirac statistics /**fair**-mee dee-**rak**/ The statistical rules for studying systems of identical fermions. It is assumed that (a) all identical particles are to be regarded as absolutely indistinguishable; and (b) fermions obey the Pauli exclusion principle, that is that no two identical fermions can have the same set of quantum numbers in a given system. The distribution of electron energies in a solid is one extremely important subject to which these rules apply. The statistics are named for the Italian physicist Enrico Fermi (1901–54) and the British mathematician and physicist Paul Dirac (1902–84). *See also* Bose–Einstein statistics; Maxwell–Boltzmann statistics; energy bands.

Fermi level The energy level in a solid at which a quantum state would be equally likely to be empty or to contain an electron. In a metal the Fermi level is in the conduction band. At absolute zero all the quantum states up to this level would be occupied and all states with higher energy would be empty. At non-zero temperatures some states below this level are empty and some above it are occupied. In insulators and semiconductors the Fermi level is in the forbidden band between the valence band and the conduction band.

When two different substances are brought into contact (for example an n-type and a p-type semiconductor) electrons flow until the Fermi levels of the two substances are at the same energy.

fermion /**fer**-mee-on/ A particle that obeys Fermi–Dirac statistics and has half-integral spin.

Elementary fermions are conserved in number in all interactions. The production of a new fermion is always accompanied by the production of an ANTIPARTICLE which is counted as *minus* the particle. The lepton is one type of fermion which appears most often in the charged form, the electron. When an electron is created in beta decay an antineutrino is also produced. This can be regarded as an uncharged antiparticle, and is counted as *minus one* lepton. The positron is regarded as a charged antiparticle to the electron. When electron–positron pairs are generated by radiation or annihilate to radiation the total number of leptons is conserved.

Similarly baryons are conserved. In a high-energy reaction a proton–antiproton pair may be generated. The proton counts as *plus one* baryon and the antibaryon as *minus one*. Thus the number remains constant on production of the pair, and similarly when the antiproton subsequently annihilates with a proton, producing pions.

See boson; elementary particles; Fermi–Dirac statistics.

fermium /**fer**-mee-ŭm/ A radioactive transuranic element of the actinoid series, not found naturally on Earth. It is produced in very small quantities by bombarding ^{239}Pu with neutrons to give ^{253}Fm (half-life 3 days). Several other short-lived isotopes have been synthesized.

Symbol: Fm; p.n. 100; most stable isotope ^{257}Fm (half-life 100.5 days).

ferrimagnetism /fe-ri-**mag**-nĕ-tiz-ăm/ A type of magnetic behavior shown by certain solids (e.g. ferrites). Ferrimagnetic materials are similar to ferromagnetic materials in behavior, but have a weaker magnetization. They contain two types of magnetic particle with unequal magnetic moments aligned antiparallel, producing a net magnetization. *See also* antiferromagnetism.

ferrite /**fe**-rÿt/ One of a class of iron compounds with weak permanent magnetism. Ferrites are important as they are electrical insulators so cannot suffer from eddy-current power loss, but they do have a permanent magnetization as a result of their ferrimagnetism. They are used as cores in high-frequency circuits.

The composition of the ferrites is $Fe_2O_3.MO$, where M is a divalent metal. One common example is magnetite, in which M is Fe^{2+}.

ferroelectric /fe-roh-i-**lek**-trik/ A type of solid material that has a spontaneous dipole moment. The name comes from the analogy with a spontaneous magnetic moment in a ferromagnetic material. Like ferromagnets, ferroelectric materials have a domain structure and undergo hysteresis. Barium titanate and Rochelle salt are examples of ferroelectric materials.

ferromagnetism /fe-roh-**mag**-nĕ-tiz-ăm/ The MAGNETISM of substances caused by a domain structure. Such substances have a large susceptibility and relative permeability. Rough values of relative permeability are: iron 1000, cobalt 50, nickel 40. *See also* Curie's law; hysteresis cycle.

fertile material Material that can be converted into fissile material by irradiation with neutrons. ^{238}U and ^{232}Th are fertile materials, being converted into ^{239}Pu (via ^{239}U) and ^{233}U (via ^{233}Th) respectively. Conversion is commonly carried out in breeder reactors. Present breeder reactors use ^{238}U as the fertile material.

FET *See* field-effect transistor.

Feynman diagram /**fÿn**-măn/ A drawing convention that represents interactions between elementary particles, which are represented as directional lines; photons are drawn as wavy lines. The diagrams are named for the American physicist Richard Feynman (1918–88).

fiber optics The use of fine transparent fibers to transmit light. The light passes along the fibers by a series of internal reflections. *Optical fibers* of this type can be used to view inaccessible objects and to carry laser signals in telecommunications.

field The concept of a field was introduced to explain the interaction of particles or bodies through space. An electric charge, for instance, modifies the space around it such that another charge in this region experiences a force. The region is an electric field (a field of force). Magnetic and gravitational fields can be similarly described.

field coil The winding in a generator or electric motor that produces the magnetic field.

field-effect transistor (FET) A solid-state electronic device with three terminals that, like the junction TRANSISTOR, is used in amplifiers. It controls the current between two terminals, the *source* and the *drain*, by the voltage at a third, the *gate*. An n-type FET consists of a single piece of n-type semiconductor, which has the source at one end and the drain at the other. A heavily-doped p-type region in the middle forms the gate. If the gate voltage is more negative than the source, electrons move from the n-type region to the p-type region leaving an area round the gate with fewer electrons to carry current. This has the effect of narrowing the conducting channel through which the source-drain current flows by an amount that depends on the source-gate voltage difference. In the FET, only one type of charge carrier – electrons in the n-type FET and holes in the p-type FET – determines the current and it is therefore known as a *unipolar* transistor. In the *bipolar* junction transistor, both positive and negative charge carriers contribute to the other current.

field emission (**cold emission**) The release of electrons from a surface as a result of a strong external electric field. Very high electric fields are necessary; these are obtained at sharp points.

field-emission microscope (**FIM**) A type of electron microscope which uses field emission from a metal point at very high voltage to investigate atoms in the surface of the material. Electrons emitted from the point are accelerated to a conducting fluorescent screen, on which they form a magnified image of the metal tip.

field ionization The ionization of atoms at a surface caused by a high electric field. It is used in the *field-ion microscope*, which operateson a similar principle to the field-emission microscope.

field lens The lens farthest from the eye in a compound eyepiece. *Compare* eye lens.

field magnet An electromagnet or permanent magnet that provides the magnetic field in an electric machine, such as a motor or generator.

field strength, electric *See* electric field strength.

fifth force *See* interaction.

filament A fine thread of metal, glass, etc.

filter 1. A device that passes only certain frequencies of radiation. A red filter, for example, is a piece of translucent material that transmits red light, absorbing other wavelengths.
2. An electronic circuit that transmits currents within a certain range of frequencies, while reducing (attenuating) other frequencies.

filter pump A simple type of laboratory vacuum pump in which there is a narrow nozzle in the middle of the pump through which water is forced. There is a reduction in pressure near the nozzle (*see* Bernoulli effect), causing air to flow in from a side tube that connects the pump to the vessel that is being evacuated. The water and air are forced through the bottom of the pump.

fine structure Closely spaced lines seen at high resolution in a spectral line or band. Fine structure may be caused by vibrational motion of the molecules or by electron spin. *Hyperfine structure*, seen at very high resolution, is caused by interactions between the spin of the atomic nucleus and the orbiting electrons affecting the possible energy levels of the atom.

fine structure constant Symbol α. A constant that characterizes the electromagnetic interaction at the quantum level. It is defined by $\alpha = e^2/\hbar c$, where e is the charge of an electron, $\hbar$ is the Dirac constant, i.e. the Planck constant divided by 2π, and c is the speed of light in a vacuum. It has been suggested that the value of α has changed during the evolution of the Universe but at present there is not conclusive evidence for this.

fissile material /**fiss**-ăl/ Material that undergoes nuclear fission, sometimes spontaneously but usually when irradiated by neutrons. Examples are ^{235}U, ^{239}Pu, and ^{233}U. Fissile material is the fuel of nuclear reactors and the explosive material of nuclear weapons.

fission /**fish**-ŏn/ *See* nuclear fission.

fission products Isotopes that result from NUCLEAR FISSION, usually having about the same mass.

fission reactor *See* nuclear reactor.

fission-track dating A method of measuring the age of glasses and other minerals that depends on the tracks made in these solids by fragments from the spontaneous fission of contained uranium. The density of tracks in the material depends on the uranium content, the age of the material, and any fading of the tracks. After counting the tracks, the uranium content is estimated by irradiating the materials with neutrons to induce fissions. The number of extra tracks produced depends on the uranium content, which can then be estimated, giving the age since solidification.

fixed point *See* International Practical Temperature Scale; temperature scale.

flash point The lowest temperature at which a volatile substance produces enough flammable vapor to produce a flash when a flame is applied.

flavor *See* quark.

F-layer *See* Appleton layer.

Fleming's rules A method of remembering the relationship between the directions of current, magnetic field, and motion in electric motors and generators. Fleming's *left-hand rule* applies to motors. If the left hand is held with the thumb and first two fingers all mutually at right angles, the thu*m*b shows *m*otion (force); the *f*irst finger, *f*ield; and the se*c*ond finger, *c*urrent. Fleming's *right-hand rule* is similar but with the fingers of the right hand showing the operation of an electrical generator. The rules were formulated by the British physicist and electrical engineer Sir John Ambrose Fleming (1849–1945).

flint glass *See* optical glass.

flip-flop *See* bistable circuit.

floating objects, law of *See* flotation; law of.

flotation, law of An object floating in a fluid displaces its own weight of fluid. This follows from Archimedes' principle for the special case of floating objects. (A floating object is in equilibrium, its only support coming from the fluid. It may be totally or partly submerged.)

fluctuations Random differences between the average value of a quantity and the actual values found. Fluctuations are important in statistical mechanics, especially the theory of phase transitions and Brownian motion, and in quantum mechanics.

fluid A substance that flows; i.e. a liquid or a gas.

fluidics /floo-**id**-iks/ The study and use of fluid flow through pipes in an analogous way to the flow of electric current through circuits. Fluidic circuits may be useful in circumstances in which there are high magnetic fields or radiation levels, which would affect electrical circuitry.

fluidity The property of a fluid that enables it to flow easily; it is the opposite of VISCOSITY.

fluidization /floo-id-ă-**zay**-shŏn/ A method of passing a gas through a finely powdered solid so that is acquires the characteristics of a fluid (so that it flows, can be pumped, etc.).

fluid mechanics The study of the mechanics of fluids, i.e. liquids and gases. The study of fluids that are at rest is called *fluid statics* and the study of fluids in motion is called *fluid dynamics*. The specific study of liquids at rest is called *hydrostatics*. Fluid dynamics is divided into *hydrodynamics*, which is the study of motion in liquids, and *aerodynamics*, which is the study of motion in gases.

fluorescence /floo-ŏ-**ress**-ĕns/ The absorption of energy by atoms, molecules, etc., followed by immediate emission of electromagnetic radiation as the particles make transitions to lower energy states. Fluorescence is a type of LUMINESCENCE in which the emission of radiation does not persist after the exciting source has been removed. The excited states have very short lifetimes.

Fluorescence, like luminescence, can be produced in a number of ways. Examples are by bombardment with electrons and by absorption of other electromagnetic radiation. In the case of electromagnetic radiation the emitted radiation often has lower frequency than the absorbed radiation. A particular example is the absorption of ultraviolet radiation followed by emission of visible radiation. This effect is used in fluorescent paints and in 'fabric brighteners' added to detergents. The inside coating of fluorescent tubes is another example of a material that converts ultraviolet radiation (from the discharge) into visible radiation.

fluorescent lamp /floo-ŏ-**ress**-ĕnt/ A lamp consisting of an electric glow discharge tube, the inner surface of which is coated with phosphor. The gas in the tube (often mercury vapor) gives visible light and ultraviolet. The ultraviolet radiation generates further visible light in the phosphor by fluorescence. The fluorescence increases the visible output and also, by a suitable choice of phosphor, it is possible to obtain a combination of wavelengths visually equivalent to white light.

fluorine /**floo**-ŏ-reen, -rin/ A slightly greenish-yellow highly reactive gaseous element belonging to the halogens. It is slightly more abundant than chlorine, accounting for about 0.065% of the Earth's crust. Fluorine is the most reactive and most electronegative element known; in fact it reacts directly with almost all the elements, including the rare gas xenon.

Symbol: F; b.p. −188.14°C; m.p. −219.62°C; d. 1.696 kg m^{-3} (0°C); p.n. 9; r.a.m. 18.99840.32.

fluoroscope /**floo**-ŏ-rŏ-skohp/ An instrument that makes x-rays, or other non-visible radition, visible by means of a fluorescent screen.

flux In general, a flow or apparent flow.
1. A flow of particles per unit area.
2. The density of lines of force in a field. *See* magnetic flux.
3. *See* luminous flux.

flux density *See* magnetic flux density.

fluxmeter /**fluks**-mee-ter/ An instrument for measuring magnetic flux. The most common type uses a small movable exploring coil (a *search coil*), which is placed in the field and then removed quickly. The induced current can be measured by a ballistic galvanometer. *See also* Earth inductor.

flywheel A large heavy wheel (with a large moment of inertia) used in mechanical devices. Energy is used to make the wheel rotate at high speed; the inertia of the wheel keeps the device moving at constant speed, even though there may be fluctuations in the torque. A flywheel acts as an 'energy-storage' device.

FM *See* frequency modulation.

f-number For an optical instrument, especially a camera, the ratio of the focal distance (f) to the aperture diameter (d):

$$\text{f-number} = f/d$$

If the ratio is 16 the f-number is written f16, f:16, or f/16.

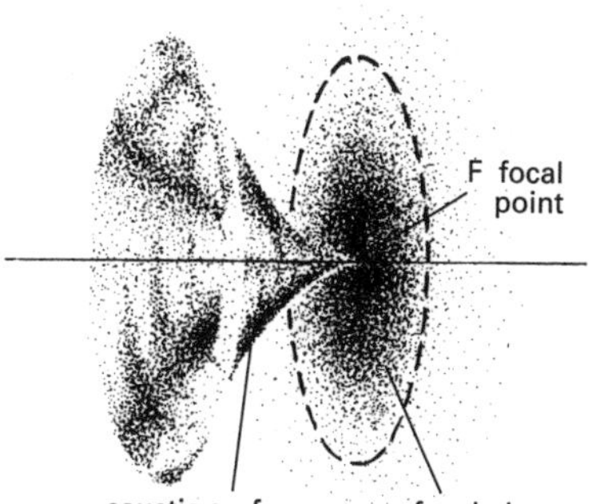

Focal point

focal distance (**focal length**) A measure of the power of a lens or mirror to converge a parallel beam. The focal distance of a lens or mirror is the distance between the pole and the focal point. Normally, positive values relate to converging lenses and mirrors, and negative values relate to diverging ones, (real is positive, virtual is negative convention). The New Cartesian convention is less common. It gives positive focal distances to diverging mirrors and converging lenses, and negative values for converging mirrors and diverging lenses.

For a refracting surface:

$$f = n_1r/(n_2 - n_1)$$

where n is refractive constant and r is radius of curvature.

For a reflecting surface:

$$f = r/2$$

For a thin lens in air:

$$1/f = (n - 1)(1/r_1 - 1/r_2)$$

For two thin lenses in contact:

$$1/f = 1/f_1 + 1/f_2$$

focal length *See* focal distance.

FOCAL POINTS FOR LENSES AND MIRRORS

	Focal point	Position	Focal distance
converging lens	real	far side of lens	positive
diverging lens	virtual	near side of lens	negative
converging mirror	real	in front of mirror	positive
diverging mirror	virtual	behind mirror	negative

focal plane The plane perpendicular to the axis centered on the focal point of a lens or mirror.

focal point (**focus**; **principal focus**) A point through which rays close and parallel to the axis pass or appear to pass after reflection or refraction. A reflector has one focal point. As radiation can enter a lens from either side, a symmetrical lens has a focal point on each side, at the same distance from the lens. Only rays close and parallel to the axis pass through the focal point. These are called *paraxial* rays. Rays parallel to the axis but not close to it focus somewhere on the caustic surface. Rays close to the axis but not parallel to it focus somewhere on the focal plane. *See illustration on page 97.*

focal power *See* diopter; power.

focus (*pl.* **focuses** or **foci**) **1.** *See* focal point.
2. An image is in focus if it is sharp rather than blurred (out of focus). Only then do the set of rays from a point on the object all pass through a single point.
3. A point through which each member of a set of rays passes or appears to pass after reflection or refraction. Paraxial rays focus at the focal point; other sets may focus on the focal plane or on the caustic surface. Others again, from the surface of an object, focus on the image plane to form the image.

foot Symbol: ft The base unit of length in the f.p.s. system (one third of a yard). It is equal to 0.304 8 meter.

foot lambert An obsolete unit of luminance equal to the luminance of a surface that reflects or emits one lumen per square foot. It is equal to 3.426 259 candelas per square meter.

foot pound A f.p.s. unit of energy or work equal to the work required to move a one-pound mass one foot in the direction of the applied force. It is equal to 1.355 818 joules.

forbidden band *See* energy bands.

force Symbol: F That which tends to change an object's momentum. Force is a vector; the unit is the newton (N). In SI, this unit is so defined that:

$$F = \mathrm{d}(mv)/\mathrm{d}t$$

(from Newton's second law). *See* Newton's laws of motion.

forced convection *See* convection.

forced oscillation The oscillation of a system or object subjected to an external periodic force. The oscillations have the same frequency as the applied force. The amplitude depends greatly upon the difference between the driving frequency and the frequency of free oscillation of the system. *Compare* free oscillation. *See also* resonance.

force meter An instrument for directly measuring force (in newtons). It resembles a spring balance.

force ratio (**mechanical advantage**) For a simple machine, the ratio of the output force (load) to the input force (effort). There is no unit; the ratio is, however, often given as a percentage. It is quite possible for force ratios far greater than one to be obtained. Indeed the design of many machines is to make this so, to make a small effort overcome a large load.

However efficiency cannot be greater than one; a large force ratio implies a large distance ratio. *See also* efficiency; machine.

forces, parallelogram (law) of *See* parallelogram of vectors.

forces, triangle (law) of *See* triangle of vectors.

formula (*pl.* **formulas** or **formulae**) **1.** A general rule, often represented by symbols. For example, $A = \pi r^2$ is a formula giving the area of a circle in terms of the radius.
2. A representation of a compound (or molecule) using the chemical symbols for the elements.

Foucault pendulum /foo-**koh**/ A simple pendulum consisting of a heavy bob on a long string. The plane of vibration rotates slowly over a period of time as a result of the rotation of the Earth below it. The apparent force causing this movement is the so-called Coriolis force. The pendulum is named for the French physicist Jean Bernard Léon Foucault (1819–68).

Fourier methods /**foo**-ree-ay/ Methods for analyzing periodic functions in which the periodic function is represented as a sum of sine waves in which the frequencies of all the waves are simple multiples of a fundamental frequency. There are many important applications of Fourier methods in physics, such as the theory of sound and the analysis of patterns in x-ray crystallography. This analytical approach is named for the French mathematician and physicist Jean Baptiste Joseph Fourier (1768–1830).

fovea /**foh**-vee-ă/ (*pl.* **foveae**) An area about half a millimeter across in the center of the yellow spot of the human retina. Here the cones are most dense and there are no rods. It is the part with the greatest visual acuity but has low sensitivity to dim light. *See* eye; yellow spot.

fps units A system of units based on the foot, pound, and second. It has been replaced in scientific use by SI units.

frame of reference A set of coordinate axes by which the position of any object may be specified as it changes with time. The origin of the axes and their spatial direction must be specified at every instant of time for the frame to be fully determined.

francium /**fran**-see-ŭm/ A radioactive element of the alkali-metal group. It is found on Earth only as a short-lived product of radioactive decay, occurring in uranium ores in minute quantities. A large number radioisotopes of francium are known.

Symbol: Fr; m.p. 27°C; b.p. 677°C; p.n. 87; most stable isotope ^{223}Fr (half-life 21.8 minutes).

Fraunhofer diffraction /**frown**-hoh-fer/ Diffraction observed with incident parallel light. In Fraunhofer diffraction the wavefronts are parallel. Although a special case of FRESNEL DIFFRACTION, it is far more important in most practical cases. Thus it is used to explain single – and multiple – slit patterns, as well as those produced by circular holes. The Fraunhofer diffraction is named for the German optician and physicist Josef von Fraunhofer (1787– 1826).

Fraunhofer lines The dark lines in the spectrum of light from the Sun, caused by the absorption of particular wavelengths by certain elements in its cooler outer regions. The wavelengths of these lines are used as reference points in specifying quantities that vary with wavelength, e.g. refractive constants.

free electron An electron that can move freely from one atom or molecule to another under the influence of an external electric field. The moving free electrons constitute an electric current, as in conducting metals or electrons moving in a thermionic tube.

free energy A measure of the ability of a system to do useful work. *See* Gibbs function; Helmholtz function.

free fall Movement of a body in a gravitational field with no other forces acting on it. *See* acceleration of free fall; weightlessness.

free oscillation (**free vibration**) An oscillation at the natural frequency of the system or object. Thus a pendulum can be forced to swing at any frequency; it will swing freely at only one, which depends on its length. *Compare* forced oscillation. *See also* resonance.

free surface energy *See* surface tension.

free vibration *See* free oscillation.

freezing The change from the liquid to the solid state that occurs when energy is transferred from a substance. For pure substances this happens at a characteristic temperature known as the *freezing point* (or *melting point*). This depends on pressure, so is usually measured at standard atmospheric pressure. Impurities generally lower the freezing point.

freezing mixtures Two or more substances mixed together to produce a low temperature. A mixture of salt and ice in water is a common example.

freezing point *See* freezing.

Frenkel defect /**frenk**-ĕl/ *See* defect.

frequency Symbol: f, ν The number of cycles per unit time of an oscillation (e.g. a pendulum, vibrating system, wave, alternating current, etc.). The unit is the hertz (Hz).

Angular frequency (pulsatance) are related to frequency by $\omega = 2\pi f$. The unit is the radian per second (rad s^{-1}).

frequency modulation (FM) A type of modulation in which a carrier wave is made to carry the information in a signal (audio or visual) by fluctuations in the frequency of the carrier. The variation of the carrier frequency is proportional to the frequency of the signal, while the amplitude of the carrier remains constant. Frequency modulation is superior to amplitude MODULATION because a wider band of frequencies may be transmitted with less interference and noise, although it is limited to line-of-sight distances. *See also* carrier wave.

Fresnel biprism /fray-**nel**/ *See* biprism, Fresnel's.

Fresnel diffraction Diffraction observed when either source or screen (or both) are close to the diffractor. In Fresnel diffraction, the wavefronts are not plane (as in Fraunhofer diffraction) and analysis is difficult. The approach is useful in explaining (for instance) diffraction around a circular obstacle. Fresnel diffraction is named for the French physicist Augustin Jean Fresnel (1788–1827).

Fresnel lens A type of lens with one surface cut in steps so that transmitted light is refracted just as if by a much thicker (and heavier and more expensive) conventional lens. Lighthouse lamp lenses have long been of this type. Very cheap plastic Fresnel lenses now have many uses, for example in overhead projectors, flashlamps, etc. The angle of each step is made to produce the desired effect.

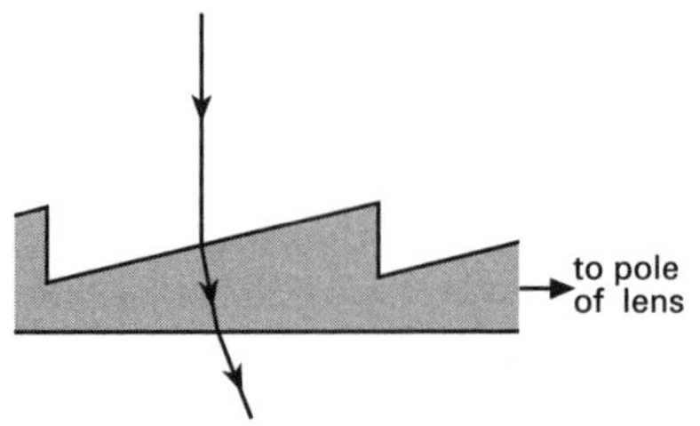

Fresnel lens

Fresnel zones *See* half-period zones.

friction Friction is the name given to various forces that oppose the relative motions of bodies.

Fluid friction is usually caused by turbulent flow for large systems and high speeds, and by viscosity for low speeds and small systems, for example thin films of lubricant.

When two solid surfaces are in contact each exerts a force of friction on the other, preventing sliding, up to a limiting value, the force of *static* or *limiting* friction. If sliding occurs there are dissipative forces of *kinetic friction*. When a body rolls on a flat surface its motion is opposed by *rolling friction*.

The *laws of friction* are:

1. The frictional forces are independent of the apparent area of contact for the same normal force.
2. The forces of limiting, kinetic, and rolling friction are proportional to the normal forces.

These laws are fairly accurate for limiting and kinetic friction but less so for rolling friction.

The coefficient of limiting friction is defined by

$$\mu_L = F/S$$

where F is the limiting value of the force of friction and S is the normal (support) force. μ_L is very nearly constant for given surfaces and typically has values from 0.5 to 2.

The coefficient of kinetic friction is defined by

$$\mu_K = F_1/S$$

where F_1 is the force of friction when the bodies are sliding at constant speed. μ_K is generally less than μ_L for the same surfaces and is independent of speed, except at high speeds.

The coefficient of rolling friction is defined by

$$\mu_R = F_R/S$$

where F_R is the force opposing the motion of the rolling body. Generally μ_R is much less than μ_L and μ_K for the same materials. It increases slightly with increase of S and depends in a complicated way upon speed. It is smaller for rollers or wheels of large radius.

Friction generally is caused by imperfect elasticity in the substances interacting. In the case of hard materials such as metals the real area of contact is much less than the apparent area. The bodies interact by microscopic irregularities (*asperities*), which occur even on those surfaces which appear smooth on optical examination or touch. The asperities are subject to large stresses and therefore undergo plastic flow or even weld together. Lubricants keep the surfaces apart and so prevent the asperities from touching. For softer substances the real areas of contact are larger and friction involves elastic hysteresis in the surface layers.

Rolling friction is caused mainly by elastic hysteresis, for example in the case of a car tire on hard ground it is caused by hysteresis in the rubber. Lubrication has very little effect on rolling friction since there is generally no slip between the surfaces.

fringes The light and dark bands obtained by interference or diffraction of light.

frost 1. Hoarfrost or white frost: a deposit of needle-like ice crystals on the ground or other exposed surfaces. It is formed by direct condensation of water vapor on sur-

Slip about to occur:
$\mu_s = F/R = P/W$
μ_s is coefficient of sliding friction

Slip at steady speed:
$\mu_k = F_1/R = P_1/W$
μ_k is coefficient of sliding (kinetic) friction
$P_1 < P \quad \mu_k < \mu_s$

Slip on a slope:
$\mu = F/R = \tan\theta$
(This can be applied either to static or to sliding friction)

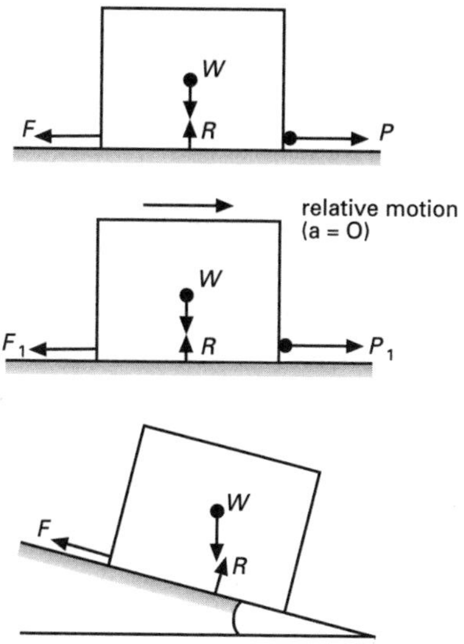

Friction

SOME FUNDAMENTAL CONSTANTS

Quantity	*Magnitude*	*Unit*
speed of light	$2.997\,925 \times 10^{8}$	$m\ s^{-1}$
Planck constant	$6.626\,069 \times 10^{-34}$	J s
Boltzmann constant	$1.380\,625\,5 \times 10^{-23}$	$J\ K^{-1}$
Avogadro constant	$6.022\,142 \times 10^{23}$	mol^{-1}
mass of proton	$1.672\,622 \times 10^{-27}$	kg
mass of neutron	$1.674\,927 \times 10^{-27}$	kg
mass of electron	$9.109\,383 \times 10^{-31}$	kg
charge of proton or electron	$\pm 1.602\,176\,5 \times 10^{-19}$	C
specific charge of electron	$-1.758\,820 \times 10^{11}$	$C\ kg^{-1}$
molar volume at s.t.p.	$2.241\,40 \times 10^{-2}$	$m^3 mol^{-1}$
Faraday constant	$9.648\,534 \times 10^{4}$	$C\ mol^{-1}$
triple point of water	273.16	K
absolute zero	–273.15	°C
permittivity of vacuum	$8.854\,187\,8 \times 10^{-12}$	$F\ m^{-1}$
permeability of vacuum	$4\pi \times 10^{-7}$	$H\ m^{-1}$
Stefan constant	$5.670\,400 \times 10^{-8}$	$W\ m^{-2}K^{-4}$
molar gas constant	8.314 472	$J\ mol^{-1}K^{-1}$
gravitational constant	$6.674\,2 \times 10^{-11}$	$N\ m^2kg^{-2}$

faces below the freezing point. *See also* dew.
2. Glazed frost: a thin clear layer of ice formed on a cold surface by impact of water droplets, or by the freezing of a layer of water.

fuel cell A type of cell in which fuel is converted directly into electricity. In one form, hydrogen gas and oxygen gas are fed to the surfaces of two porous nickel electrodes immersed in potassium hydroxide solution. The oxygen reacts to form hydroxyl (OH^-) ions, which it releases into the solution, leaving a positive charge on the electrode. The hydrogen reacts with the OH^- ions in the solution to form water, giving up electrons to leave a negative charge on the other electrode. Large fuel cells can generate tens of amperes. Usually the e.m.f. is about 0.9 volt and the efficiency around 60%.

full-wave rectifier An electrical device that ensures that current flows almost entirely in one direction by using two pairs of diodes. In this way, every half-cycle contributes to the current. The signal produced in this way can be smoothed by using a capacitor.

fundamental The simplest way (mode) in which an object can vibrate. The fundamental frequency is the frequency of this vibration. The less simple modes of vibration are the higher *harmonics*; their frequencies are higher than that of the fundamental. *See also* quality of sound.

fundamental constants (universal constants) Quantities that do not change under any known circumstances. Examples are the speed of light in a vacuum and the charge on an electron. It has been speculated that the values of the fundamental constants could change over the history of the universe. There is some tentative evidence that this is the case in the FINE STRUCTURE CONSTANT. At the time of writing, this evidence cannot be regarded as firmly established.

fundamental interval *See* temperature scale.

fundamental particles *See* elementary particles.

fundamental units The base units of a measurement system, especially those that deal with force and motion. Usually these are the units of length, mass, and time, plus at least one electrical unit. In SI, the fundamental units are the meter, the kilogram, the second, and the ampere. *See also* base unit.

fuse A protective device in an electric circuit consisting of a low-melting wire that melts and breaks the circuit if too much current flows through it. *See also* circuit breaker.

fusion 1. The interaction of light nuclei such that heavier nuclei are produced. There are two distinct classes of fusion reactions.

In the Sun and other stars fusion of protons to form helium nuclei takes place slowly by processes involving *weak interactions* (*see* carbon cycle; proton-proton chain reaction). On Earth nuclei of heavy isotopes of hydrogen can be made to react by *strong interactions*, as follows:

$${}^{2}_{1}\text{H} + {}^{2}_{1}\text{H} \rightarrow {}^{3}_{2}\text{He} + {}^{1}_{0}\text{n}$$
$${}^{2}_{1}\text{H} + {}^{2}_{1}\text{H} \rightarrow {}^{3}_{1}\text{H} + {}^{1}_{1}\text{H}$$
$${}^{3}_{1}\text{H} + {}^{2}_{1}\text{H} \rightarrow {}^{4}_{2}\text{He} + {}^{1}_{0}\text{n}$$

In each case two particles are produced, one heavier and the other lighter than the original ones. The large energy release appears as kinetic energy of the products. More energy is released by the subsequent reactions of the neutrons with nearby materials.

In both classes of reaction the nuclei have to approach so closely that the short-range nuclear field of force can act. This is opposed by the long-range electric repulsion of the positive charges. This is overcome by the substance being at a very high temperature (10^7–10^8 K) so that the nuclei have sufficient kinetic energies to make enough close collisions for the interactions to occur. Hence such fusion processes are called *thermonuclear reactions*. At such high temperatures matter is in the form of a plasma of free nuclei and free electrons; there are no atoms.

Fusion can be initiated by the explosion of an atomic bomb using nuclear fission. A hydrogen bomb contains a quantity of heavy isotopes of hydrogen which are caused to fuze by the detonation of a fission bomb. Controlled fusion reactions for operating electrical generating stations are the subject of much research. The gas is heated to form a plasma at high temperatures by electric discharges, or possibly by laser radiation. The plasma is contained by intense magnetic fields. Such plasmas are found to be subject to many kinds of instability and there are large problems to be overcome before fusion power becomes a commercial proposition.

2. *See* melting.

fusion reactor *See* nuclear reactor.

G

gadolinium /gad-ŏ-**lin**-ee-ŭm/ A ductile malleable silvery element of the lanthanoid series of metals. It occurs in association with other lanthanoids. Gadolinium is used in alloys, magnets, and in the electronics industry.

Symbol: Gd; m.p. 1313°C; b.p. 3266°C; r.d. 7.9 (25°C); p.n. 64; r.a.m. 157.25.

gain A ratio measuring the efficacy of an electronic system, usually an amplifier or an antenna. The voltage gain of an amplifier is the ratio of the output voltage developed across a load to the input voltage. The gain of a power amplifier is the ratio of the output power to the input power.

galactic dynamics The mathematical study of the dynamics of galaxies. Galactic dynamics is based on the Newtonian theory of gravity. An example of galactic dynamics is the orbit of a galaxy about the center of gravity of the galaxy cluster to which the galaxy belongs.

galaxy A collection of a large number of stars (up to 100 billion) and their satellites existing as a discernable unit held together by gravity. The galaxy to which our Sun (and the Earth) belong is called the *Milky Way Galaxy*, or, often, just the *Galaxy* (capital letter 'G').

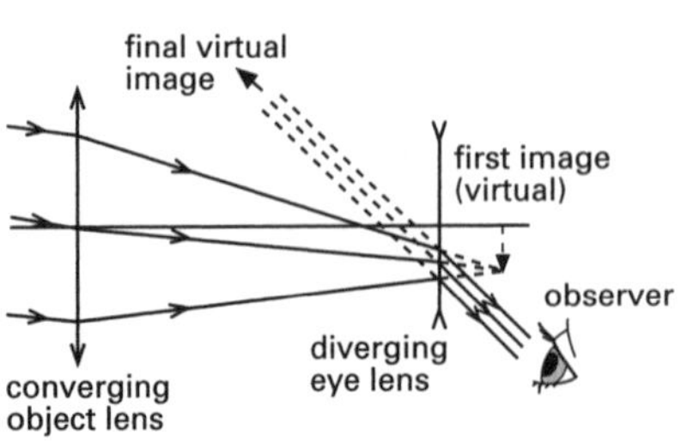

Galilean telescope

Galilean telescope /gal-ă-**lee**-ăn/ A type of refracting telescope having a converging objective and a diverging eyepiece. The Galilean arrangement produces an upright final image. However the field of view is smaller than that of the Keplerian telescope (which gives an inverted image). The arrangement is used in cheap binoculars (opera glasses). The normal adjustment of this telescope has the intermediate image in the focal planes of both lenses. These are therefore separated by a distance $f_O - f_E$ (New Cartesian sign convention). This was the first kind of telescope to be used in astronomy. With its aid, the Italian astronomer and physicist Galileo (1564–1642) discovered the four largest moons of Jupiter in 1610. *See also* refractor.

gallium /**gal**-ee-ŭm/ A soft silvery low-melting metallic element. Gallium is used in low-melting alloys, high-temperature thermometers, and as a doping impurity in semiconductors. Gallium arsenide is a semiconductor used in light-emitting diodes and in microwave apparatus.

Symbol: Ga; m.p. 29.78°C; b.p. 2403°C; r.d. 5.907 (solid at 20°C), 6.114 (liquid); p.n. 31; r.a.m. 69.723.

galvanic cell /gal-**van**-ik/ *See* cell.

galvanometer /gal-vă-**nom**-ĕ-ter/ An instrument for measuring small direct currents. *See* moving-coil instrument; tangent galvanometer. *See also* ballistic galvanometer.

gamma /gam-ă/ (γ) The ratio of the principal specific thermal capacity of a gas at constant pressure to that at constant volume (c_p/c_v). *See* adiabatic change; sound; specific thermal capacity.

gamma decay A type of radioactive decay in which gamma rays are emitted by the specimen. Gamma decay occurs when a nuclide is produced in an excited state, gamma emission occurring by transition to a lower energy state. It can occur in association with alpha decay and beta decay.

gamma radiation Properly, gamma radiation is a form of electromagnetic radiation emitted by atomic nuclei, but often the term is used loosely for radiations with similar (or higher) quantum energies derived from other sources.

Gamma rays from nuclei have wavelengths typically in the range 10^{-10} m to 10^{-13} m, with quantum energies in the range 10 keV to 10 MeV (10^{-15} J to 10^{-12} J). Homogeneous gamma radiation is absorbed exponentially in matter, causing ionization by secondary electrons. This ionization is the basis of methods of detecting the radiation. In living cells it can cause chemical changes that can be harmful.

Penetration of matter by gamma radiation depends very much upon the wavelength, and the atomic numbers of the atoms. For radiations of the shorter wavelengths shields of lead several centimeters thick are needed, whereas some long wavelength rays may be absorbed almost totally in thin foils.

Nuclei emit gamma radiation when they return to the ground state after excitation. The excited states have lifetimes that are usually of the order of a picosecond (10^{-12} s) or less, but some states are known with much longer lifetimes. The emission of alpha or beta rays, or the capture of electrons by the nucleus, often leaves the nucleus in excited states, as does bombardment with nuclear particles – particularly neutrons. Gamma emission does not change the proton or neutron numbers of the nucleus.

gamma-ray burst A dramatic cosmic event in which a powerful source of gamma rays briefly appears in the sky (for times ranging from less than a second to a few minutes). It is thought that gamma-ray bursts occur because of the formation of a BLACK HOLE, either by the merging of two neutron stars or the gravitational collapse of a single large star.

gamma rays Streams of gamma radiation

gas *See* states of matter.

gas constant (**universal molar gas constant**) Symbol: R The universal constant 8.314 472 J mol^{-1}K^{-1} appearing in the equation of state for an ideal gas.

gas-cooled reactor A nuclear reactor in which cooling of the core is achieved by circulation of a gas. Early reactors were cooled by air. Magnox reactors are cooled by high-pressure carbon dioxide. Advanced *gas-cooled reactors* (*AGR*) are also cooled by carbon dioxide. The *high-temperature gas-cooled reactor* (*HTGR*) is cooled by helium.

gas equation The equation of state for an ideal gas. *See* gas laws.

gas laws Laws relating the temperature, pressure, and volume of a fixed mass of gas. The main gas laws are Boyle's law and Charles' law. The laws are not obeyed exactly by any real gas, but many common gases obey them under certain conditions, particularly at high temperatures and low pressures. A gas that would obey the laws over all pressures and temperatures is a *perfect* or *ideal gas*. Boyle's and Charles' laws can be combined into an equation of state for ideal gases:

$$pV_m = RT$$

where V_m is the molar volume and R the molar gas constant. For n moles of gas

$$pV_m = nRT$$

An ideal gas is defined as one that obeys Boyle's law at all temperatures and pressures. It is realized by extrapolating results obtained for real gases to zero pressure. If pV_m is plotted against p at a given temperature, where V_m is the molar volume, the value $(pV_m)_{p=0}$ is used to define the ideal gas temperature T

$$(pV_m)_{p=0} = RT$$

where the molar gas constant R is the same for all gases. Real gases often conform closely to Boyle's law over a considerable range of p and T, and the equation of state for this range can be written

$$pV_m = RT$$

More exact equations are needed when there are significant departures from Boyle's law. *See* van der Waals' equation.

gas thermometer A thermometer that uses a fixed mass of gas as its working substance. It can take two forms. A *constant volume gas thermometer* depends on the direct proportion relationship between the pressure of a fixed mass of gas and its absolute temperature at constant volume. A *constant pressure gas thermometer* is dependent on the similar relationship between the volume and temperature of the gas.

Constant volume gas thermometers may use helium, hydrogen, or nitrogen. The typical arrangement is a bulb connected to a U-tube containing mercury. As the temperature changes the heights of mercury are adjusted such that the volume of gas in the bulb is constant. The pressure is measured by differences in the heights of the mercury. Corrections can be made for expansion of the bulb, temperature of the mercury, and nonideal behavior of the gas.

gate **1.** A signal that enables a circuit to function.

2. An electrode in a semiconductor device to which a biasing voltage is applied to modify the conductivity. *See* field-effect transistor.

3. A circuit that has one or more inputs but only one output. This is a digital gate as used in logic circuits.

4. A circuit that admits an input for only a specified time, producing an output proportional to the input, or to some other function of the input. *See also* semiconductor.

gauss /gows/ Symbol: G The unit of magnetic flux density in the c.g.s. system. It is equal to 10^{-4} tesla.

Gauss's law A result in electrostatics stating that the total electric flux at right angles to a closed surface with electric charges inside is proportional to the algebraic sum of these charges. It is named for the German mathematician Carl Friedrich Gauss (1777–1855), who discovered the law in the 1830s.

Gay-Lussac's law *See* Charles' law.

gear In a machine, a toothed wheel that meshes with another toothed wheel to transfer rotation from one shaft to another. Generally the shafts are parallel, but bevel gears meshing at right angles change the direction of rotation through 90°. The ratio of the numbers of teeth on two meshing gear wheels is the gear ratio. If they are different, the second (driven) shaft rotates at a different speed from the drive shaft.

Geiger counter /gÿ-ger/ (**Geiger-Müller counter**) A device for detecting ionizing radiation to provide a count of individual particles and photons. The construction is a gas-filled tube with a cylindrical metal cathode and a thin axial wire anode. The potential difference applied between the two depends on the design, but is usually between about 400 and 2000 volts. The gas may be low-pressure air or argon with ethanol vapor. Low-energy particles can enter through a thin 'window', often of mica. Passage of an alpha or beta particle or a photon of gamma radiation through the gas produces ions, which move towards the electrodes. The electrons accelerating towards the central anode acquire sufficient kinetic energy to ionize other gas molecules, initiating an 'avalanche' of electrons, which produces a measurable pulse at the anode. The pulse of current can be used to operate counting equipment. Some Geiger tubes contain a small amount of halogen vapor to quench the avalanche rapidly so that the tube returns quickly to a state in which it is ready to detect another pulse. Several thousand particles per second can be counted in this way. The device is named for the German physicist Hans Wilhelm Geiger (1882–1945), who invented it in 1912. The design was improved by the German physicist Walther Müller (1905–79) in 1928.

Geiger–Nuttall law /**nut**-awl/ A relationship between the decay constant γ of an alpha emitter and the kinetic energy E of the alpha particle emitted. The rule is that E is directly proportional to log γ.

Geissler tube /**gÿs**-ler/ An electric discharge tube consisting of a glass tube with sealed-in electrodes and containing traces of gas at very low pressure. When a high voltage is applied to the electrodes, the gas ionizes and light is emitted from excited atoms and ions (fluorescence). The tube is named for the German physicist Johann Heinrich Wilhelm Geissler (1815–79).

gel /jel/ A semi-rigid COLLOID.

general theory *See* relativity.

generator A large machine for converting mechanical energy into electrical energy. It has two sets of coil windings. One (the stator) is fixed and the other (the rotor) rotates. The relative motion of the two sets of coils induces an electric current in them. This is the reverse of a motor. In an electrical power station, generators are usually turned by steam turbines or water turbines. The turbine turns the generator, often called an alternator, which produces alternating current of a fixed frequency. Diesel motors are used to turn smaller generators of the type kept for emergency use in hospitals, etc. *See also* dynamo; electric motor.

geomagnestism /jee-ŏ-**mag**-nĕ-tiz-ăm/ *See* Earth's magnetism.

geometrical optics *See* optics.

geophysics The branch of physics concerned with the Earth and its interior. It is often taken to include also meteorology and oceanography.

geothermal energy Energy that comes from within the Earth. Examples of sources of geothermal energy include volcanoes, hot springs, and geysers. In some parts of the World, it is possible to utilize sources of geothermal energy.

germanium /jer-**may**-nee-ŭm/ A hard brittle gray metalloid element. Most germanium is recovered during zinc or copper refining as a by-product. Germanium was extensively used in early semiconductor devices but has now been largely superseded by silicon. It is used as an alloying agent, catalyst, phosphor, and in infrared equipment.

Symbol: Ge; m.p. 937.45°C; b.p. 2830°C; r.d. 5.323 (20°C); p.n. 32; r.a.m. 72.61.

getter /**get**-er/ A chemical substance used for removing residual gases after a vacuum has been produced by pumping. Getters are usually metals that combine with oxygen or nitrogen to produce a very high vacuum.

GeV Gigaelectronvolts; 10^9 electronvolts. Often *BeV* is used in the USA.

ghost A faint image near the image required, caused by radiation that has taken a different path. In the case of traditional 'silvered' glass mirrors, the main images are caused by light from the backing of the glass. Ghost images, formed by reflection from the front surface of the glass, may be a tenth as bright as the main images. Other 'ghost' images may arise following reflections inside the glass.

Ghost effects can also arise with non-visible electromagnetic radiations. Thus, ghost television images can appear on the screen if some radio waves reach the antenna after reflection by large nearby objects. This effect can be used to obtain a value for the speed of radio waves.

Gibbs function (**Gibbs free energy**) Symbol: G A thermodynamic function defined by $G = H - TS$, where H is the enthalpy, T the thermodynamic temperature, and S the entropy. The function is useful for specifying the conditions of chemical equilibrium for reactions for constant temperature and pressure (for which G is a minimum). The Gibbs function is named for the American scientist and mathematician Josiah Willard Gibbs (1839–1903). *See also* free energy.

giga- Symbol: G A prefix denoting 10^9. For example, 1 gigahertz (GHz) = 10^9 hertz (Hz).

gilbert The e.g.s. unit of magnetomotive force, equal to 0.795 775 ampere.

glass A solid substance in which there is a non-regular arrangement of atoms. Glasses are thus not crystalline – their atoms have a fairly random arrangement and they soften over a range of temperature, showing no definite melting point. They can be regarded as supercooled liquids. *See also* optical glass.

glass electrode An half-cell electrode consisting of a platinum wire in a bulbous thin glass membrane containing dilute acid, used to measure hydrogen-ion concentration (as in pH measurements).

global warming *See* greenhouse effect.

glow discharge A silent discharge of electricity through a gas at low pressure. Usually luminous, it consists of dark and light bands between the electrodes.

gluon /**gloo**-on/ A massless spin-1 particle that is responsible for transferring the strong interactions between QUARKS in quantum chromodynamics, just as photons transfer the electromagnetic interactions in quantum electrodynamics. Isolated gluons have never been observed.

gold A transition metal that occurs native. It is unreactive and is very ductile and malleable. Gold is used in jewelry, often alloyed with copper, and in electronics and colored glass. Pure gold is 24 carat; 9 carat indicates that 9 parts in 24 consist of gold.

Symbol: Au; m.p. 1064.43°C; b.p. 2807°C; r.d. 19.320 (20°C); p.n. 79; r.a.m. 196.96654.

gold-leaf electroscope *See* electroscope.

governor A mechanical device to control the speed of rotation of a machine. A simple governor consists of two masses attached to a shaft so that as the speed of rotation of the shaft increases, the masses move farther outward from the center of rotation, while still remaining attached to the shaft. As the weights move outward they operate a control that reduces the fuel or energy input to the machine. As they reduce speed and move inward they increase the fuel or energy input. Thus, on the principle of negative feedback, the speed of the machine is kept constant under varying conditions of load.

grad *See* vector calculus.

grade A unit of angle equal to 1/100 of a right angle. $1^g = 0.9°$.

Graham's law (of diffusion) Gases diffuse at a rate that is inversely proportional to the square root of their density. Light molecules diffuse faster than heavy molecules. The principle is used in the separation of isotopes. The law is named for the Scottish chemist Thomas Graham (1805–69).

gram Symbol: g A unit of mass defined as 10^{-3} kilogram.

gram-atom *See* mole.

gram-molecule *See* mole.

grand unified theory (GUT) A theory that attempts to unify the strong, weak, and electromagnetic INTERACTIONS into a single theory. Several different GUTs have been proposed. These all postulate that at high energies the interactions merge into a single interaction, with the STANDARD MODEL emerging from the GUT as a result of spontaneous symmetry breaking associated with the HIGGS MECHANISM. The energies at which the interactions come together in GUTs has been calculated to be about 1015 GeV. This is much higher than any conceivable accelerator of the future but was attained in the very EARLY UNIVERSE. For that reason GUTs are of relevance to the early Universe; for example, they provide a solution to the problem of matter–antimatter asymmetry in the Universe.

GUTs suggest that very massive bosons called *x-bosons* mediate interactions between quarks and leptons. This leads to the main prediction of GUTs, which is that protons decay to positrons and pions, with baryon number ceasing to be a conserved quantity. The lifetime of the proton is predicted to be much larger than the age of the Universe. Nevertheless, the probabilistic nature of quantum mechanics means that if sufficiently many protons are present in an observed sample then some should decay each year. Extensive experiments have been performed in attempts to detect proton decay. These have indicated that if proton decay does occur then the lifetime of the proton must be longer than the predictions of GUTs. This finding rules out many GUTs. However, if GUTs are combined with SUPERSYMMETRY to give *supersymmetric grand unified theories (SUSY GUTs)*, then such theories predict a lifetime about a thousand times longer than the predictions of nonsupersymmetric GUTs. It is hoped that experiments in the next few years will be able to test this prediction of SUSY GUTs.

Most GUTs correctly predict that neutrinos have nonzero masses. Many also predict the existence of MAGNETIC MONOPOLES and/or COSMIC STRINGS for topological reasons but there is no experimental evidence for such entities.

GUTs were first proposed in the mid-1970s by Howard Georgi, Sheldon GLASHOW, and others and combined with supersymmetry in the early 1980s.

graticule /**grat**-ă-kyool/ A grid of squares used with the eyepieces of many optical instruments to provide easier measurement and/or reference.

grating *See* diffraction grating.

gravimeter /gră-**vim**-ĕ-ter/ An instrument for measuring the Earth's gravitational field (the force of gravity), usually by detecting the effect of a changing field on a suspended or vibrating weight.

gravitation The concept originated by Isaac Newton around 1666 to account for the apparent motion of the Moon about the Earth, the essence being a force of attraction, called gravity, between the Moon and the Earth. Newton used this theory of gravitation to give the first satisfactory explanations of many diverse physical facts, such as Kepler's laws of planetary motion, the ocean tides, and the precession of the equinoxes. *See also* Newton's law of universal gravitation.

gravitational acceleration *See* acceleration of free fall.

gravitational collapse The collapse of a body that occurs because of the gravitational attraction between all the particles of the body. Gravitational collapse happens when there is insufficient opposing force to counteract the gravitational force. For example, the gravitational collapse of a star occurs when the nuclear fuel of the star is exhausted. The end point of this type of gravitational collapse is a WHITE DWARF, a NEUTRON STAR, or a BLACK HOLE, depending on the original mass.

gravitational constant Symbol: G The constant of proportionality in the equation that expresses NEWTON'S LAW OF UNIVERSAL GRAVITATION: $F = Gm_1m_2/r^2$, where F is the gravitational attraction between two point masses, m_1 and m_2, separated by a distance r. The value of G is $6.674\,2 \times 10^{-11}$ N m^2 kg^{-2}. It is regarded as a fundamental constant, although it has been suggested that the value of G may be changing slowly with time owing to the expansion of the Universe.

gravitational field The region of space in which one body attracts other bodies as a result of their mass. Bodies on or near the surface of the Earth are influenced by the Earth's gravitational field. To escape from this field a body has to be projected outward with a certain velocity (the *escape velocity*). The strength of the gravitational field at a point on the Earth's surface is given by the ratio force/mass, which is equivalent to the acceleration of free fall. At a point, it is defined as GM/r^2, where G is the gravitational constant, M the mass of the Earth, and r the distance between the center of the Earth and the point in question. The standard value of the acceleration of free fall is 9.806 65 m s^{-2}, but it varies with altitude (i.e. with r^2).

gravitational interaction *See* interaction.

gravitational lens A massive heavenly body that has a high enough gravitational field to bend the direction of light rays passing close to its surface. This type of light bending is predicted to occur in general relativity theory and has been observed in certain circumstances.

gravitational mass The mass of a body as measured by the force of attraction between masses, the value being given by Newton's law of universal gravitation. Inertial and gravitational masses are equal in a uniform gravitational field. *See also* inertial mass.

gravitational red shift *See* red shift.

graviton /**grav**-ă-ton/ A fundamental particle postulated in order to explain the very weak gravitational interaction in a way that fits in with quantum mechanics. Gravitons have zero rest mass, spin two and travel at the speed of light between the two interacting masses. None has yet been observed.

gravity The gravitational pull of the Earth (or other celestial body) on an object. The force of gravity acting on a mass is identical to weight.

gravity, center of *See* center of mass.

gravity waves Waves that are the analog in general relativity theory of electromagnetic waves in classical electrodynamics. Like electromagnetic waves, gravity waves propagate at the speed of light. Gravity waves are far more difficult to detect than electromagnetic waves and, despite a number of experiments, have never been detected directly. However, their existence can be inferred from the analysis of the period of a binary pulsar system (*see* binary star).

gray Symbol: Gy. The derived SI unit of absorbed IONIZING RADIATION dose; equivalent to an absorption per unit mass of one joule per kilogram of irradiated material. It is named for the British physician L. Harold Gray (1905–65). *See* rad; sievert.

greenhouse effect A heating effect that occurs in the atmosphere of planets such as the Earth as a result of the presence of certain gases in the atmosphere that absorb infrared radiation. Radiation from the Sun, mainly ultraviolet and visible radiation, passes through the atmosphere and is absorbed at the Earth's surface. Most of this energy is re-radiated by the surface of the Earth in the form of infrared radiation and radiation at these wavelengths is absorbed in the atmosphere by certain compounds (*greenhouse gases*). This raises the temperature of the atmosphere and consequently the Earth itself. It is generally believed that this effect may be the cause of *global warming*, a measurable increase in average global temperatures with time that has occurred over a period of years. The main greenhouse gas is carbon dioxide, which is produced by burning fossil fuels (petroleum, natural gas, and coal). There is considerable international concern about the effects of global warming on the environment and international action is being taken to reduce so-called carbon emissions from the burning of coal, gas, gasoline, etc. The term takes its name from the effect in a greenhouse, where the glass allows incedent light and some ultraviolet radiation through, but absorbs the infrared radiation re-emitted from surfaces within the greenhouse.

Gregorian telescope /grĕ-**gor**-ee-an, -**goh**-ree-/ An early type of reflecting telescope similar to the Cassegrainian arrangement, except that both mirrors were converging. The telescope is named for the Scottish astronomer and mathematician James Gregory (1638–75). *See also* reflector.

grid **1.** An open wire mesh interposed between the anode and cathode of a thermionic tube. *See also* thermionic tube; triode.
2. The system of distributing electrical energy nationally at high voltage.

grounding (**earthing**) The connection of an electrical conductor to a large conducting body, such as the Earth, which is used as the zero in the scale of electrical potential. In electrical appliances, any metal casings are usually grounded for safety

ground state The lowest energy state of an atom, molecule, or other system. *Compare* excited state.

ground wave A radio wave that travels close to the surface of the Earth between the transmitter and the receiver. *Compare* sky wave.

group A mathematical structure that is used to study symmetry. Formally, a group G is a collection of *elements* a, b, c, etc., with the elements being related to each other by certain rules.
(1) For any two members a and b of G the product ab must also be a member of G.
(2) For any three elements a, b, c, there is an associative relation $a(bc) = (ab)c$.
(3) There is an element called the *identity element* (sometimes called the *unit element*), usually denoted e, such that $ae = ea = a$ for all elements a.
(4) Each element a has an inverse, denoted a^{-1}, which is also a member of G.

Two elements a, b of a group G are said to *commute* if $ab = ba$. If $ab = ba$ for all pairs a, b of elements of G then G is said to be an *Abelian group*. If this is not the case then G is said to be a *non-Abelian group*. If a group has a finite number of elements then the group is said to be a *discrete group*. For example, the set of reflection and rotation symmetries of shapes that molecules can have, such as triangles and tetrahedra, gives discrete point groups. If the group has an infinite number of continuous elements then the group is said to be a *continuous group*. For example, the set of rotations about a fixed axis forms a continuous group called the *rotation group*.

Because of the connection between symmetries and conservation laws there is a close relation between group theory and conserved quantities. For example, the rotation group is closely associated with angular momentum. Group theory is of particular importance in quantum mechanics.

group velocity If a wave motion has a phase velocity that depends on wavelength, the disturbance of a progressive wave travels with a different velocity from the phase velocity. This is called the *group velocity*. The group velocity is the velocity with which the group of waves travels. It is given by $U = c - \lambda dc/d\lambda$, where c is the phase velocity. The group velocity is the one that is usually obtained by measurement. If there is no dispersion, as for electromagnetic radiation in free space, the group and phase velocities are equal.

GUT *See* grand unified theory.

gyromagnetic ratio /jÿ-roh-mag-**net**-ik/ Symbol g For an atom or nucleus, the ratio of the magnetic moment to its angular momentum.

gyroscope /**jÿ**-rŏ-skohp/ A rotating object that tends to maintain a fixed orientation in space. For example, the axis of the rotating Earth always points in the same direction toward the Pole Star (except for a small PRECESSION). A spinning top or a cyclist are stable when moving at speed because of the gyroscopic effect. Practical applications are the navigational gyrocompass and automatic stabilizers in ships and aircraft.

H

hadron /**had**-ron/ A type of ELEMENTARY PARTICLE. The hadrons are subdivided into baryons and the mesons. The hadrons are distinguished from the leptons by their type of interaction.

hafnium /**haf**-nee-ŭm/ A transition metal found in zirconium ores. Hafnium is difficult to work and can burn in air. It is used in control rods for nuclear reactors and in certain specialized alloys and ceramics.

Symbol: Hf; m.p. 2230°C; b.p. 5197°C; r.d. 13.31 (20°C); p.n. 72; r.a.m. 178.49.

Hagen's formula /**hah**-gĕnz/ *See* Poiseuille's formula.

half-adder An electronic logic circuit that is used to perform addition of binary arithmetic, i.e. the four possibilities: 0 + 0, 0 + 1, 1 + 0, 1 + 1. There are several ways of combining logic gates to get a half-adder.

half cell A metal electrode in contact with a solution of the metal ions. In general there will be an e.m.f. set up between metal and solution, depending on the tendency of the element to form ions in solution. The e.m.f. cannot be measured directly since setting up a circuit results in the formation of another half cell. *See* electrode potential.

half-life (**half-life period**) Symbol: $T_{1/2}$ The time taken for half the nuclei of a sample of a radioactive nuclide to DECAY. The half-life of a nuclide is a measure of its stability. (Stable nuclei can be thought of as having infinitely long half-lives). Half-lives of known nuclear species range from less than a picosecond to thousands of millions of years. If N_0 is the original number of nuclei, the number remaining at the end of one half-life is $N_0/2$, at the end of two half-lives is $N_0/4$, etc.

half-period zones (**Fresnel zones**) The circular bands on a spherical wavefront required in the analysis of Fresnel diffraction. Each band is half a wavelength close to, or further from, the point of observation than its neighbor. Their radii are $\sqrt{(nd\lambda)}$, where n (the order of the zone) is an integer, and d is the distance from the observation point to the center of the wavefront.

half-wave antenna (dipole antenna) An antenna that is used to transmit and receive radio signals, consisting of two metal rods connected to a circuit, with the length of each rod being $\lambda/4$, where λ is the wavelength of the radio wave being transmitted or received.

half-wave plate *See* retardation plate.

half-width *See* monochromatic radiation.

Hall effect When an electric current is passed through a conductor and a magnetic field is applied at right angles, a potential difference is produced between two opposite surfaces of the conductor. The direction of the potential gradient is perpendicular to both the current direction and the field direction. It is caused by deflection of the moving charge carriers in the magnetic field. The size and direction of the potential difference gives information on the number and type of charge carriers. The effect is named for the American physicist Edwin Herbert Hall (1855–1938). *Compare* Nernst effect.

halo A rainbow-colored ring, sometimes

white, seen around the Moon or Sun. It is caused by the refraction of light by tiny ice crystals in the Earth's atmosphere.

hard (high) vacuum *See* vacuum.

hardness The resistance of a solid substance to scratching or local deformation by indentation.

harmonic One of the possible simple (sinusoidal) components of a complex waveform or vibration. The *first harmonic* is the fundamental (frequency *f*); the *second harmonic* has twice the fundamental frequency (2*f*), and so on. A waveform may contain frequencies other than these harmonics. *See also* partials; quality of sound.

harmonic motion A regularly repeated sequence that can be expressed as the sum of a set of sine waves. Each component sine wave represents a possible simple harmonic motion. The complex vibration of sound sources (with fundamental and overtones), for instance, is a harmonic motion, as is the sound wave produced. *See also* simple harmonic motion.

harmonic series In sound, a sequence in which each sound is a harmonic of a fundamental note. Mathematically, it is the sequence 1 + 1/2 + 1/3 + 1/4 + ...

hassium /**hass**-ee-ŭm/ A transactinide element formed artificially.

Symbol: Hs; p.n. 108; most stable isotope ^{265}Hs (half-life 2×10^{-3}s).

heat Symbol: *Q* The name often given to energy that is transferred from regions of high temperature to those of lower temperature. The energy transferred when a system changes temperature is equal to the product of the temperature change, the mass of the object, and its specific thermal capacity. The energy transferred when a sample changes state is the product of mass and specific latent thermal capacity. *See also* internal energy; kinetic theory.

heat capacity *See* thermal capacity.

heat engine A device that does work on being supplied by heat from a high temperature source. Ideally, a working substance is taken through a cycle of operations in which heat is taken from a furnace and waste heat is discharged into the surroundings. By the first law of thermodynamics, the work done by the engine is equal to the difference between the heat supplied and the heat discharged. In real heat engines the working substance may be expelled at the end of a cycle and then replaced. The theory of ideal heat engines has played a large part in the development of thermodynamics. *See also* Carnot cycle; efficiency; heat pump; Otto cycle; refrigerator.

heat exchanger A device for transferring heat from one fluid to another. Typically, a heat exchanger consists of a series of tubes through which one fluid flows. These are surrounded by the other hotter or cooler fluid. A car radiator is an example of a heat exchanger for transferring energy from the coolant to the surroundings.

heat pump A device, similar to a refrigerator, that continually extracts heat from a body and discharges heat at a higher temperature. By the second law of thermodynamics, this is possible only because work is done upon the system, usually by an electric motor. The heat discharged is equal to the heat taken from the cooler body plus the work done.

Heat pumps are used for space heating. The hotter body is then the building being heated, while the colder body is the atmosphere, the soil, or water such as a river. Unlike an ordinary electric heater the heat pump provides heat at a much higher rate than the electric supply does work. However, it involves much higher capital costs, which may only be justified in a severe climate. *See also* heat engine; refrigerator.

Heaviside layer /**hev**-ee-sÿd/ (**E-layer; Kennelly–Heaviside layer**) A region of the IONOSPHERE at an altitude of 90–150 km that reflects medium-frequency radio waves. The existence of the Heaviside layer was predicted by and named for the English physicist and electrical engineer Oliver

Heaviside (1850–1925) and independently by the British–American electrical engineer Arthur Edwin Kennelly (1861–1939). *See also* Appleton layer.

heavy hydrogen *See* deuterium.

heavy water *See* deuterium.

hectare Symbol: ha A metric unit of area equal to 10 000 square meters.

hecto- Symbol: h A prefix denoting 10^2. For example, 1 hectometer (hm) = 10^2 meters (m).

Heisenberg's uncertainty principle /**hȳ**-zĕn-bergz/ (**principle of indeterminacy**) According to quantum mechanics the laws of physics can control only the *probability* of certain events or values, and are not deterministic. Heisenberg's principle expresses approximately this inherent uncertainty in physical laws. It can be expressed by the relations:

$$\Delta x \Delta p_x \geq h/4\pi \sim h/2\pi$$
$$\Delta t \Delta E \geq h/4\pi \sim h/2\pi$$

Here h is the Planck constant, Δx is the inherent uncertainty in the x-coordinate of the position of a particle, Δp_x is the uncertainty in the x-coordinate of the momentum (similar rules apply to the y and z components), ΔE is the uncertainty in the energy, and Δt the uncertainty in the time. The uncertainties refer to the inherent nature of the laws and not to limitations of experimental method. They are expressed as root-mean-square deviations.

For example, consider the second relation. An excited state of an atom may typically last 10^{-9} s. Any measurement of the excitation energy must in principle be made within this time, so the uncertainty Δt cannot be greater than about 10^{-9} s. Hence the uncertainty ΔE cannot be less than about $(h/2\pi) \div 10^{-9} \sim 10^{-25}$ joule. Since the excitation energy will be about 10^{-18} joule its value can only be determinate to within about one in 10^7, however refined the measurements are. The German physicist Werner Heisenberg (1901–76) formulated his uncertainty principle in 1927.

helium /**hee**-lee-ŭm/ A colorless monatomic gas; the first member of the rare gases (group 18 of the periodic table). Helium has the electronic configuration $1s^2$ and consists of a nucleus of two protons and two neutrons (equivalent to an α-particle) with two extra-nuclear electrons. It has an extremely high ionization potential and is completely resistant to chemical attack of any sort. The gas accounts for only $5.2 \times 10^{-4}\%$ of the atmosphere; up to 7% occurs in some natural gas deposits. Helium is the second most abundant element in the universe, the primary process on the Sun being nuclear fusion of hydrogen to give helium. Helium is recovered commercially from natural gas in both the USA and countries of the former USSR and it also forms part of ammonia plant tail gas if natural gas is used as a feedstock. Its applications are in fields in which inertness is required and where the cheaper alternatives, such as nitrogen, are too reactive; for example, high-temperature metallurgy, powder technology, and as a coolant in nuclear reactors. Helium is also used for balloons (it is less dense than air) and for low-temperature physics research.

Helium is unusual in that it is the only known substance for which there is no triple point (i.e., no combination of pressure and temperature at which all three phases can co-exist). This is because the interatomic forces, which normally participate in the formation of solids, are so weak that they are of the same order as the zero-point energy. At 2.2 K helium undergoes a transition from liquid helium I to liquid helium II, the latter being a true liquid but exhibiting superconductivity and an immeasurably low viscosity (*superfluidity*). The low viscosity allows the liquid to spread in layers a few atoms thick, described by some as 'flowing uphill'.

Helium also has an isotope. ^{3}He is formed in nuclear reactions and by decay of tritium. This also undergoes a phase change at temperatures close to absolute zero.

Symbol: He; m.p. 0.95 K (pressure); b.p. 4.216 K; d. 0.1785 kg m^{-3} (0°C); p.n. 2; r.a.m. 4.002602.

Helmholtz coils /**helm**-holts/ A pair of identical coils designed so that when a current is carried the magnetic field between them is fairly uniform. The two coils are set parallel at a distance equal to the radius of each. They are connected in series so that each carries the current in the same direction. Helmholtz coils are named for the German scientist Hermann Ludwig Ferdinand von Helmholtz (1821–94).

Helmholtz function (**Helmholtz free energy**) Symbol: F A thermodynamic function defined by $F = U - TS$, where U is the internal energy, T the thermodynamic temperature, and S the entropy. It is a measure of the ability of a system to do useful work in an isothermal process. *See also* free energy.

henry /**hen**-ree/ Symbol: H The SI unit of inductance, equal to the inductance of a closed circuit that has a magnetic flux of one weber per ampere of current in the circuit. $1\ \mathrm{H} = 1\ \mathrm{Wb\ A^{-1}}$.

The unit is named for the American physicist Joseph Henry (1797–1878).

Henry's law The mass of gas dissolved by a liquid is proportional to the pressure of the gas, at constant temperature and as long as the gas does not react with the liquid. The law is named for the British chemist William Henry (1774–1836).

hertz /herts/ Symbol: Hz The SI unit of frequency, defined as one cycle per second (s^{-1}). Note that the hertz is used for regularly repeated processes, such as vibration or wave motion. An irregular process, such as radioactive decay, for which the SI unit is the becquerel, would have units that could also be expressed as s^{-1} (per second). The unit is named for the German physicist Heinrich Rudolf Hertz (1857–94).

Hertzian waves /**hert**-see-ăn/ An obsolete name for radio waves.

heterodyne /**het**-ĕ-rŏ-dÿn/ A technique in radio reception that makes use of the phenomenon of beats (*see* beats). A locally generated radio wave is superimposed on the incoming wave to bring it within audio range. In a *superheterodyne* receiver, the intermediate frequency is supersonic and is amplified before being demodulated.

HF *See* high frequency.

hidden variables *See* Bell's paradox.

Higgs boson A hypothetical particle with zero spin postulated in explaining the unification of the electromagnetic interaction and the weak interaction. So far it has not been detected but it is expected to be discovered as higher-energy accelerators are constructed. The *Higgs field* is a field associated with this particle. The particle is named for the British theoretical physicist Peter Ware Higgs (1929–) and others in the mid-1960s.

Higgs mechanism The mechanism by which the W BOSON and the Z BOSON and, more generally, all massive particles, acquire their mass. In the mid 1960s Peter HIGGS, and independently Robert Brout and François Englert, predicted the existence of a massive spin-zero particle is now known as the *Higgs boson*. The Higgs boson is associated with a scalar field, now known as the *Higgs field*.

In the WEINBERG–SALAM MODEL the W boson and the Z boson which mediate the weak interaction acquire their masses by the Higgs mechanism. Higgs bosons have not been found experimentally, although thorough searches up to 130 GeV have been conducted. It is thought that fermions in the STANDARD MODEL acquire their masses by the Higgs mechanism but it has not been possible to implement this suggestion.

high frequency (**HF**) A radio frequency in the range between 30 MHz and 3 MHz (wavelength 10 m=100 m).

high-temperature gas-cooled reactor *See* gas-cooled reactor.

high-temperature superconductivity Superconductivity occurring at higher temperatures (above 100K), found in certain mixed yttrium-barium-copper oxides. It has a different mechanism from normal superconductivity but so far no full theory has been developed.

high tension (H.T.) High voltage.

hole *See* semiconductor.

holmium /**hohl**-mee-ŭm/ A soft malleable silvery element of the lanthanoid series of metals. It occurs in association with other lanthanoids. It has few applications.

Symbol: Ho; m.p. 1474°C; b.p. 2695°C; r.d. 8.795 (25°C); p.n. 67; r.a.m. 164.93032.

holography /hŏ-**lôg**-ră-fee/ A method of recording (usually photographically) a three-dimensional image of an object. Normally laser light is used, but other radiations (including sound) can give holograms. The object is illuminated with laser light and the reflected light from the object is combined with direct light from the source, to give an interference pattern on a photographic plate. A three-dimensional image of the object is reconstructed by illuminating the interference pattern with the original light. The pattern includes information about phase and direction as well as intensity and color.

homopolar generator /hoh-mŏ-**poh**-ler, hom-ŏ-/ (**Faraday disk**) A direct-current generator. It consists of a metal disk rotating in a magnetic field at right angles to the plane of the disk. The e.m.f. is induced between the center of the disk and the edge.

Hooke's law The principle that if a body is deformed the strain produced is directly proportional to the applied stress. A graph of stress against strain for a material obeying Hooke's law is a straight line. When the stress is removed, the material returns to its original dimensions. Above a certain stress the material ceases to obey Hooke's law and the graph becomes non-linear. The point at which this occurs is the *proportional limit* of the material. The law is named for the English scientist Robert Hooke (1635–1703). *See also* elasticity.

horizontal intensity Symbol: B_H The strength of the Earth's magnetic field in a horizontal direction at a given point on or near the surface. The value is changing with time. *See also* Earth's magnetism; magnetic variation.

horsepower Symbol: hp **1.** Any of various f.p.s. units of power either equal to or nearly equal to 746 watts (W) in SI derived units. Horsepower units typically were used to measure mechanical power, the most important such unit being equal to 550 foot-pounds of force per second, or 745.7 W.
2. An m.k.s. unit of power known as the *metric horsepower*, equal to 75 kilogram meters of force per second. It is equal to 735.5 W.

hot-wire instrument An electrical measuring instrument in which the current is passed through a fine resistance wire, causing a rise in temperature. This may be measured by a thermocouple, or by the 'sag' in the wire caused by expansion. Hot-wire instruments can be used for alternating of currents without a rectifier. The deflection is roughly proportional to the square of the current.

hour Symbol: h A unit duration of time equal to 60 minutes or 3600 seconds, or 1/24 of a mean solar day.

H.T. High tension (voltage).

HTGR High-temperature gas-cooled reactor. *See* gas-cooled reactor.

Hubble constant Symbol *H*. The quantity which appears as a constant in HUBBLE'S LAW.

Hubble's law A law stating that the velocity v with which a distant galaxy is receding from the Earth is proportional to the distance d from us. This law can be stated in the form $v = Hd$, where H is the *Hubble constant*. Hubble's law can be explained in terms of the description of an expanding Universe by general relativity theory. The Hubble constant has an inverse called the *Hubble time* which is a measure of the age of the Universe. The law is named for the American astronomer Edwin Hubble (1889–1953), who found it empirically in 1929.

hue A pure color in the visible spectrum, relating to one visible wavelength. Any hue can be mixed with a certain other hue to produce the sensation of white. If a hue is mixed with white, the result is a tint. *See also* complementary colors.

humidity The amount of water vapor in the air. ABSOLUTE HUMIDITY, that is, the mass of water vapor per unit volume of air, is one way of expressing humidity. However, condensation and evaporation also depend on the air temperature, and so the RELATIVE HUMIDITY at a given temperature is more often used. Another measure is *specific humidity*, the mass of water vapor per unit mass of air.

hundredweight A unit of mass equal to 100 lbs (45.359 kg). The *long hundredweight* of 112 lbs (50.802 kg) was used in the UK (and in the Canadian province of Newfoundland).

Huygens' construction /**hÿ**-gĕnz/ A method of constructing a subsequent wavefront from an existing one. Each point on the original wavefront is thought of as a secondary point source, emitting radiation of the same frequency as the source, with a speed depending on the medium. The new wavefront is the surface tangential to all the *secondary wavelets* produced by those point surfaces in the forward direction. The construction can easily be used to show the effects of rectilinear propagation, reflection, and refraction. Huygens' principle also gives good explanations for interference and diffraction. The construction is named for the Dutch scientist Christiaan Huygens (1629–95).

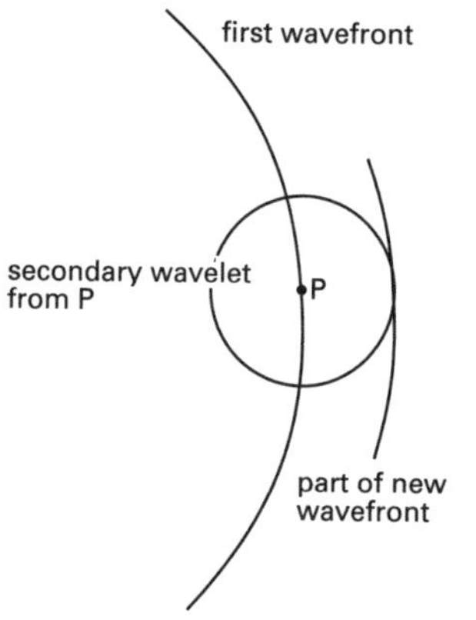

Huygens' construction of wavefronts

hybrid IC *See* integrated circuit.

hydraulic press A machine in which forces are transferred by way of pressure in a fluid. In this case (and in the related hydraulic braking system and hydraulic jack) the force the user exerts (effort) is less than the force the machine exerts (load): the force ratio is greater than one. The force ratio F_2/F_1 is equal to A_2/A_1. This machine is not very efficient; frictional effects are large.

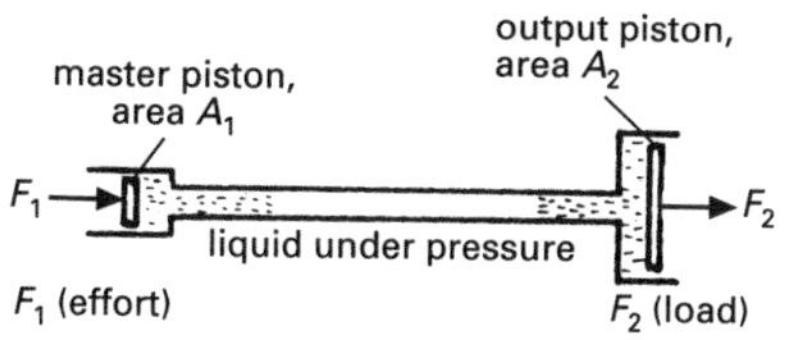

Hydraulic press principle

hydraulics The branch of physics (and engineering) concerned with the properties of fluids at rest and in motion.

hydrodynamics /hÿ-droh-dÿ-**nam**-iks/ *See* fluid mechanics.

hydrogen /**hÿ**-drŏ-jĕn/ A colorless gaseous element; the least dense and most abundant element in the Universe and the ninth most abundant element in the Earth's crust and atmosphere (by mass). It occurs principally in the form of water and petroleum products; traces of molecular hydrogen are found in some natural gases and in the upper atmosphere. Hydrogen occupies a unique position among the elements as hydrogen atoms are the simplest of all atoms. The hydrogen atom consists of a proton (positive charge) with one extranuclear electron ($1s^1$).

Atomic nuclei possess the property of 'spin' and for diatomic molecules there exists the possibility of having the spins of adjacent nuclei aligned (*ortho*) or opposed (*para*). Because of the small mass of hydrogen, these forms are more important in hydrogen molecules than in other diatomic molecules. The two forms are in equilibrium with parahydrogen dominant at low temperatures, rising to 75% orthohydrogen at room temperatures. Although chemically identical the melting point and boiling point of the para form are both about 0.1° lower than the 3:1 equilibrium mixture.

Natural hydrogen in molecular or combined forms contains about one part in 2000 of deuterium, D, an isotope of hydrogen that contains one proton and one neutron in its nucleus.

Symbol: H; m.p. 14.01 K; b.p. 20.28 K; d. 0.089 88 kg m^{-3} (0°C); p.n. 1; r.a.m. 1.0079.

hydrogen electrode An electrode based on hydrogen, and assigned zero electrode potential, so that other elements may be compared with it. The hydrogen is bubbled over a platinum electrode, coated in 'platinum black', in 1M acid solution. Hydrogen is adsorbed on the platinum black, enabling the equilibrium, $H(g) = H^+(aq) + e^-$, to be set up. *See also* electromotive series.

hydrometer /hÿ-**drom**-ĕ-ter/ An instrument for measuring the relative density of a fluid, usually a liquid. The most common type is a weighted glass bulb that, when floated in a liquid, shows the liquid's relative density by the depth of immersion. There is usually a calibrated scale marked on the glass.

hydrophone /**hÿ**-drŏ-fohn/ A type of transmitter or microphone (transducer) designed for use underwater with sound or ultrasound. *See* echolocation; sonar.

hydrosol /**hÿ**-drŏ-sol, -sohl/ A solution of a COLLOID in water.

hydrostatics /hÿ-drŏ-**stat**-iks/ The study of fluids (liquids and gases) in equilibrium.

hygrometer /hÿ-**grom**-ĕ-ter/ An instrument for measuring humidity. In a *chemical hygrometer*, air is drawn through a drying agent. Saturated air at the same temperature is then drawn through an identical sample of drying agent. The ratio of the different amounts of water absorbed indicates relative humidity. A *wet and dry bulb hygrometer* consists of two ordinary thermometers, one used normally and the other with a bulb kept moist by a damp wick. The cooling of the wet bulb by evaporation depends on the humidity of the air surrounding it.

hygroscope /**hÿ**-grŏ-skohp/ A device that indicates the humidity of the air, often in the form of a substance that changes color in the presence of moisture. *See also* hygrometer.

hyperbola /hÿ-**per**-bŏ-lă/ A conic with an eccentricity greater than 1. The hyperbola has two branches and two axes of symmetry. An axis through the foci cuts the hyperbola at two *vertices*. The line segment joining these vertices is the *transverse axis* of the hyperbola. The *conjugate axis* is a line at right angles to the transverse axis through the center of the hyperbola. A chord through a focus perpendicular to the transverse axis is a *latus rectum*.

In Cartesian coordinates the equation:

$$x^2/a^2 - y^2/b^2 = 1$$

represents a hyperbola with its center at the origin and the transverse axis along the x-axis. $2a$ is the length of the transverse axis. $2b$ is the length of the conjugate axis. This is the distance between the vertices of a different hyperbola (the *conjugate hyperbola*) with the same asymptotes as the given one. The foci of the hyperbola are at the points $(ae,0)$ and $(-ae,0)$, where e is the eccentricity. The asymptotes have the equations:

$$x/a - y/b = 0$$
$$x/a + y/b = 0$$

The equation of the conjugate hyperbola is

$$x^2/a^2 - y^2/b^2 = -1$$

The length of the latus rectum is $2b^2/ae$. A hyperbola for which a and b are equal is a rectangular hyperbola:

$$x^2 - y^2 = a^2$$

If a rectangular hyperbola is rotated so that the x- and y-axes are asymptotes, then its equation is

$$xy = k$$

where k is a constant.

See also conic.

hyperfine structure /**hÿ**-per-fÿn/ *See* fine structure.

hypermetropia /hÿ-per-mĕ-**troh**-pee-ă/ *See* hyperopia.

hyperons /**hÿ**-pĕ-ronz/ A group of elementary particles classified as baryons. They are heavier than the nucleons (proton and neutron) and have very short lifetimes. *See also* elementary particles.

hyperopia /hÿ-pĕ-**roh**-pee-ă/ (**hypermetropia**) Long sight caused if the distance between the eye lens and the back of the eyeball is too short. Rays from a distant object would focus behind the retina if the eye were fully relaxed. The eye must accommodate to allow clear vision of distant objects. This means that light from close objects cannot be focused onto the retina. Longsighted people have a distant near point. The defect is corrected with converging spectacle or contact lenses.

hypsometer /hip-**som**-ĕ-ter/ An instrument used for calibrating thermometers at the steam temperature. The thermometer bulb is placed above pure water boiling at a known pressure, and therefore a known temperature. *See* temperature scale.

hysteresis /his-tĕ-**ree**-sis/ In general, an apparent lag of an effect behind whatever is causing it. *Magnetic hysteresis* is the behavior of ferromagnetic materials as they are magnetized and demagnetized. The flux density, B, lags behind the external field strength H. *See* hysteresis cycle.

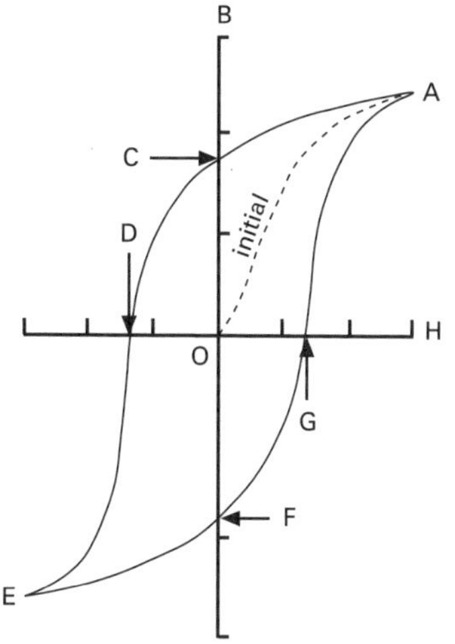

Hysteresis cycle

hysteresis cycle A closed loop obtained by plotting the flux density, B, of a ferromagnetic substance against the magnetizing field strength, H. The substance is first brought to magnetic saturation from an unmagnetized state – this produces curve OA (see illustration). As the field strength is taken through one cycle of reductions, reversals, and increases, the curve follows the path ACDEFGA. This is known as a hysteresis loop.

The area of the loop equals the energy loss in taking the sample once through the cycle. This is known as *hysteresis loss* and depends on the substance.

OC (or OF) represents the *remanence*, the magnetic flux density remaining in the specimen when the saturating field is removed. OD (or OG) represents the *coercivity*, the value of the magnetizing field strength needed to reduce this remaining flux density to zero. All these details are readily explained by the domain behavior of ferromagnetics.

I

IC *See* integrated circuit.

Iceland spar *See* calcite.

ice point The freezing point of water saturated with dissolved air, at standard atmospheric pressure. It was used as a standard fixed point in thermometry for over 200 years but has now been replaced by the triple point of pure water. The ice point (273.15 K) is 0.01 K lower than the triple point because of the pressure difference and the dissolved air.

See also International Practical Temperature Scale.

ideal gas *See* gas laws; kinetic theory.

ideal solution A (hypothetical) solution that exactly obeys RAOULT'S LAW.

illumination /i-loo-mă-**nay**-shŏn/ Symbol: E A measure of the visible-radiation energy reaching a surface in unit time. Once called 'intensity of illumination', it is measured in lux (lx). One lux is an illumination of one lumen per square meter. *See also* photometry.

image The point or region from which rays from an object appear to have come after reflection or refraction. Rays actually pass through a *real image* (like that produced on a screen by the LENS of a reflex camera or slide projector). Rays do not actually pass through a *virtual image* (like that produced behind a make-up MIRROR or by a magnifying lens).

Image position is measured in terms of image distance (v) – the distance between the image and the lens or mirror. It relates to object distance (u) and focal distance (f) by the equation:

$$1/f = 1/v + 1/u$$

(in the real is positive, virtual is negative convention).

Truly sharp images are formed only with rays of a single wavelength passing fairly close and parallel to the principal axis of the lens or mirror.

To describe an image, three other factors are used in addition to its distance from the lens or mirror.

(1) Is it real or virtual?
(2) Is it upright or inverted – the same way up as the object or the other way up?
(3) How large is it compared to the object – magnified, same size, or diminished?

See also magnification.

image converter An electron tube that converts invisible images (those formed by 'illumination' with radiation outside the visible spectrum, such as infrared) into visible images.

image intensifier A type of image converter that produces a brighter image than that of the original visible image.

image plane The plane perpendicular to the axis centered on the image.

immersion objective A type of high-power microscope objective whose effectiveness is obtained by filling the space between lens and microscope slide with a suitable liquid. Traditionally oil is used, hence the common name *oil-immersion objective*. However, sugar solution is preferable. The liquid increases the effective aperture of the objective because of its high refractive constant. In turn this maximizes the resolution obtainable.

impedance /im-**pee**-dăns/ Symbol: Z A

measure of the opposition of a circuit to an alternating current. Impedance is the resultant of total REACTANCE X and resistance R:

$$Z^2 = R^2 + X^2$$

See also alternating current circuit; phasor.

imperfections Entities that violate the structure of a perfect crystal. Defects and impurities are examples of imperfections.

Imperial units The f.p.s. system of rationalized weights and measures established by the British Weights and Measures Act of 1824, during the reign of George IV, and subsequently adopted throughout Britain's colonies and possessions. Although Imperial units for quantities of length and weight (or mass) remained essentially the same as their predecessors, those for volume and capacity differed significantly. In particular the Imperial gallon was made equal to the volume of 10 pounds of water at 62 degrees Fahrenheit, or 4.546 09 cubic decimeters (dm^3) in SI units. This value was 20% larger than the old English wine gallon, equal to 3.785 412 dm^3, that had long ago been adopted in the former colonies which had become the United States. Further, the Imperial quart was divided into 40 fluid ounces (equal to 1.136 522 dm^3) whereas the U.S. quart was divided into 32 fluid ounces (equal to 0.946 353 dm^3), making the U.S. fluid ounce larger. In the Imperial system the bushel is equal to 8 Imperial quarts, and dry and liquid measures are the same, unlike the situation in the United States.

impulse (**impulsive force**) A force acting for a very short time, as in a collision. If the force is constant the impulse is $F\delta t$; if it is a variable force the impulse is the integral of this over the short time period. An impulse is equal to the change of momentum that it produces.

impulse noise *See* noise.

impulsive force *See* impulse.

incandescence /in-can-**dess**-ĕns/ The radiation of visible light by a surface at a high temperature. Radiant electric heaters and the filament lamp are domestic examples of incandescent sources. *See also* black body.

inch A unit of length in f.p.s. systems. It is equal in SI units to 25.4 millimeters.

incidence, angle of *See* angle of incidence.

inclination (**angle of dip**) Symbol: δ The angle between the Earth's magnetic field and the horizontal at a given point on the Earth's surface. It is measured with an inclinometer and changes to some extent with time. *See also* Earth's magnetism; magnetic variation.

inclined plane A type of simple machine. Effectively a plane at an angle, it can be used to raise a weight by movement up an incline.

Both distance ratio and force ratio depend on the angle of inclination. Efficiency can be fairly high if friction is kept low. The screw and the wedge are both examples of inclined planes.

inclinometer /in-klă-**nom**-ĕ-ter/ (**dip circle**) An instrument used to measure inclination (the angle of dip). A magnetic needle is pivoted so that it is free to move in a vertical plane in front of a circular scale. The instrument is leveled and then rotated through a horizontal axis until the needle is vertical. The needle is now under the influence only of the vertical component of the Earth's field. When the dip circle is turned through 90° (into the meridian) the needle lines up along the Earth's magnetic field. Then the angle between the needle and the horizontal is the inclination at that point.

indium /**in**-dee-ŭm/ A soft silvery metallic element. It is found in minute quantities, primarily in zinc ores, and is used in alloys, in several electronic devices, and in electroplating.

Symbol: In; m.p. 155.17°C; b.p. 2080°C; r.d. 7.31 (25°C); p.n. 49; r.a.m. 114.818.

induced current Electric current in a conductor caused by e.m.f. set up by a changing magnetic field surrounding it. *See* electromagnetic induction.

induced magnetism The creation of MAGNETISM in a ferromagnetic material by placing it in a magnetic field. This causes the material's magnetic domains to become aligned. The inducing field may come from another permanent magnet or from an electromagnet. *See* domain.

inductance /in-**duk**-tăns/ A measure of the e.m.f. produced in a circuit as a result of the magnetic effect of a changing electric current, either in that circuit or in another. *See* mutual induction; self-induction.

induction /in-**duk**-shŏn/ In general, the production of an effect by interaction with a field. **1.** The production of magnetic order in a material by an external field. *See* magnetic induction.
2. The separation of electric charge in a material by an external electric field. *See* electrostatic induction.
3. The production of an e.m.f. (or resulting current) in a conductor in a changing magnetic field. *See* electromagnetic induction. *See also* mutual induction; self-induction.
4. *See* magnetic flux density.

induction coil A device that produces a series of high-voltage pulses by means of electromagnetic induction. It consists of a coil of insulated wire with only a few turns, wound on an iron core and surrounded by another with many more turns. When the current in the first coil is interrupted suddenly, a large e.m.f. is induced in the second. A pulsed current in the first coil induces a large pulsed e.m.f in the second.

induction heating (**eddy-current heating**) The temperature rise that occurs in conductors in a varying magnetic field. It is caused by the induced eddy currents inside the block of material. Induction heating can be used for heating metals in a furnace. Some stoves work by heating the metal pots in this way. However, induction heating also results in power losses from electrical machines, although usually this can be minimized with careful design.

induction motor An alternating-current ELECTRIC MOTOR in which the changing magnetic field of one coil, connected to the power supply, induces a current in another winding not connected to the supply. The magnetic force between the two coils turns the motor. Unlike most other electric motors, no brushes are needed to connect moving and stationary electrical parts. This eliminates sparking and therefore large motors are often of this type. *See also* synchronous motor.

inductor /in-**duk**-ter/ (**choke**) A circuit component that has a significant inductance. *See* inductance.

inelastic collison /in-i-**las**-tik/ A collision for which the restitution coefficient is less than one. In effect, the relative velocity after the collision is less than that before; kinetic energy is not conserved in the collision, even though the system may be closed. Some of the kinetic energy is converted into internal energy. *See also* restitution; coefficient of.

inertia /i-**ner**-shă/ An inherent property of matter implied by Newton's first law of motion: 'a body continues in a state of rest or constant velocity unless acted on by an external force'. An object automatically opposes a change of motion by reason of its inertia. *See also* inertial mass; Newton's laws of motion.

inertial mass /i-**ner**-shăl/ The mass of an object as measured by the property of inertia. It is equal to the ratio force/acceleration when the object is accelerated by a constant force. In a uniform gravitational field, it is equal to GRAVITATIONAL MASS, since all objects have the same gravitational acceleration at the same place.

inertial system A FRAME OF REFERENCE in which an observer sees an object that is free of all external forces to be moving at constant velocity. The observer is called an *inertial observer*. Any frame that moves with constant velocity and without rotation relative to an inertial frame is also an inertial frame. NEWTON'S LAWS OF MOTION are valid in any inertial frame (but not in an accelerated frame), and the laws are therefore independent of the velocity of an inertial observer.

inflation *See* early Universe.

information theory The branch of mathematics that analyzes the concept of information mathematically. Information theory has been used extensively in several branches of physics, particularly in connection with the concept of entropy. It has been suggested that the fundamental principles of physics can be expressed in terms of information theory.

infrared /in-frā-**red**/ (IR) Electromagnetic radiation past the red end of the visible spectrum. The range of wavelengths is approximately 0.7 micrometer to 1 millimeter. Infrared is sometimes called *thermal radiation*; most of the energy radiated by surfaces below about 6000 K is infrared. Many materials transparent to visible light are opaque to infrared, including glass. Rock salt, quartz, germanium, or polyethylene prisms and lenses are suitable for use with infrared. Infrared radiation close to visible light in wavelength can be detected by thermometers with blackened bulbs and with photographic film. Otherwise a thermocouple or a bolometer can be used.

Infrared radiation is produced by movement of charges on the molecular scale; i.e. by vibrational or rotational motion of molecules. *See also* electromagnetic spectrum.

infrasound /in-frā-**sownd**/ Vibrations in a medium with frequencies below about 16 hertz. The ear distinguishes such vibrations as a series of pulses, rather than a continuous sound.

instantaneous value /in-stăn-**tay**-nee-ŭs/ The value of a varying quantity (e.g. acceleration, current, power, etc.) at a particular instant in time.

insulation, electrical The use of nonconductors to coat or separate electrical conductors (for safety or prevention of short circuits). Rubber, PVC, and ceramics are common insulators.

insulation, thermal Any means of preventing or reducing the transfer of thermal energy. Lagging of hot-water tanks and foam-filled cavity walls are examples. *See also* insulator; thermal.

insulator, electrical A material with a very high electrical resistivity. Most nonmetallic materials are good electrical insulators. *Compare* conductor (electrical), semiconductor.

insulator, thermal A material that does not readily transmit heat. Many insulators, such as asbestos, are porous or fibrous solids that trap small pockets of air within them. Gases do not conduct heat well, and convection is prevented because the air cannot flow. *See also* conductor (thermal).

integral /**in**-tĕ-grăl, in-**teg**-răl/ The result of integrating a function. *See* integration.

integrated circuit (IC) A circuit that incorporates numerous components into one unit. In a *monolithic IC* a single chip of silicon is the base or substrate onto which all the individual components are integrated during manufacture. *Hybrid ICs* consist of one or more monolithic ICs mounted on a substrate, or several components similarly mounted and interconnected. After manufacture neither type can be dismantled.

integration The continuous summing of change in a function, $f(x)$, over an interval of the variable x. It is the inverse process of differentiation in calculus, and its result is known as the *integral* of $f(x)$ with respect to x. An integral:

$$\int f(x)dx$$

can be regarded as the area between the curve and the x-axis, between the values x_1 and x_2. It can be considered as the sum of a number of column areas of width Δx and heights given by $f(x)$. As Δx approaches zero, the number of columns increases infinitely and the sum of the column areas approaches the area under the curve. The integral of velocity is distance. Integrals between definite *limits* are known as *definite integrals*. An *indefinite integral* is one without limits. The result of an indefinite integral contains a constant – the *constant of integration*. For example,

$$\int x dx = x^2/2 + C$$

where C is the constant of integration. A table of integrals is given in the Appendix. *See also* differentiation.

intensity A measure of the rate of energy transfer by radiation. The unit of radiant intensity is the watt per square steradian (W sr^{-1}). **1.** The intensity of visible radiation is related to brightness. In photometry special units are used because it is necessary to consider the sensitivity of the eye to different visible radiations. Thus the intensity of a light source – sometimes called luminous intensity – is measured with the candela. The intensity of illumination of a surface is measured in lux. The intensity of visible radiation itself is measured in lumens.
2. The intensity of sound relates to the sensation of loudness, which also depends on frequency. It is inversely proportional to the square of the distance from the source and is measured in watts per square meter. The *intensity level* of a sound is the intensity relative to an agreed standard. *See* decibel.

intensity, electric *See* electric field strength.

intensity, magnetic *See* magnetic field strength.

intensive variable A quantity in a system for which the value of that quantity does not depend on the size of that system. Examples of intensive variables include pressure, temperature, and density. For example, if two systems with the same temperature are brought together to form a larger system, the temperature does not change. *Compare* extensive variable.

interaction A mutual effect between two or more systems or bodies, so that the overall result is not simply the sum of the separate effects. There are four separate interactions distinguished in physics:
1. *Gravitational interaction* The weakest of the four, about 10^{40} times weaker than the electromagnetic interaction. It is an interaction between bodies or particles on account of their mass, and operates over long distances. *See* Newton's law of universal gravitation.
2. *Electromagnetic interaction* The interaction between charged bodies or particles (stationary or moving). It falls off with the square of distance and operates over all distances. *See also* Coulomb's law.
3. *Strong interaction* An interaction between hadrons, about 100 times greater than the electromagnetic interaction. It operates at very short range (up to around 10^{-15} m) and is the force responsible for holding nucleons together in the atomic nucleus.
4. *Weak interaction* An interaction about 10^{10} times weaker than the electromagnetic interaction. It can occur for both leptons and hadrons and is the interaction in beta decay.

So far it has not proved possible to formulate a unified field theory for all four types of interaction, although there has been some success in unifying the electromagnetic and weak interactions.

interface The common surface (boundary of contact) between two adjacent PHASES in a system. Either or both of the surfaces may be a gas, liquid, or solid.

interference When similar waves with a regular phase relationship pass through the same region they are said to *interfere*. The resultant displacement at any point is the sum of the individual displacements, taking into account their directions and phases. The waves emerge from the overlapping region unaffected. The combination of such coherent waves is to be contrasted with the combination of incoherent waves, for example light from two lamps falling on a surface, where the resultant intensity is just the sum of the separate intensities.

There are two simple cases. If the WAVES travel in the same or very nearly the same direction, interference fringes are produced. If waves of comparable intensity travel in opposite directions stationary waves are produced. In all other cases, a complex system of interference fringes and stationary waves is caused, which does not admit simple analysis or description.

Interference fringes can be observed with all types of waves. They can be understood most easily by considering waves from just two sources but the ideas can then be generalized to more complicated systems. The two sources must give waves of the same frequency and with a regular phase relationship. This is easily achieved using sources of sound, ripples or radiowaves but in optics it is normally necessary to derive the two sources from one original source by division of wavefront or division of amplitude. The first method is used in Young's double slit, Fresnel's biprism, and Lloyd's mirror. The second is used in Newton's rings, the air wedge, and thin films.

Interference fringes are normally observed at a distance from the sources that is very large compared with their separation. Thus the waves travel in almost exactly the same direction and, if the sources are equal, they have equal intensity. There are then maxima of intensity (light fringes) where the path difference is zero or an integral number of wavelengths, and minima (dark fringes) where the path difference is an odd number of half wavelengths.

See also wave.

interferometry /in-ter-fĕ-**rom**-ĕ-tree/ The technique of producing interference patterns and using these for accurate wavelength measurement, measuring small distance changes, testing flat surfaces, etc. Devices, such as etalons, used for producing and using interference patterns are *interferometers*.

intermolecular forces /in-ter-mŏ-**lek**-yŭ-ler/ Forces of attraction between molecules or neutral atoms. *See* van der Waals' forces.

internal combustion engine A type of HEAT ENGINE in which fuel is burned within the engine. For example, in a normal automobile engine, combustion of gasoline, diesel, or some other fuel within the cylinder causes expansion of gas, which moves the piston in the engine and performs work. In contrast, a STEAM ENGINE is an *external combustion engine*. In this case the combustion takes place outside the body of the engine and steam at high pressure is passed to the cylinder.

internal conversion A process in which the nucleus of an atom decays from an excited state into the ground state but, instead of releasing the available energy as electromagnetic radiation (a photon), transfers it by electromagnetic coupling to a bound electron in the atom (creating an ion, which may then itself emit an electron or a gamma-ray photon). *See also* Auger effect.

internal energy Symbol: U The energy of a system that is the total of the kinetic and potential energies of its constituent particles (e.g. atoms and molecules). If the temperature of a substance is raised, by transferring energy to it, the internal energy increases (the particles move faster). Similarly, work done on or by a system results in an increase or decrease in the internal energy. The relationship between heat, work, and internal energy is given by the first law of thermodynamics. Sometimes the internal energy of a system is loosely spoken of as 'heat' or 'heat energy'. Strictly, this is incorrect; heat is the transfer of energy as a result of a temperature difference.

internal friction Another term for viscosity. *See* viscosity.

internal resistance Resistance of a source of electricity. In the case of a cell, when a current is supplied, the potential difference between the terminals is lower than the e.m.f. The difference (i.e. e.m.f. – p.d.) is

INTERNATIONAL PRACTICAL TEMPERATURE SCALE		
Fixed point	*TK*	°C
triple point of hydrogen	13.81	–259.34
hydrogen with vapor pressure 25/76 atmosphere	17.042	–256.108
boiling temperature of hydrogen	20.28	–252.87
boiling temperature of neon	27.102	–246.048
triple point of oxygen	54.361	–218.789
boiling temperature of oxygen	90.188	–182.962
triple point of water	273.16	0.01
boiling temperature of water	373.15	100
melting temperature of zinc	692.73	419.58
melting temperature of silver	1235.08	961.93
melting temperature of gold	1337.58	1064.43

Note: the scale given above is based on standard interpolation procedures to be used experimentally for measurements between the fixed points. The methods employed are:

–259.35° to 630.74°C	resistance measurement with a platinum resistance thermometer
630.74°C to 1064.43°C	e.m.f. measurement with a thermocouple made of platinum and platinum/10% rhodium alloy
above 1064.43°C	radiation measurement and use of Planck radiation law

proportional to the current supplied. The internal resistance (r) is given by:

$$r = (E - V)/I$$

where E is the e.m.f., V the potential difference between the terminals, and I the current.

international candle A former unit of luminous intensity, approximately equal to 1.018 3 CANDELA. It was originally defined in terms of the light emitted per second by a specified electric lamp, but was superseded by the candela in 1948.

International Practical Temperature Scale (IPTS) A scale of temperatures (1968) based on a number of fixed points, with agreed methods of measuring temperatures over particular ranges. It is used for practical measurements of thermodynamic temperature.

interstitial /in-ter-**stish**-ăl/ *See* defect

intrinsic semiconductor *See* semiconductor.

inverse Compton effect An increase in energy of a photon when it is scattered by an electrically charged particle such as an electron. In this process the electrically charged particle loses energy. It is thought that this effect may be important in some processes in astronomy.

inverse square law A relationship in which the effect of a source is inversely proportional to the square of the distance from the source. It applies to energy radiated from a point source and also to fields from point sources.

inversion temperature *See* Joule–Kelvin effect.

inverter gate (**NOT gate**) *See* logic gate.

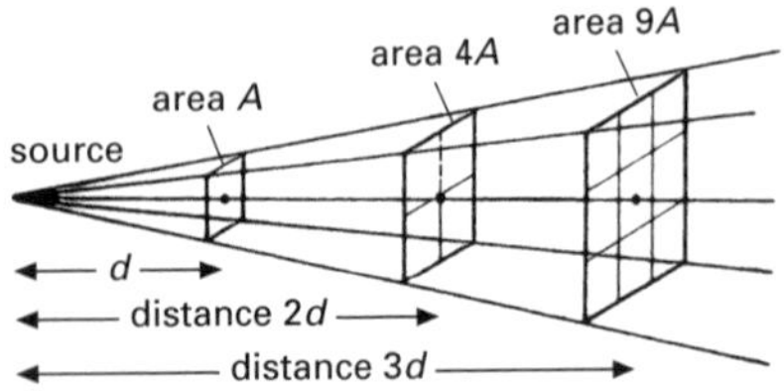

Inverse square law

inverting prism *See* erecting prism.

iodine /ÿ-ŏ-dÿn, -deen/ A dark-violet volatile solid element belonging to the halogens.

Symbol: I; m.p. 113.5°C; b.p. 184°C; r.d. 4.93 (20°C); p.n. 53; r.a.m. 126.90447.

ion /ÿ-on, -ŏn/ A charged particle consisting of an atom, or group of atoms, that has either lost or gained electrons. Sodium chloride (salt), for example, is made up of positive sodium ions – atoms that have each lost an electron – and negative chlorine ions – atoms that have each gained an electron. *See* ionization.

ion engine A device consisting of a source for producing ions and an electromagnetic field for accelerating the ions in a particular direction. In practice, an ion engine can be used in a low-pressure environment to produce propulsion by reaction forces associated with the rapidly moving ions. It is also necessary to expel oppositely charged particles to the ions from the engine to ensure charge neutrality. Ion engines have been investigated as possible power sources for spacecraft.

ionic radius The effective radius of an ion in a compound, most usually determined by comparing the interionic distances between the ion concerned and various other (oppositely charged) ions in crystals.

ionization /ÿ-ŏ-ni-**zay**-shŏn/ The process of producing ions. There are several ways in which ions may be formed from atoms or molecules. In certain chemical reactions ionization occurs by transfer of electrons; for example, sodium atoms and chlorine atoms react to form sodium chloride, which consists of sodium ions (Na^+) and chloride ions (Cl^-). Certain molecules can ionize in solution; acids, for example, form hydrogen ions as in the reaction

$$H_2SO_4 \rightarrow 2H^+ + SO_4^{2-}$$

Ions can also be produced by ionizing radiation; i.e. by the impact of particles or photons with sufficient energy to break up molecules or detach electrons from atoms: $A \rightarrow A^+ + e^-$. Negative ions can be formed by capture of electrons by atoms or molecules: $A + e^- \rightarrow A^-$.

ionization chamber A chamber in which ionizing radiation can be detected and/or measured. It contains a pair of electrodes between which is maintained a potential difference. Passage of ionizing radiation through the chamber produces electrons and positive ions, which separate and move to the electrode of opposite sign. There is thus a small electric current between the electrodes, which relates to the intensity of the radiation.

ionization potential Symbol: I The energy required to remove an electron from an atom or molecule to form a positive ion.

ionizing radiation Radiation such that the individual particle or quantum has sufficient energy to ionize substances. Electrons with kinetic energy just greater than the ionization potential will cause ionization, but other particles (e.g. molecular positive ions) require higher energies. Gamma-rays and x-rays ionize indirectly by means of the electrons they eject from substances by the photoelectric effect or Compton scattering. Short-wavelength ultraviolet quanta may ionize individual molecules by the photoelectric effect, but the ejected electrons have insufficient kinetic energy to cause further ionization unless an electric field is applied.

At ordinary intensities electromagnetic radiation with quantum energy below the ionization potential cannot cause ionization. But by focusing laser beams it is possible to ionize matter – even when the individual photons have energy less than a hundredth of the ionization potential. Such radiations are not ordinarily classified as ionizing radiations.

Ionizing radiations can be dangerous to health and precautions should be taken with their use.

ionosphere /ÿ-**on**-ŏ-sfeer/ A region in the Earth's upper atmosphere containing ions and free electrons. It is formed by ionization by ultraviolet radiation from the Sun.

It is divided into three layers: D-layer (50–90 km), E-layer (90–150 km), and F-layer (150–1000 km). Radio transmission can occur by reflection from the ionosphere.

ion pair A pair of ions of opposite charge produced, for example, by an ionizing radiation: $AB \rightarrow A^+ + B^-$.

IPTS *See* International Practical Temperature Scale.

IR *See* infrared.

iridium /i-**rid**-ee-ŭm/ A white transition metal that is highly resistant to corrosion. It is used in electrical contacts, in spark plugs, and in jewelry.

Symbol: Ir; m.p. 2410°C; b.p. 4130°C; r.d. 22.56 (17°C); p.n. 77; r.a.m. 192.217.

iris /**ÿ**-ris/ An arrangement able to vary the amount of light that enters an optical instrument. In the mammalian eye, this is a circular muscle that changes the size of the pupil. In many optical instruments, a similar effect is obtained with a diaphragm. In either case the aperture is varied.

iris diaphragm *See* diaphragm.

iron /**ÿ**-ern/ A transition element occurring in many ores. It is used in the manufacture of steel and various other alloys.

Symbol: Fe; m.p. 1535°C; b.p. 2750°C; r.d. 7.874 (20°C); p.n. 26; r.a.m. 55.845.

irradiance /i-**ray**-dee-ăns/ Symbol: E The rate of energy reaching unit area of a surface; i.e. the radiant flux per unit area. Unlike illumination, irradiance is not restricted in use to visible radiation. The unit is the watt per square meter ($W\ m^{-2}$).

The *solar constant* is the irradiance of the Earth by the Sun. The value is 1.353 $kW\ m^{-2}$. When applying this figure to a consideration of solar power, allowance must be made for atmospheric absorption and reflection, and for the angle of the Sun from the zenith.

irreversible change /i-ri-**ver**-să-băl/ *See* reversible change.

isobars /**ÿ**-sŏ-barz/ **1.** Two or more nuclides that have the same nucleon numbers but different proton numbers.
2. Lines joining points of equal pressure.

isoclinic line /ÿ-sŏ-**klin**-ik/ An imaginary line on the Earth's surface joining points with the same inclination (angle of dip). The *magnetic equator* (or *aclinic line*) is the line that joins points of zero inclination. *See also* Earth's magnetism.

isodynamic line /ÿ-sŏ-dÿ-**nam**-ik/ An imaginary line on the Earth's surface joining points at which the Earth's total magnetic field strength is the same. *See also* Earth's magnetism.

isoentropic process /ÿ-sŏ-**trop**-ik/ A thermodynamic process in which the entropy of the system stays constant throughout the change.

isogonal /ÿ-**sog**-ŏ-năl/ *See* isogonic line.

isogonic line /ÿ-sŏ-**gon**-ik/ (**isogonal**) An imaginary line on the Earth's surface joining points of equal declination. Isogonic lines run roughly East to West. *Agonic lines* are isogonic lines that join points having zero declination. *See also* Earth's magnetism.

isolated system *See* closed system.

isomers, nuclear *See* nuclear isomers.

isotherm /**ÿ**-sŏ-therm/ A line on a chart or graph joining points of equal temperature. *See also* isothermal change.

isothermal change /ÿ-sŏ-**therm**-ăl/ A process that takes place at constant temperature. If work is done on a body in such a way as would otherwise increase its temperature (for example by compressing it) it is possible to keep the temperature constant by heat transfer from the body. Conversely a body may expand doing external work, while heat is transferred to it to prevent the temperature from falling. In the case of an ideal gas, for which the internal energy depends only on the temperature, the heat transferred from the substance must be exactly equal to the work done on it. For real substances the work and heat may not balance in an isothermal process since the internal energy may depend also on the volume.

Although no real process can be perfectly isothermal there are many phenomena that approximate very closely to it. Perfectly reversible isothermal volume changes of an ideal gas are assumed in thermodynamic analyses; for example, Kelvin based his thermodynamic temperature scale upon the behavior of an ideal gas in a Carnot cycle.

Isothermal changes are distinct from adiabatic changes, in which no heat enters or leaves the system so any work done on or by it changes the internal energy and hence the temperature. *See* adiabatic change.

isotones /ÿ-sŏ-tohnz/ Two or more nuclides that have the same neutron numbers but different proton numbers.

isotopes /ÿ-sŏ-tohps/ Two or more species of the same element differing in their mass numbers because of differing numbers of neutrons in their nuclei. The nuclei must have the same number of protons (an element is characterized by its proton number). Isotopes of the same element have very similar chemical properties (the same electron configuration), but differ slightly in their physical properties. An unstable isotope is termed a *radioactive isotope* or *radioisotope*. For example, potassium has 3 naturally occurring isotopes of mass numbers 39, 40, and 41 respectively.
${}^{39}_{19}K$ has 19 protons and 20 neutrons
${}^{40}_{19}K$ has 19 protons and 21 neutrons
${}^{41}_{19}K$ has 19 protons and 22 neutrons
${}^{40}_{19}K$ is radioactive with a half-life of 1.3×10^9 years.

At least eight other isotopes of potassium can be produced by nuclear reactions but all are highly unstable (radioactive).

isotope separation The process of separating the different isotopes of an element according to their atomic masses. Most methods depend on small differences in physical properties between isotopes.

On the laboratory scale isotopes of elements can be separated in the mass spectrometer in which the paths of nuclei depend on the ratio of charge to mass. Similar methods have been used on a larger scale. The difference in rates of diffusion of the volatile compounds ^{238}uranium hexafluoride and ^{235}uranium hexafluoride has been used on a large scale to separate the isotopes of uranium. The ultracentrifuge process is also used. $^{235}UF_6$ tends to stay near the axis of the centrifuge and the $^{238}UF_6$ tends to drift to the perimeter. Slight separation occurs at each stage.

Early methods of isolating deuterium used electrolysis of water. Deuterium ions are discharged at a slightly slower rate; thus the residue becomes progressively richer in deuterium during electrolysis. The process that is currently used is a chemical exchange between water and hydrogen sulfide.

isotopic mass /ÿ-sŏ-**top**-ik/ (**isotopic weight**) The mass number of given isotope of an element.

isotopic number The difference between the number of neutrons in an atom and the number of protons.

isotopic weight *See* isotopic mass.

isotropy /ÿ-**sot**-rŏ-pee/ A medium is isotropic if the value of a measured physical quantity does not depend on the direction. *Compare* anisotropy.

J

jansky Symbol: Jy /**jan**-ski/ A m.k.s.a. unit used to indicate the strength of radio signals detected during astronomical observations. One jansky is equal to a power of 0.01 yoctowatt (yW) per square meter (m) per hertz (Hz) (1 Jy = 0.01 yW m^{-2} Hz^{-1}).

Joly steam calorimeter /**joh**-lee/ An apparatus for measuring the specific thermal capacity of a gas at constant volume. Two identical copper globes are suspended from the arms of a balance – one is evacuated and the other filled with the sample. The two globes are surrounded by a container into which steam is passed. More condensation occurs on the filled globe because of the higher thermal capacity. The specific heat capacity (C_v) of the gas can be calculated from

$$mC_v(\theta_2 - \theta_1) = m_sL$$

where m is the mass of gas, θ_2 the temperature of the steam, θ_1 the initial temperature of the globes, m_s the mass of steam condensed, and L the specific latent thermal capacity of evaporation of water. The calorimeter is named for the Irish physicist John Joly (1857–1933).

Josephson effect /**joh**-zĕf-sŏn/ Any of various electrical effects observed with pairs of superconductors. *Josephson junctions* (between semiconductors) are finding various applications in investigating fundamental constants, the phenomenon of electron tunneling, and in very high-speed computers. The effect is named for its discoverer, the British physicist Brian Josephson (1940–).

joule /jool/ Symbol: J The SI unit of energy and work, equal to the energy required or work done when the point of application of a force of one newton moves an object one meter in the direction of action of the force. 1 J = 1 N m. The joule is the unit of all forms of energy. The unit is named for the British physicist James Prescott Joule (1818–89).

Joule–Kelvin effect /jool **kel**-vin/ (**Joule–Thomson effect**) A temperature change that occurs when a thermally insulated gas expands through a small hole or a porous barrier into a region of lower pressure. If the gas is initially above a certain temperature, the *inversion temperature* of the gas, the temperature rises. If it is initially below the inversion temperature it cools. It can be shown that an ideal gas would not show this effect.

The Joule–Kelvin effect is used in the method of liquefying gases invented by Linde (1895). The gas, which has been cooled slightly by passing through the hole, flows back over the surface of the input pipe, thereby cooling the gas approaching the hole. Thus the temperature drops steadily until liquefaction takes place. The method can be used for liquefying any gas, but in the important cases of hydrogen and helium the gases must first be cooled to below their inversion temperatures, which are far below room temperature. Helium must be cooled by liquid hydrogen, which is itself initially cooled by liquid nitrogen.

The Joule–Kelvin effect is essentially thermodynamically *irreversible* and should not be confused with adiabatic expansion of a gas in a cylinder or turbine, which can approximate to an ideal *reversible* process. The Joule–Kelvin effect was discovered by Joule and William Thomson (subsequently Lord Kelvin) in the 1850s (*see* kelvin).

Joule's equivalent Symbol: *J* A constant, 4.1855×10^7 ergs per calorie (15°), relating former units of 'heat' energy to units of mechanical energy. It arose from early experiments by Joule showing that mechanical work always relates to an equivalent quantity of 'heat'. The constant is also called the *mechanical equivalent of heat.*

In SI units, work, heat, and all forms of energy are measured in joules. The value of J (4.1855 joules per calorie) is the factor defining the calorie.

Joule's laws **1.** The electrical work done per unit time (the power) when a current I flows through a resistor of resistance R with a potential difference V is given by

$$\text{power} = VI = RI^2 = V^2/R$$

In the steady state this gives the rate of emission of heat by the resistor. These relationships follow from the modern definitions of potential difference and resistance, but Joule discovered them experimentally in terms of arbitrarily defined quantities.

2. The internal energy of an ideal gas depends only on the temperature, i.e. it is independent of pressure and volume if the temperature is held constant. Joule originally expressed the law for real gases on the basis of insensitive methods. Later he showed that the temperature of a real gas falls slightly on expanding into a vacuum with perfect insulation. Since in such a case no work is done and no heat flows the internal energy is constant, and for an ideal gas the temperature would be constant.

Joule–Thomson effect *See* Joule–Kelvin effect.

J-particle Another name for a psi particle. *See* fundamental particles.

junction rectifier *See* rectifier.

junction transistor *See* transistor.

K

kaon /**kay**-on/ (**K-meson**) A type of meson. There are four types of meson: positively charged, negatively charged, and two types of neutral particle. They are strongly interacting bosons of zero spin with several alternative modes of decay. *See also* elementary particles.

keepers Pieces of 'soft' iron placed between the poles of permanent magnets when stored to help retain the magnetism. Bar magnets are placed in pairs with opposite poles next to each other and keepers are placed across each end. The magnet's internal field opposes its external field; this causes a demagnetizing effect. When keepers are used, they become magnetized with an internal field in the same direction as that of the magnet. Thus the demagnetizing effect is reduced. *See also* magnetic induction.

kelvin Symbol: K The SI base unit of thermodynamic temperature. It is defined as the fraction 1/273.16 of the thermodynamic temperature of the triple point of water. Zero kelvin (0 K) is absolute zero. One kelvin is the same as one degree on the Celsius scale of temperature. The unit is named for British physicist William Thomson, later made Lord Kelvin (1824–1907).

Kelvin temperature scale The TEMPERATURE SCALE that has its origin at absolute zero, with 273.16 K as the triple point of water. *See* absolute temperature; kelvin.

Kennelly–Heaviside layer /**ken**-ĕ-lee **hev**-ee-sÿd/ *See* Heaviside layer.

Keplerian telescope /kep-**leer**-ee-ăn/ (**astronomical telescope**) The most common type of refracting telescope arrangement, consisting of converging objective and eyepiece. Unlike the Galilean telescope, it provides an inverted image and has a greater length. However the field of view is larger and image quality is higher.

The angular magnification is given by $(f_O + f_E)$; the lens separation is $(f_O + f_E)$. For terrestrial use an inverting lens can be included between the objective and the eyepiece. This is placed so that it is a distance $2f_I$ (f_I is its focal distance) from the first image formed by the objective. An erect image is formed a distance $2f$ behind the inverting lens, arranged to be at the principal focus of the eyepiece. The magnification is not affected but the distance from objective to eyepiece is now $f_O + f_E + 4f$. The telescope is named for the German astronomer Johannes Kepler (1571–1630). *See also* refractor.

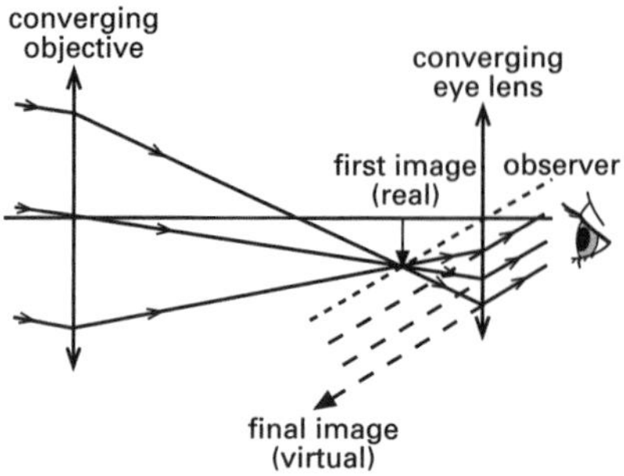

Keplerian telescope

Kepler's laws /**kep**-lerz/ The laws of planetary motion deduced in about 1610 by Johannes Kepler using astronomical observations made by Tycho Brahe:
(1) the planets describe elliptical orbits with the Sun at one focus of the ellipse.

(2) the line between a planet and the Sun sweeps out equal areas in equal times.
(3) the square of the period of a planet's orbit is proportional to the cube of the semi-major axis of the ellipse.

Kerr effect /ker/ The appearance of birefringence in certain isotropic substances when placed in a strong electric field. The effect is proportional to the square of the field strength. It is used in *Kerr cells*, which can switch light radiation extremely rapidly (often within a few picoseconds). Benzene is an example of a substance showing the Kerr effect. It is named for the British physicist John Kerr (1824–1907).

keV kiloelectronvolts; 10^3 electronvolts.

kilo- Symbol: k A prefix denoting 10^3. For example, 1 kilometer (km) = 10^3 meters (m).

kilocalorie /**kil**-ŏ-kal-ŏ-ree/ An energy unit equal to 1000 calories, sometimes called the *Calorie* (with a capital C), as used by dietitians.

kilogram /**kil**-ŏ-gram/ Symbol: kg The SI base unit of mass, equal to the mass of the international prototype of the kilogram, which is a cylinder of platinum–iridium kept at Sèvres in France. It is the only SI base unit not defined in terms of fundamental constants.

kilogram-force Symbol: kgf A m.k.s. unit of gravitational force equal to that acting on a mass of one kilogram at sea level. In SI derived units it is equal to 9.806 65 newtons. The alternate term *kilopond* has been used to avoid potential confusion of the kilogram-force with the kilogram.

kilopond Symbol: kp *See* kilogram-force.

kilowatt-hour /**kil**-ŏ-wot/ Symbol: kWh A unit of energy, usually electrical, equal to the energy transferred by one kilowatt of power in one hour. It is the same as the Board of Trade unit (BTU) and has a value of 3.6×10^6 joules.

kinematics /kin-ĕ-**mat**-iks/ The study of the motion of objects. *See also* dynamics.

kinematic viscosity Symbol: v The ratio of a fluid's VISCOSITY to its density.

kinetic energy /ki-**net**-ik/ Symbol: T The work that an object can do because of its motion. For an object of mass m moving with velocity v, the kinetic energy is given by $mv^2/2$. This gives the work the object would do in coming to rest. The rotational kinetic energy of an object of moment of inertia I and angular velocity ω is given by $I\omega^2/2$.

See also energy.

kinetic equation A type of equation used to describe the way in which a system changes with time in statistical mechanics. The first example, known as the *Boltzmann equation*, was put forward by the Austrian physicist Ludwig Boltzmann (1844–1906) in 1872. Kinetic equations are often used to calculate transport quantities, such as thermal or electrical conductivity.

kinetic friction *See* friction.

kinetic theory (of gases) The molecules or atoms of a gas are in continuous random motion and the pressure (p) exerted on the walls of a containing vessel arises from the bombardment by these fast moving particles. These have average speeds, at normal pressures and temperatures, of around one kilometer per second. When the temperature is raised these speeds increase; so consequently does the pressure. If more particles are introduced or the volume is reduced there are more particles to bombard unit area of the walls and the pressure also increases. When a particle collides with the wall it experiences a rate of change of momentum, which is equal to the force exerted. For a large number of particles this provides a steady force per unit area (or pressure) on the wall.

Following certain additional assumptions, the kinetic theory leads to an expression for the pressure exerted by an ideal gas. The assumptions are:

1. The particles behave as if they are hard smooth perfectly elastic points.
2. They do not exert any appreciable force on each other except during collisions.
3. The volume occupied by the particles themselves is a negligible fraction of the volume of the gas.
4. The duration of each collision is negligible compared with the time between collisions.

By considering the change in momentum on impact with the walls it can be shown that $p = \rho c^2/3$ where ρ is the density of the gas and c is the root-mean-square speed of the molecules. The mean square speed of the molecules is proportional to the absolute temperature:

$$Nmc^2 = RT$$

See also degrees of freedom; equipartition of energy.

Kirchhoff's law (of radiation) /**kersh**-hofs/ For a given wavelength the emissivity of a surface in a particular direction is equal to the absorptance for radiation incident from that direction. The law is named for the German physicist Gustav Robert Kirchoff (1824–87).

Kirchhoff's laws A set of rules for calculating unknown currents, resistances, and voltages in an electric circuit. They are:
1. The algebraic sum of the currents at any point in any circuit is zero. For example, if 6 amperes enter a three-way junction through one wire, then 6 amperes must leave through the other two. A current flowing away from a junction has an opposite sign to one flowing toward the junction.
2. The algebraic sum of the e.m.f.s round any closed loop in any circuit is equal to the sum of the products of current and resistance around the loop. For example, in a circuit with e.m.f. E, current I, and resistances R and r, $E = Ir + IR$.

klystron /**klis**-tron, **klÿs**-tron/ A cavity resonator used for generating microwave radiation. It is a velocity-modulated electron beam tube, originally used in radar equipment. *See also* magnetron.

K-meson *See* kaon.

knife edge A sharp wedge used as a fulcrum or support, as in a balance. The sharp edge minimizes the area of contact between moving parts, thereby reducing the friction between them. Knife edges are made of hard material such as agate.

krypton /**krip**-ton/ A colorless odorless monatomic element of the rare-gas group, known to form unstable compounds with fluorine. It occurs in minute quantities (0.001% by volume) in air. Krypton is used in fluorescent lights.

Symbol: Kr; m.p. –156.55°C; b.p. –152.3°C; d. 3.749 (0°C) kg m^{-3}; p.n. 36; r.a.m. 83.80.

Kundt's tube /kûnt/ A device for showing stationary waves in a gas (or liquid). A horizontal rod, clamped at its midpoint, has a flat disk (E) on one end. The disk just fits into the bore of a glass tube, which is closed at the other end with a flat surface (R). Fine powder is sprinkled along the inside of the tube. The rod is stroked to produce longitudinal vibrations; sound waves are generated by the disk, travel down the tube, and are reflected at the closed end. The position of the disk is changed (changing the length of the column of gas) until standing waves are produced. The dust then vibrates strongly and settles in regularly-spaced heaps along the tube. These are pressure antinodes (i.e. particle-velocity nodes).

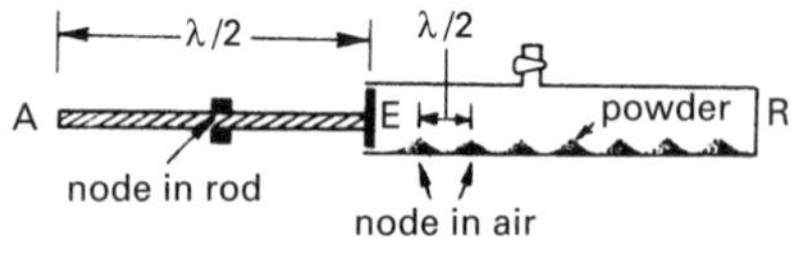

Kundt's tube

L

label A stable or radioactive nuclide used to investigate some process, such as a chemical reaction. It is also sometimes called a tracer. *See* radioisotope.

lag The time or angle by which one periodic quantity is delayed with respect to another.

Lagrangian point /lă-**gran**-jee-ăn, -**grahn**-/ A point in the plane of a two-body system, such as the Earth and the Moon, in which small bodies, such as artificial satellites, can be trapped because the gravitational fields of the two bodies are equal at that point. The idea was first put forward by the French astronomer Joseph Louis Lagrange in 1772.

lambda particle /**lam**-dă/ *See* elementary particles.

lambda point The temperature at which liquid helium I becomes the superfluid helium II. *See* helium.

lambert /**lam**-bert/ The c.g.s. unit of luminance defined as that of a perfectly diffusing surface which reflects one lumen per square centimeter. In SI units it is equal to 3183.099 candelas per square meter. It is approximately 3.18×10^3 Cd m^{-2}. The unit is named for the German mathematician, physicist, astronomer, and philosopher Johann Heinrich Lambert (1728–77).

Lambert's law **1.** First proposed in 1760; then restricted to visible light, it is now used with all radiations. The law concerns the rate of absorption of radiation as it travels deeper into a medium. It states that equal thicknesses of the medium absorb equal proportions of the incident radiation. In other words, the intensity I of the transmitted radiation falls off exponentially with distance d in the medium:

$$I = I_0 \exp{-\alpha d}$$

Here I_0 is the intensity of the initially incident radiation, and α is the *linear absorption coefficient* of the medium. As well as depending on the medium, α varies with wavelength.

2. In photometry, the fact that the luminous intensity of a diffuse surface varies with angle of view:

$$I_0 = I_0 \cos\theta$$

Here, I_0 is the intensity along the normal, while I is that along a line at angle θ to the normal. The principle is often called *Lambert's cosine law.*

laminar flow /**lam**-ă-ner/ Steady flow in which the fluid moves past a surface in parallel layers of different velocities. *Compare* turbulent flow. *See also* Poiseuille's formula.

laminated iron /**lam**-ă-nay-tid/ A piece of iron constructed in thin layers that are separated by electrical insulator. Laminated iron cores are used in many electric machines. The laminations are used to reduce eddy currents caused by a changing magnetic field.

lanthanum /**lan**-thă-nŭm/ A soft ductile malleable silvery metallic element that is the first member of the lanthanoid series. It is found associated with other lanthanoids in many minerals, including monazite and bastnaesite. Lanthanum is used in several alloys (especially for lighter flints), as a catalyst, and in making optical glass.

Symbol: La; m.p. 921°C; b.p. 3457°C; r.d. 6.145 (25°C); p.n. 57; r.a.m. 138.9055.

Laplace equation /lă-**plass**, **-plahss** lah-/ An important equation of mathematical physics, which can be written as $\nabla^2\phi = 0$, where ϕ is a scalar function such as the gravitational or electric potential of a system and ∇^2 is the LAPLACIAN. The Laplace equation is a special case of the POISSON EQUATION. The equation was first put forward by the French mathematician Pierre Simon Laplace (1749–1827). It is used in many branches of physics, including fluid mechanics, electromagnetic theory, and gravity.

Laplacian /lă-**plass**-ee-ăn/ Symbol∇^2. A mathematical operator defined by $\partial^2/\partial x^2 + \partial^2/\partial y^2 + \partial^2/\partial z^2$. The Laplacian occurs in many equations of mathematical physics including the LAPLACE EQUATION and the POISSON EQUATION.

large-scale structure Structure in the Universe that involves large numbers of objects, as in galaxies and clusters of galaxies. It is thought that the inhomogeneous distribution of matter into such structure is a consequence of quantum fluctuations in the early Universe.

Larmor precession /**lar**-mor/ The precession in the orbital motion of an electron about the nucleus of an atom resulting from the application of a small external magnetic field. The precession is named for the Irish physicist Sir Joseph Larmor (1857–1942).

laser /**lay**-zer/ (*l*ight *a*mplification by *s*timulated *e*mission of *r*adiation) A device able to produce a beam of radiation with unusual properties. Generally the beam is: coherent (the waves are in phase); monochromatic (the waves are of effectively the same wavelength); parallel; and intense (carrying a great deal of energy).

There are innumerable applications of such beams in communications, engineering, science, and medicine. Laser action is obtained in a volume of suitable material (solid, liquid, or gas) into which energy is passed at a high rate.

The input energy excites the active particles to a higher energy state W_2 from which they return to a comparatively stable state W_1 above the ground state, W_0. They accumulate there, forming a *population inversion*. Passing photons of energy (W_1–W_0) stimulate decay to the ground state. The photons emitted travel in phase with, and in the same direction as, those that stimulated their production.

In practice, reflecting surfaces are used at each end of the device – one totally reflecting and the other partially reflecting. The radiation is reflected backward and forward, building up an intense beam, which is emitted through the partially reflecting surface.

In solid lasers, such as the ruby laser, the population inversion is produced by an intense external light source. Generally it is pulsed. The wavelength is 694.3 nm. In gas lasers, a discharge is used. The carbon dioxide laser gives a wavelength of 10.5 μm. Helium-neon lasers produce a number of separate wavelengths, including 1.153 μm. Gas lasers can produce continuous (rather than pulsed) beams.

See also holography.

latent heat /**lay**-tĕnt/ The energy transferred when a substance changes its state from solid to liquid, liquid to gas, solid to gas, etc. For example, when a liquid boils the latent heat taken in is partly used to overcome attractive forces between the molecules, and partly to do work in expanding against atmospheric pressure. The *specific latent heat* (specific latent thermal capacity) is the energy transferred in converting unit mass from one state to another at the same temperature. For instance, the specific latent thermal capacity of fusion (solid-liquid) or the specific latent thermal capacity of evaporation (liquid-gas). The units are joules per kilogram (J kg^{-1}). Latent heats are standard enthalpies for these processes.

lateral inversion *See* perversion.

lattice *See* crystal.

lattice energy The energy required to separate an ion from a crystal to an infinite distance. It is measure of the strength of an ionic bond.

law of flotation *See* flotation; law of.

law of magnetic poles (**law of magnetism**) The rule describing the forces between nearby poles – like poles repel each other; unlike poles attract each other.

law of moments *See* moment.

lawrencium /lor-**en**-see-ŭm/ A radioactive transuranic element of the actinoid series, not found naturally on Earth. Several very short-lived isotopes have been synthesized by bombarding ^{252}Cf with boron nuclei or ^{249}Bk with ^{18}O nuclei.

Symbol: Lr; p.n. 103; most stable isotope ^{262}Lr (half-life 261 minutes).

laws of conservation *See* conservation of mass and energy; constant energy (law of), constant (linear) momentum (law of), constant mass (law of), constant angular momentum (law of).

laws of electromagnetic induction *See* electromagnetic induction.

laws of friction *See* friction.

laws of reflection *See* reflection; laws of.

laws of refraction *See* refraction; laws of.

laws of thermodynamics *See* thermodynamics.

LCR circuit *See* alternating-current circuit.

LDR *See* light-dependent resistor.

lead /led/ A dense, dull, gray, soft metallic element; the end product of radioactive decay series. It occurs in small quantities in a wide variety of minerals but only a few are economically important. Lead is used in (lead–acid) accumulators, alloys, radiation shielding, and water and sound proofing. It is also used in the petrochemical, paint, and glass industries.

Symbol: Pb; m.p. 327.5°C; b.p. 1830°C; r.d. 11.35 (20°C); p.n. 82; r.a.m. 207.2.

lead-acid accumulator A type of electrical accumulator used in vehicle batteries. It has two sets of plates: spongy lead plates connected in series to the negative terminal and lead oxide plates connected to the positive terminal. The material of the electrodes is held in a hard lead-alloy grid. The plates are interleafed. The electrolyte is dilute sulfuric acid.

The e.m.f. when fully charged is about 2.2 V. This falls to a steady 2 V when current is drawn. As the accumulator begins to run down, the e.m.f. falls further. During discharge the electrolyte becomes more dilute and its relative density falls. To recharge the accumulator, charge is passed through it in the opposite direction to the direction of current supply. This reverses the cell reactions, and increases the relative density of the electrolyte (*c.* 1.25 for a fully charged accumulator).

The electrolyte contains hydrogen ions (H^+) and sulfate ions (SO_4^{2-}). During discharge, H^+ ions react with the lead (IV) oxide to give lead (II) oxide and water

$$PbO_2 + 2H^+ + 2e^- \rightarrow PbO + H_2O$$

This reaction takes electrons from the plate, causing the positive charge. There is a further reaction to yield soft lead sulfate:

$$PbO + SO_4^{2-} + 2H^+ \rightarrow PbSO_4 + H_2O + 2e^-$$

Electrons are released to the electrode, producing the negative charge.

During charging the reactions are reversed:

$$PbSO_4 + 2e^- \rightarrow Pb + SO_4^{2-}$$

$$PbSO_4 + 2H_2O \rightarrow PbO_2 + 4H^+ + SO_4^{2-} + 2e^-$$

Le Chatelier principle /lă-sha-**tel**-yay/ If a constraint is applied to any system in EQUILIBRIUM, the system will tend to adjust itself to minimize the effect of the constraint. The principle is usually applied to cases of chemical equilibrium. The principle is named for the French chemist Henri Louis Le Chatelier (1850–1936).

Leclanché cell /lă-**klahn**-shay/ A primary voltaic cell consisting, in its 'wet' form, of a carbon-rod anode and a zinc cathode, with a 10–20% solution of ammonium chloride as electrolyte. Manganese(IV) oxide mixed with crushed carbon in a porous bag or pot surrounding the anode acts as a depolarizing agent. The dry form (*dry cell*) is widely used for flashlight batteries, transistor radios, etc. It has a mixture of ammonium chloride, zinc chloride, flour, and gum forming an electrolyte paste. Sometimes the dry cell is arranged in layers to form a rectangular battery, which has a longer life than the cylinder type. The cell is named for the French engineer Georges Leclanché (1839–82).

LED /el-ee-**dee**/ *See* light-emitting diode.

LEED Low-energy electron diffraction. *See* electron diffraction.

Lees' disk An apparatus for measuring the thermal conductivity of a poor conductor. The sample is a thin flat disk of known area (A) and thickness (l) placed between two disks of brass (or other good conductor), each having a thermometer in a hole in its side. The apparatus is suspended by strings and the top disk is raised in temperature. Under steady conditions energy passes at a steady rate and the temperatures of the top and bottom metal disks are θ_2 and θ_1. The bottom conductor loses energy by radiation: to measure the rate of loss of energy, the bottom slab is warmed to the temperature θ_2, the insulator placed on top, and a cooling curve plotted so that that rate of temperature fall can be obtained at the lower temperature θ_1. Then k is found from the equation

$$kA(\theta_2 - \theta_1)/l = mc\mathrm{d}\theta/\mathrm{d}t$$

where m is the mass of the bottom slab, c its specific thermal capacity, and $\mathrm{d}\theta/\mathrm{d}t$ its rate of temperature fall obtained from the cooling curve.

The apparatus can be modified for use with liquids, these being contained in a insulating ring between the two metal disks. It is named for the English physicist Charles Herbert Lees (1864–1952).

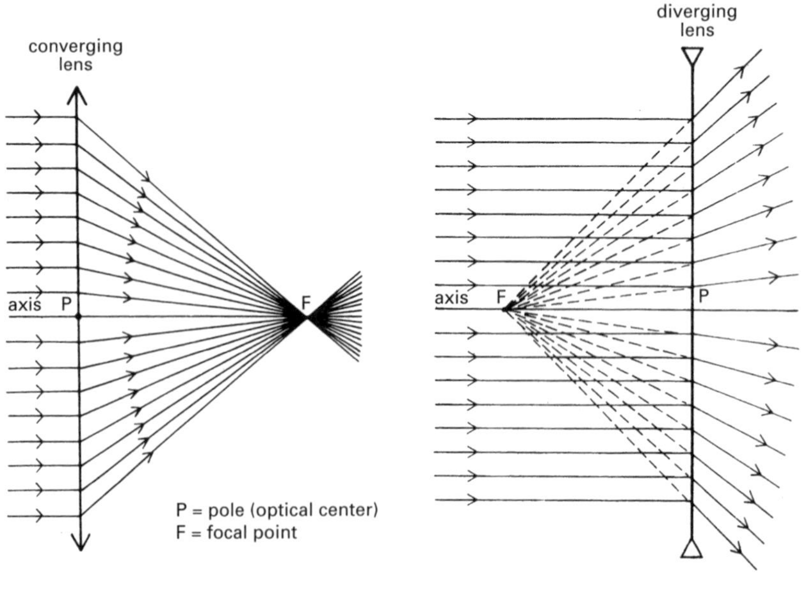

The paths of light rays through converging and diverging lenses

left-hand rule *See* Fleming's rules.

lens An optical component that refracts rays passing through, and either converges or diverges them. With visible radiation, glass lenses in air are normally used; however lenses of any transparent substance would work. A lens is best described by its effect on a parallel beam. Converging lenses make it convergent; diverging lenses make it divergent. Focal distance is the usual measure of the power of a lens to converge a parallel beam. Parallel rays are refracted by converging and diverging lenses as shown in the diagram. Strictly, such focusing occurs only with rays parallel to and close to the principal axis. Rays from an object will be refracted so that they appear to come from somewhere else – the image. Two examples are given overleaf to show how ray diagrams are drawn.

The object distance (u) and the image distance (v) relate to the focal distance (f) thus:

$1/f = 1/v + 1/u$ (in the real is positive, virtual is negative sign convention).

The formula applies to real or virtual focal points, objects, and images. All the above can be adapted to lenses for use with nonvisible radiations, including pressure waves and electron beams.

Lenz's law /lents-iz/ The direction of an induced e.m.f. is such that it opposes the change that produces it. For example, if a permanent bar magnet is moved toward a coil that forms part of a complete circuit, then current flows so that the magnetic field around the coil repels the magnet. If this were not the case, no work would need to be done to cause charge to flow. The law is named for the Estonian-born German physicist Heinrich Friedrich Emil Lenz (1805–65).

lepton /lep-ton/ A type of elementary particle including the electron, the muon, and the neutrino. Leptons are distinguished from hadrons by their type of interaction. *See also* elementary particles.

Leslie's cube /lez-leez/ A cube-shaped can with the four sides painted different colors or given different finishes (polished, rough, etc.). It is filled with hot water and used in experiments on the effect of surface on emissivity. The cube is named for the Scottish scientist Sir John Leslie (1766–1832).

lever A class of machine. Levers are rigid objects able to turn around some point (pivot or fulcrum). The force ratio and the distance ratio depend on the relative posi-

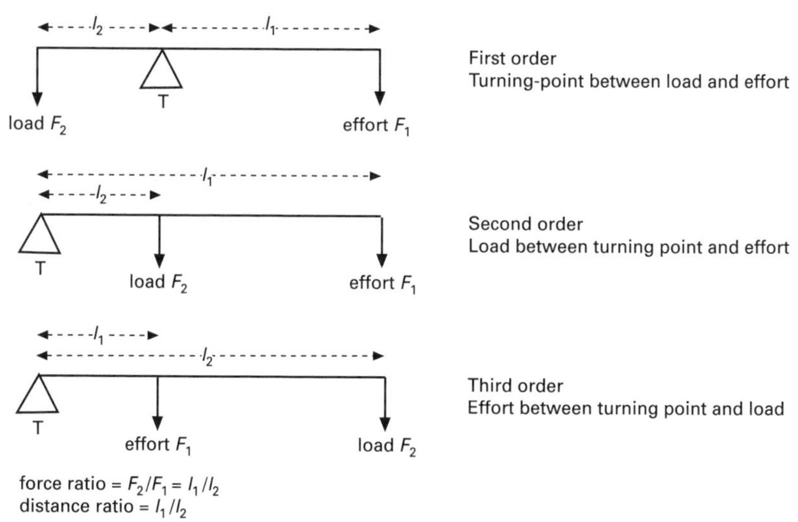

Lever

OBJECT AND IMAGE POSITIONS FOR CONVERGING LENS

Object position	*Image nature*	*Image position*
infinity	real	F or F′
beyond 2F or 2F′	real	between F′ (F) and 2F′ (2F)
2F or 2F′	real	2F′ or 2F
between 2F (2F′) and F (F′)	real	beyond 2F′ or 2F
F or F′	virtual	infinity
between F (F′) and P	virtual	between infinity and P

OBJECT AND IMAGE POSITIONS FOR DIVERGING LENS

Object position	*Image nature*	*Image position*
infinity	virtual	F or F′
beyond 2F or 2F′	virtual	between F or F′ and P
2F or 2F′	virtual	between F or F′ and P
between 2F (2F′) and F (F′)	virtual	between F or F′ and P
F or F′	virtual	between F or F′ and P
between F (F′) and P	virtual	between F or F′ and P

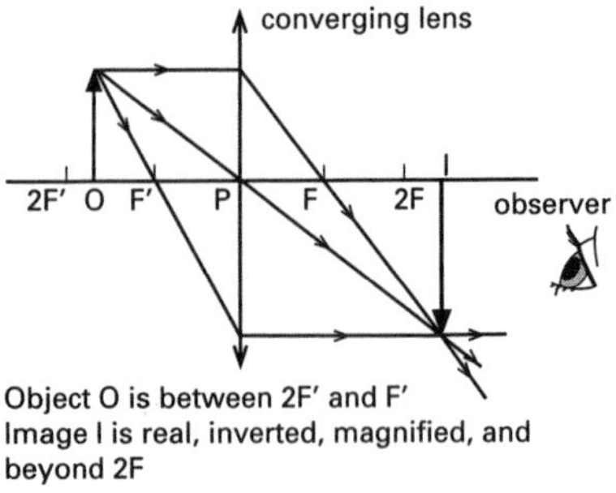

Ray diagram showing a real (inverted) image formed from a real object

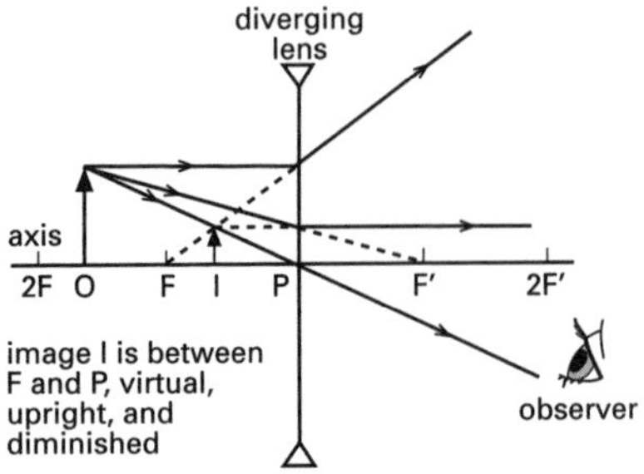

Ray diagram showing a virtual (erect) image formed from a real object

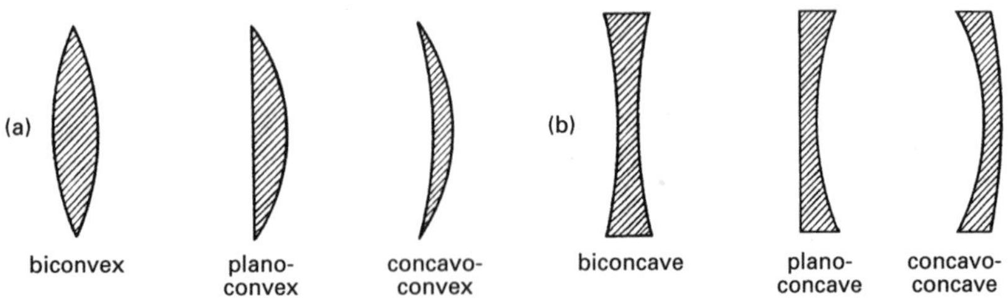

Types of lens

tions of the pivot, the point where the user exerts the effort, and the point where the lever applies force to the load. There are three types (orders) of lever.

Levers can have high efficiency; the main energy losses are by friction at the pivot, and by moving the lever itself.

levorotatory /lee-voh-**roh**-tă-tor-ee. -toh-ree/ *See* optical activity.

Leyden jar /**lÿ**-d'n/ An early type of capacitor made from a glass jar with metal foil on its inner and outer surfaces. The Leyden jar (sometimes spelt Leiden jar) is named for the Dutch town of Leyden, where it was invented.

LF *See* low frequency.

lift A force exerted by a fluid (gas or liquid) on an object in its flow. It acts upward at right angles to the direction of the flow, and is therefore the force that enables birds and winged aircraft to fly. *See also* airfoil; drag.

light (**visible radiation**) A form of electromagnetic radiation able to be detected by the human eye. Its wavelength range is between about 400 nm (far 'red') and about 700 nm (far 'violet'). The boundaries are not precise as individuals vary in their ability to detect extreme wavelengths; this ability also declines with age.

Light is produced by surfaces at temperatures above about 900 K. Below about 6000 K, however, the majority of the radiation emitted is infrared. The standard household filament lamp works at about 2800 K; at this temperature only a few percent of the emitted radiation is visible. 'Cold' sources, such as certain chemical reactions, glowworms, lasers, and discharge tubes, are more efficient in this sense.

These produce light by specific transitions between electronic energy levels in atoms and molecules. Their spectra are therefore not continuous, but sets of lines.

light-dependent resistor (**LDR**) A resistor whose resistance decreases when it is illuminated. LDRs are made from semiconductors such as cadmium sulfide and selenium.

light-emitting diode (**LED**) A semiconductor DIODE, made from certain materials (e.g. gallium arsenide), in which light is emitted in response to the forward-bias current. The light results from the recombination of electrons and positive holes, with a transition to a lower energy state. *See also* transistor.

lightning An electrical discharge in the atmosphere between charged clouds and the Earth, or between the clouds.

lightning conductor A thick metal conductor (usually copper) that runs from the top of a high building to the ground (where the end is buried). Its function is to conduct any lightning strikes safely to earth.

light-year Symbol: ly A unit of distance used in astronomy, defined as the distance that light travels through space in one year. It is approximately equal to $9.460\,528 \times 10^{15}$ meters.

limit cycle A closed curve in phase space to which a system evolves. A limit cycle is a type of ATTRACTOR characteristic of oscillating systems.

limiting friction *See* friction.

linac /**lin**-ak/ *See* linear accelerator.

THE RANGE OF COLORS IN THE VISIBLE SPECTRUM

Wavelength range (nm)	*Color*
400–420	violet
420–450	indigo
450–500	blue
500–550	green
550–600	yellow
600–650	orange
650–700	red

Note that perception of different colors varies between individuals.

Linde process /**lin**-dĕ/ A method of liquefying gases by compression followed by expansion through a nozzle. The temperature falls by the Joule–Kelvin effect. The cooled gas is used to reduce the temperature of the compressed gas. Eventually the temperature falls below the boiling point and the gas liquefies. The gas must start this process below its inversion temperature. Hydrogen can be liquefied by this method if it is first cooled below its inversion temperature using liquid air. Helium is cooled below its inversion temperature by using liquid hydrogen boiling under reduced pressure. The process is named for the German engineer Karl von Linde (1842–1934). *See also* cascade process.

linear accelerator (**linac**) A device for accelerating charged particles in a straight line. It consists of a series of cylindrical electrodes of increasing length, alternate ones of which are connected together and between which an alternating voltage is applied. A charged particle is accelerated to enter the first cylinder. By the time it emerges the voltages of the cylinders have reversed, and it is now repelled from the first cylinder and attracted towards the second, so accelerating and gaining energy each time it crosses a gap between cylinders. The cylinders have to be made progressively longer as the particles speed up so that the particles arrive at the gaps at the correct point in the voltage cycle.

A more advanced type of linear accelerator used for electrons and protons has a traveling radio-frequency electromagnetic wave in a waveguide. The particles are carried by the electric component of the wave. *See also* tandem generator.

linear charge density *See* charge density.

linear density *See* density.

linear expansivity *See* expansivity.

linear momentum *See* momentum.

linear momentum, conservation of *See* constant (linear) momentum (law of).

linear motor An ELECTRIC MOTOR in which the 'rotor' is rectangular and travels along rails (which act as the 'stator'). It is a type of INDUCTION MOTOR, used for such applications as opening automatic doors.

line defect *See* defect.

lines of force Imaginary lines drawn to represent a force field. They show the direction of the field at each point; their closeness represents the field strength (the closer the lines, the greater the strength). In the case of magnetism, a line of force shows the path a free N-pole would take in the field.

Lines of force are a useful convention for showing fields; in magnetic fields they are the original basis of magnetic flux. They have, however, no real existence.

line spectrum A SPECTRUM composed of a number of discrete lines corresponding to single wavelengths of emitted or absorbed radiation. Line spectra are produced by atoms or simple (monatomic) ions in gases. Each line corresponds to a change in electron orbit, with emission or absorption of radiation.

line width The width of a spectral line in wavelength terms. The usual measure is the half-width of the line. *See* monochromatic radiation.

liquefaction /lik-wĕ-**fak**-shŏn/ A change of state to a liquid. The term is often used for the conversion of gases to their liquid state.

liquid *See* states of matter.

liquid barometer *See* barometer.

liquid crystal A compound that, although liquid at room temperature, shows behavior normally expected only from solid crystalline substances. Some liquid crystals change orientation in an electric field and are used in liquid crystal displays (LCDs); others change color with temperature, and are used for simple thermometers. *See also* Kerr effect.

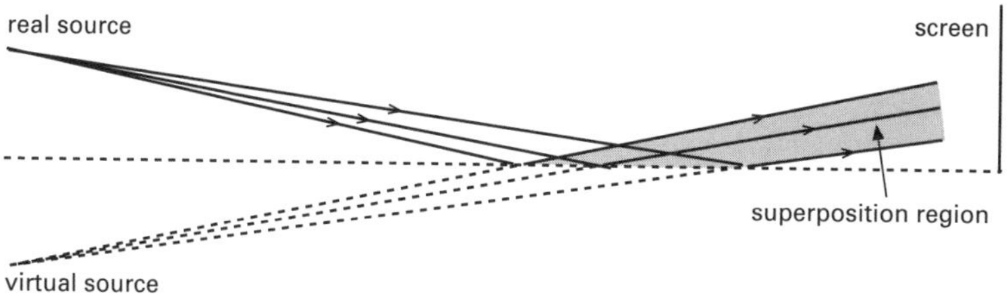

Lloyd's mirror

liquid crystal display (LCD) A type of digital display, used in calculators, watches, and other electronic equipment, based on the properties of LIQUID CRYSTALS.

liquid-drop model A model used in nuclear physics in which the interaction between nucleons is regarded as being analogous to the intermolecular forces between water molecules in a drop of water. The liquid-drop model can be used to describe certain nuclear phenomena and is particularly successful in describing nuclear fission. More complete quantitative analyses of nuclear physics require the combination of the liquid-drop model with the nuclear SHELL MODEL.

Lissajous' figures /**lee**-sa-zhooz/ Patterns obtained by combining two simple harmonic motions in different directions. They can be demonstrated with an oscilloscope, deflecting the spot with one oscillating signal along one axis and with another signal along the other axis. A variety of patterns are produced depending on the frequencies and phase differences. They are named for the French physicist Jules Antoine Lissajous (1822–80).

liter /**lee**-ter/ Symbol: L or l A unit of volume defined as one cubic decimeter or 10^{-3} metre^3. The name is not recommended for precise measurements. Formerly, the liter was defined as the volume of one kilogram of pure water at 4°C and standard pressure. On this definition, 1l = 1000.028 cm^3.

lithium /**lith**-ee-ŭm/ A light silvery moderately reactive metal; it is a rare element accounting for 0.0065% of the Earth's crust.

Symbol: Li; m.p. 180.54°C; b.p. 1347°C; r.d. 0.534 (20°C); p.n. 3; r.a.m. 6.941.

Lloyd's mirror An arrangement in which light meets a mirror at grazing incidence and is reflected to interfere with the direct radiation. INTERFERENCE of the reflected light with the direct light produces fringes. The device is effectively a two-source arrangement. The apparatus is named for the Irish physicist Humphrey Lloyd (1800–81).

logic gate An electronic circuit than can be used to perform simple logical operations. Examples of such operations are

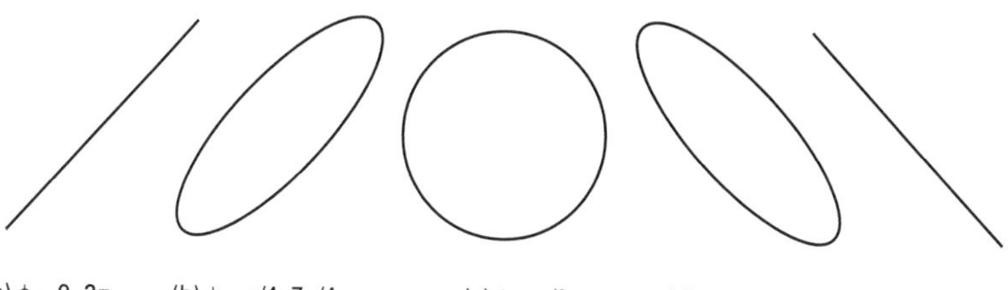

Lissajous' figures

'and', 'either–or', 'not', 'neither–nor', etc. Logic gates operate on high or low input and output voltages. Binary logic circuits, those that switch between two voltage levels (high and low), are widely used in digital computers. The *inverter gate* or *NOT gate* simply changes a high input to a low output and vice versa. In its simplest form, the *AND gate* has two inputs and one output. The output is high if and only if both inputs are high. The *NAND gate* (not and) is similar, but with the output negated; that is a low output if and only if both inputs are high. The *OR gate* has a high output if one or more of the inputs are high. The *exclusive OR gate* has a high input only if one of the inputs, but not more than one, is high. The *NOR gate* has a high output only if all the inputs are low. Logic gates are constructed using transistors, but in a circuit diagram they are often shown by symbols that denote only their logical functions.

See also bistable circuit; multivibrator.

long hundredweight *See* hundredweight.

longitudinal wave A wave motion in which the vibrations in the medium are in the same direction as the direction of energy transfer. Sound waves are an example of longitudinal waves. *Compare* transverse wave.

long sight *See* hyperopia.

Lorentz–Fitzgerald contraction /lo-**rents** fits-**je**-răld/ According to classical theory, space contained an imponderable fluid *ether* providing an absolute reference frame with respect to which bodies could be shown to be moving or at rest. The failure of the Michelson–Morley experiment to show motion with respect to the ether was explained by Lorentz and Fitzgerald by the hypothesis that the length of any object contracts in the direction of its motion through the ether by the factor $\sqrt{(1 - v^2/c^2)}$, where v is its speed in the ether and c the speed of light. This was assumed to be a real physical change in the object and various experiments were done to detect it, with negative results.

In the theory of relativity there is no absolute motion, the concept of an ether that provides a standard of rest is rejected. Hence the negative result of the Michelson–Morley experiment is expected and there is no justification for the Lorentz–Fitzgerald hypothesis. In this theory the length of a body as measured by any observer with respect to whom the body has a velocity v differs from the length as measured by an observer at rest with respect to the body by the same factor as assumed by Lorentz and Fitzgerald. This is purely a mathematical result used by a particular observer (observers with different uniform relative velocities would obtain different values) and does not represent any physical change in the body. The contraction is named for the Dutch physicist Hendrik Antoon Lorentz (1853–1928) and the Irish physicist George Francis Fitzgerald (1851–1901).

Lorentz force An aspect of the motor effect; the force on a charge Q, moving at velocity v across a magnetic field B.

$$f = BQv\sin\theta$$

θ is the angle between v and B.

loudness The sensation that a sound produces in a listener corresponding to its intensity (i.e. the energy flow per unit area). For a given frequency, the greater the intensity, the greater the loudness. Loudness also depends on the frequency of the sound. *See* phon.

loudspeaker A device for converting an electrical signal into sound. Most loudspeakers consist of a small coil, through which the signal flows, attached to the center of a diaphragm or cone. The coil is free to move in an annular gap between the poles of a magnet. The changing current in the coil causes it to vibrate in the magnetic field and the cone sends out these vibrations as sound waves.

lowering of vapor pressure The lowering of the vapor pressure of a liquid that occurs when a solute is dissolved in the liquid. The reduction is proportional to the number of molecules of solute dissolved per unit volume of liquid solvent, i.e. the lowering of vapor pressure is one of the COLLIGATIVE PROPERTIES of matter.

low frequency (LF) A radio frequency in the range between 300 kHz and 30 kHz (wavelength 1 km-10 km).

low tension (L.T.) Low voltage.

lumen (*pl.* **lumens** or **lumina**) /**loo**-mĕn/ Symbol: lm The SI unit of luminous flux, equal to the luminous flux emitted by a point source of one candela in a solid angle of one steradian. 1 lm = 1 cd sr.

luminance /**loo**-mă-năns/ Symbol: L_v A measure of the brightness of an extended source (one that cannot be considered a point). In a given direction, it is the luminous intensity per unit area projected at right angles to the direction. The unit is the candela per square meter (cd m^{-2}).

luminescence /loo-mă-**ness**-ĕns/ The emission of radiation from a substance in which the particles have absorbed energy and gone into excited states. They then return to lower energy states with the emission of electromagnetic radiation. If the luminescence persists after the source of excitation is removed it is called PHOSPHORESCENCE; if not, it is called FLUORESCENCE.

The excitation of the particles may occur by a variety of mechanisms. Absorption of other electromagnetic radiation gives *photoluminescence.* If the original excited states are produced by bombardment with electrons the phenomenon is *electroluminescence. Chemiluminescence* is luminescence produced by chemical reactions. *Bioluminescence* occurs in natural systems, e.g. glowworms and fireflies. *Radioluminescence* occurs in radioactive materials. Luminescence produced in materials by friction is called *triboluminescence.*

luminosity /loo-mă-**noss**-ă-tee/ *See* brightness; luminous intensity.

luminous exitance *See* exitance.

luminous flux Symbol: Φ_v The rate of flow of energy of visible radiation. It is the radiant flux corrected for the fact that the sensitivity of the eye is different for different wavelengths. The unit is the lumen (lm).

luminous intensity Symbol: I_v The luminous flux from a point source per unit solid angle. The unit is the candela (cd).

lunar eclipse /**loo**-ner/ *See* eclipse.

lunar time *See* time; year.

lutetium /loo-**tee**-shee-ŭm/ A silvery element of the lanthanoid series of metals. It occurs in association with other lanthanoids. It is rare and has few uses.

Symbol: Lu; m.p. 1663°C; b.p. 3395°C; r.d. 9.84 (25°C); p.n. 71; r.a.m. 174.967.

lux (*pl.* **lux**) /luks/ Symbol: lx The SI unit of illumination, equal to the illumination produced by a luminous flux of one lumen falling on a surface of one square meter. 1 lx = 1 lm m^{-2}.

Lyman series /**lÿ**-măn/ A series of lines in the ultraviolet spectrum emitted by excited hydrogen atoms. They correspond to the atomic electrons falling into the lowest energy level and emitting energy as radiation. The wavelength (λ) of the radiation in the Lyman series is given by $1/\lambda = R(1/1^2 - 1/n^2)$, where n is an integer and R is the Rydberg constant. The series is named for the American physicist Theodore Lyman (1874–1954). *See also* spectral series.

M

machine A device for transmitting power between one place and another. The user applies a force (the effort) to the machine; the machine applies a force to a load of some kind. These two forces need not be the same. In fact the purpose of the machine may well be to overcome a large load with a small effort. For any machine this relationship is measured by the *force ratio* (or mechanical advantage): this is the force applied by the machine (load, F_2) divided by the force applied by the user (effort, F_1).

The work done by the machine cannot exceed the work done to the machine. Therefore, for a 100% efficient machine:

$$\text{if } F_2 > F_1 \text{ then } s_2 < s_1$$
$$\text{and if } F_2 < F_1 \text{ then } s_2 > s_1.$$

Here s_2 and s_1 are the distances moved by F_2 and F_1 in a given time.

The relationship between s_1 and s_2 in a given case is measured by the distance ratio (or velocity ratio). This is the distance moved by the effort (s_1) divided by the distance moved by the load (s_2).

Neither distance ratio nor force ratio has a unit; neither has a standard symbol. *See also* efficiency; hydraulic press; inclined plane; lever; pulley; screw.

Mach number /makh/ The ratio of the speed of a moving object (e.g. a high-speed aircraft) to the speed of sound in the air or other medium through which the object is traveling. An aircraft passes through the 'sound barrier' as the Mach number exceeds one; at this speed the air resistance increases sharply. The number is named for the Austrian physicist Ernst Mach (1838–1916).

magic number In nuclear physics, one of several numbers of neutrons or protons that characterize especially stable atomic nuclei. They are 2, 8, 20, 28, 50 and 82. The magic numbers in nuclei are the analog of closed electron shells in atoms.

magnesium /mag-**nee**-zee-ŭm, -seee-ŭm/ A light metallic element. It has the electronic configuration of neon with two additional outer 3s electrons. The element accounts for 2.09% of the Earth's crust and is eighth in order of abundance. Magnesium metal is industrially important as a major component in lightweight alloys (with aluminum and zinc). The metal surfaces develop an impervious oxide film which protects them from progressive deterioration.

Symbol: Mg; m.p. 649°C; b.p. 1090°C; r.d. 1.738 (20°C); p.n. 12; r.a.m. 24.3050.

magnet /**mag**-nĕt/ *See* permanent magnet.

magnetic bottle /mag-**net**-ik/ A magnetic force field used to contain charged particles as in nuclear fusion experiments (where it contains the plasma). The plasma temperature may exceed 100 million degrees Celsius and any material container would vaporize instantly. The temperature is so high that all the electrons are stripped from the atoms and the matter consists entirely of charged particles, which can be restrained by a suitably arranged magnetic field.

magnetic circuit A closed path formed by magnetic lines of force. A simple example is given by a horseshoe magnet; its magnetic circuit is improved by the use of a keeper.

The concept is very important in electrical engineering; it compares directly to an electric circuit. Magnetomotive force F (unit – the ampere, A) compares to e.m.f.;

magnetic flux Φ (unit – the weber, Wb) is the analog of current; reluctance R (unit – reciprocal henry, H^{-1}) is like resistance.

magnetic compass *See* compass.

magnetic constant *See* permeability.

magnetic cooling A method of attaining very low temperatures (well below 1 K). A salt is magnetized isothermally and then demagnetized adiabatically. *See* cryogenics.

magnetic declination *See* declination.

magnetic dip *See* inclination.

magnetic dipole A pair of north-seeking and south-seeking magnetic poles a distance apart, as in a bar magnet. A loop carrying an electric current also acts as a magnetic dipole, as do many atoms.

magnetic dipole moment *See* magnetic moment.

magnetic domain *See* domain.

magnetic elements The Earth's magnetism at any point is defined by three magnetic elements, which give the strength and direction of the field at that point. They are the horizontal component of the field and the angles of inclination and declination. The horizontal intensity can be measured by an Earth inductor, the inclination by an inclinometer, and the declination by a compass.

The elements change from place to place and with time. *See also* magnetic variation.

magnetic equator *See* isoclinic line.

magnetic field A region in which a magnetic force can be observed; i.e. a small magnet or a small loop of wire carrying a current will experience a force. The strength and direction of a field can be represented by lines of force. The number of lines per unit cross-sectional area is a measure of magnetic field strength – the magnetic flux density B.

magnetic field strength (**magnetic intensity**) Symbol: H The force that would be exerted on a unit N-pole at a given point in a magnetic field. The unit is the ampere per meter (A m^{-1}).

magnetic flux Symbol: Φ The strength of a magnetic field through an area, based on the idea of the number of lines of force passing through the area. It is given by the product of magnetic flux density (B) and area. The unit is the weber (Wb).

magnetic flux density (**magnetic induction**) Symbol: B The flux per unit perpendicular area of a magnetic field; it is sometimes thought of as the number of lines of force per unit area. It is defined by the effect on a current-carrying conductor in the field. For a field B:

$$B = F/Qv$$

where F is the force on a charge Q moving perpendicular to the field with velocity v.

The unit of magnetic flux density is the tesla (T). It is equivalent to the weber per square meter (Wb m^{-2}) since

$$B = \Phi A$$

where Φ is magnetic flux and A is area. In SI the relationship between magnetic flux density and magnetic field strength is:

$$B = \mu_r \mu_0 H$$

where μ_r is the relative permeability of the medium and μ_0 the permeability of free space.

magnetic focusing The use of shaped magnetic fields to focus beams of charged particles. Applications of the technique are found in cathode-ray tubes, accelerators, and electron microscopes.

magnetic hysteresis *See* hysteresis.

magnetic induction 1. The magnetization of a substance by an external magnetic field. If a sample of magnetic material is placed near a strong magnet N- and S-poles are induced in it. The sample will then act as an induced magnet and be attracted to the other. *See also* induction.

2. *See* magnetic flux density.

magnetic meridian An imaginary line on the Earth's surface joining the two magnetic poles, and passing through the observer's position. It is the line along which a compass needle comes to rest at that point. The angle between the magnetic and geographic meridians is the declination at that point, one of the three magnetic elements.

See also Earth's magnetism.

magnetic moment (**magnetic dipole moment**) Symbol: m A measure of the strength of a magnet or current-carrying coil. It relates to the turning effect (moment) on it when in a given field. It is defined as the torque T observed in a unit field at 90° to the magnetic axis:

$$m = T/B$$

The unit is ampere meters-squared (A m^2). For a coil with N turns and area A carrying a current I:

$$m = NIA$$

In this case m is often called the *electromagnetic moment* of the coil.

magnetic monopole An elementary particle postulated but not discovered, equivalent to an isolated N- or S-pole. The monopole was postulated to provide symmetry between theories of electricity and magnetism.

magnetic north The location of the north magnetic pole, the direction toward which a magnetic compass needle points. Its position varies from year to year and also from time to time during the year. *See* poles; magnetic.

magnetic permeability *See* permeability.

magnetic pole *See* poles; magnetic.

magnetic pole strength *See* pole strength.

magnetic potential *See* magnetomotive force.

magnetic resistance *See* reluctance.

magnetic resonance imaging (MRI) A technique used in medicine for producing images of the human body (or parts of the body) which makes use of nuclear magnetic resonance. In MRI the sample is placed in a strong magnetic field and irradiated with radiofrequency electromagnetic radiation, causing hydrogen nuclei to be raised to a higher spin energy state. Reversion of these to the ground state emits radiation, which is detected and analyzed to give a three-dimensional picture of internal structures. MRI is less dangerous than x-ray diagnosis, but is also better for investigating soft tissues.

magnetic susceptibility *See* susceptibility.

magnetic variation 1. Various changes with time of the magnetic elements at a point on the Earth's surface. Secular changes are slow continuous changes. Thus the angle of declination at London has changed from zero in 1659 to 8° in 1960; the inclination at the same place has changed from 74° in 1700 to 66° in 1960. Abrupt changes in the magnetic elements are due to magnetic storms, which are related to solar activity. *See also* annual variation.

2. A term still widely used in navigation instead of *declination*.

magnetism /**mag**-nĕ-tiz-ăm/ The study of the nature and cause of magnetic force fields, and how different substances are affected by them. Magnetic fields are produced by moving charge – on a large scale (as with a current in a coil, forming an *electromagnet*), or on the small scale of the moving charges in the atoms. It is generally assumed that the Earth's magnetism and that of other planets, stars, and galaxies have the same cause. Substances may be classified on the basis of how samples interact with fields. Different types of magnetic behavior result from the type of atom. Diamagnetism, which is common to all substances, is due to the orbital motion of electrons. Paramagnetism is due to electron spin, and is a property of materials containing unpaired electrons. Ferromagnetism, the strongest effect, also involves electron spin and the alignment of magnetic moments in domains. Anti-ferromagnetism and ferrimagnetism are rarer effects involving antiparallel alignment of spin magnetic moments.

magnetism, terrestrial *See* Earth's magnetism.

magnetocaloric effect /mag-nee-toh-kă-**lor**-ik, -**loh**-rik/ The change of temperature of a sample as the external magnetic field changes. In particular, energy is stored in ferromagnetic domains; as these appear or disappear, energy is absorbed or released. A use of the effect is in adiabatic demagnetization for the production of very low temperatures (close to absolute zero).

magnetohydrodynamics /mag-nee-toh-hÿ-droh-dÿ-**nam**-iks/ (**MHD**) The study of how magnetic fields interact with conducting fluids (e.g. plasmas or liquid metals).

magnetometer /mag-nĕ-**tom**-ĕ-ter/ A form of compass used for comparing magnetic field strengths. The *deflection magnetometer* consists of a very small bar magnet, pivoted so that it is free to move in a horizontal plane. A circular scale, graduated in degrees, is placed below the magnet and a long light pointer is attached over it at 90°. The magnetometer is rotated in a magnetic field, H_1, so that the reading is 360. When another field, H_2, is placed at right angles to H_1, the needle is deflected by an angle θ. Then:

$$\tan\theta = H_1/H_2$$

The *vibration magnetometer* makes use of the (simple harmonic) vibration of a magnetic needle to measure the horizontal magnetic field strength at the needle. The field strength is inversely proportional to the square of the period.

magnetomotive force /mag-nee-toh-**moh**-tiv/ (**magnetic potential; m.m.f.**) Symbol: F The integral of the magnetic field strength around a closed path; i.e.

$$\oint H dl$$

The unit is the ampere (A). Magnetomotive force is analogous to electromotive force in electric circuits. *See also* magnetic circuit.

magneton /**mag**-nĕ-ton/ *See* Bohr magneton.

magneto-optical effects /mag-nee-toh-**op**-ti-kăl/ Changes of the optical behavior of matter when the external magnetic field changes. Examples are the Faraday effect and the Kerr effect. *See* Faraday effect; Kerr effect.

magnetoresistive effect /mag-nee-toh-ri-**zis**-tiv/ The change of electrical resistance of a sample when the external magnetic field changes.

magnetosphere / mag-**nee**-tŏ-sfeer/ A region of space surrounding a planet with a magnetic field. The size and nature of the magnetosphere depends on the magnetic field in the body. In the case of the Earth, the magnetosphere is the region in which the motion of electrically charged particles from the Sun is dictated by the magnetic field of the Earth. It is a pear-shaped region extending to a distance of about 60 000 km on the side of the Earth nearest the Sun, and to a much greater distance on the side opposite the Sun.

magnetostatics /mag-nee-toh-**stat**-iks/ A development of magnetic theory analogous to electrostatics but based on magnetic poles rather than electric charges. Although free poles are not known, the results of assuming their existence have some validity.

magnetostriction /mag-nee-toh-**strik**-shŏn/ The change in length (and section area) of a ferromagnetic rod with change in external magnetic field. The effect is the result of domain boundary changes. Such a rod will vibrate longitudinally in an alternating field (particularly at the natural frequency); magnetostriction is a common technique for producing ultrasonic radiation.

magnetron /**mag**-nĕ-tron/ A type of thermionic tube used for producing pulsed microwaves. A steady magnetic field is applied from outside the tube. The electrons emitted from a hot cathode circle under the influence of the field and generate electromagnetic oscillations in resonant cavities in the surrounding anode. Magnetrons are used in radar sets and microwave ovens. *See also* klystron.

magnification /mag-nă-fă-**kay**-shŏn/ Symbol: m The extent to which an optical system changes the apparent size of an object; i.e. the image size in comparison with the object size. It is measured by the heights perpendicular to the axis: $m = y/x$, where y is the image height and x the object height. It also depends on the distances of image and object from the center of the system: $m = v/u$, where v is the image distance and u the object distance. If the image is smaller than the object the value of m is less than 1; a negative value of m means that the image (or object) is virtual. This is for the 'real is positive, virtual is negative' sign convention; in the New Cartesian convention a negative value of m denotes an inverted image. *See also* angular magnification.

magnifying glass /**mag**-nă-fÿ-ing/ *See* microscope; simple.

magnifying power *See* angular magnification.

magnitude A logarithmic scale used to measure the brightness of celestial objects, ranging from minus values for the brightest objects to positive values for the faintest. For example, Sirius, the brightest star, has a magnitude of –1.4, while the faintest objects visible to the Hubble Space Telescope have magnitudes of +28. *Absolute magnitude* refers to the actual brightness a celestial object would have at a distance of 10 parsecs (308.57 petameters in SI units) from Earth. *Apparent magnitude* refers to the perceived brightness of an object from a vantage point on Earth, and as such depends on both the object's luminosity and its distance.

magnox /**mag**-noks/ A type of alloy containing magnesium (88%), aluminum (11%), and traces of other elements. It is used for canning the uranium rods in some types of nuclear reactors.

majority carrier *See* semiconductor.

make-and-break An arrangement by which a circuit is alternately opened and closed; for instance by movement of an armature. *See* electric bell.

malleability /mal-ee-ă-**bil**-ă-tee/ The property of a substance that allows it to be beaten or rolled into thin sheets. *Compare* ductility.

Malus's law /ma-**lyoos**, mah-/ A law describing the intensity of polarized light in a polarizing apparatus consisting of a polarizer and an analyzer. It states that when θ is the angle between the *transmission axes* of the polarizer and the analyzer, i.e. the directions along which light can pass unaffected, the intensity of polarized light is proportional to $\cos^2\theta$. This law was first stated by the French engineer Étienne Malus (1775–1812) in 1809.

manganese /**mang**-gă-neez, -nees/ A transition metal occurring naturally as oxides. Nodules found on the ocean floor are about 25% manganese. Its main use is in alloy steels made by adding pyrolusite to iron ore in an electric furnace.

Symbol: Mn; m.p. 1244°C; b.p. 1962°C; r.d. 7.44 (20%C); p.n. 25; r.a.m. 54.93805.

manometer /mă-**nom**-ĕ-ter/ A device for measuring pressure. A simple type is a U-shaped glass tube containing mercury or other liquid. The pressure difference between the arms of the tube is indicated by the difference in heights of the liquid.

maser /**may**-zer/ (*m*icrowave *a*mplification by *s*timulated *e*mission of *r*adiation) A device for producing an intense source of coherent microwave radiation. Masers, like lasers, operate by population inversion and stimulated emission.

mass Symbol: m A measure of the quantity of matter in an object. Mass is determined in two ways: the INERTIAL MASS of a body determines its tendency to resist change in motion; the GRAVITATIONAL MASS determines its gravitational attraction for other masses. The SI unit of mass is the kilogram. *See also* weight.

mass, center of *See* center of mass.

mass defect The mass equivalent of the BINDING ENERGY of a nucleus.

mass density *See* density.

mass-energy equation The equation $E = mc^2$, where E is the total energy (rest mass energy + kinetic energy + potential energy) of a mass m, c being the velocity of light. The equation is a consequence of Einstein's special theory of relativity; mass is a form of energy and energy also has mass. Conversion of rest-mass energy into kinetic energy is the source of power in radioactive substances and the basis of nuclear-power generation.

mass number *See* nucleon number.

mass spectrograph *See* mass spectrometer.

mass spectrometer An instrument for producing ions in a gas and analyzing them according to their charge/mass ratio. The earliest experiments by Francis Aston used a stream of positive ions from a discharge tube, which were deflected by parallel electric and magnetic fields at right angles to the beam. Each type of ion formed a parabolic trace on a photographic plate (a *mass spectrograph*).

In modern instruments, the ions are produced by ionizing the gas with electrons from an electron gun. The positive ions are accelerated out of this ion source into a high-vacuum region. Here, the stream of ions is deflected and focused by a combination of electric and magnetic fields, which can be varied so that different types of ion fall on a detector. In this way, the ions can be analyzed according to their mass, giving a *mass spectrum* of the material. Mass spectrometers are used for accurate measurements of relative atomic masses, for analysis of isotope abundance, and for chemical identification of compounds and mixtures.

mass spectrum *See* mass spectrometer.

matrix /may-triks, **mat-**riks/ A set of quantities arranged in rows and columns to form a rectangular array. The common notation is to enclose these in parentheses. Matrices do not have a numerical value, like determinants. They are used to represent relations between the quantities. For example, a plane vector can be represented by a single column matrix with two numbers, a 2 × 1 matrix, in which the upper number represents its component parallel to the x-axis and the lower number represents the component parallel to the y-axis. Matrices can also be used to represent, and solve, simultaneous equations. In general, an $m \times n$ matrix – one with m rows and n columns – is written with the first row:

$$A = a_{11}a_{12} \dots a_{1n}$$

The second row is:

$$a_{21}a_{22} \dots a_{2n}$$

and so on, the mth row being:

$$a_{m1}a_{m2} \dots a_{mn}$$

The individual quantities a_{11}, a_{21}, etc., are called *elements* of the matrix. The number of rows and columns, $m \times n$, is the *order* or *dimensions* of the matrix. Two matrices are equal only if they are of the same order and if all their corresponding elements are equal. Matrices, like numbers, can be added, subtracted, multiplied, and treated algebraically according to certain rules. However, the commutative, associative, and distributive laws of ordinary arithmetic do not apply. *Matrix addition* consists of adding corresponding elements together to obtain another matrix of the same order. Only matrices of the same order can be added. Similarly, the result of subtracting two matrices is the matrix formed by the differences between corresponding elements.

Matrix multiplication also has certain rules. In multiplication of an $m \times n$ matrix by a number or constant k, the result is another $m \times n$ matrix. If the element in the *i*th row and *j*th column is a_{ij} then the corresponding element in the product is ka_{ij}. This operation is distributive over matrix addition and subtraction, that is, for two matrices A and B,

$$k(A + B) = kA + kB$$

Also, $kA = Ak$, as for multiplication of numbers. In the multiplication of two matrices, the matrices A and B can only be multiplied together to form the product AB if the number of columns in A is the same as the number of rows in B. In this case they are called *conformable matrices*. If A is an $m \times p$ matrix with elements a_{ij} and B is a $p \times n$ matrix with elements b_{ij}, then their product $AB = C$ is an $m \times n$ matrix with elements c_{ij}, such that c_{ij} is the sum of the products

$$a_{ij}b_{ij} + a_{i2}b_{2j} + a_{i3}b_{3j} + \dots + a_{ip}b_{pj}$$

Matrix multiplication is not commutative, that is, $AB \neq BA$.

matter The term matter is commonly used very loosely and was formerly regarded as synonymous with substance. Since MASS is defined as the quantity of matter, strictly the term should be used to mean anything that has mass. It has been found necessary to define a distinct concept, *substance*, measured in terms of the number of particles in a system. The measured mass of a body depends on the relative motion of the body and the observer but substance is invariant. *See also* mole; relativistic mass.

maximum-and-minimum thermometer A thermometer designed to record the maximum and minimum temperatures during a period of time. It usually consists of a graduated capillary tube at the base of which is a bulb containing ethanol. Within the capillary is a thin thread of mercury with a steel index at each end. When the temperature rises, the alcohol expands, thus forcing the mercury up the tube, carrying the upper steel index. Therefore the position of the upper index marks the maximum temperature reached, and similarly the position of the lower index, the minimum temperature attained.

maxwell /maks-wĕl/ Symbol: Mx A unit of magnetic flux used in the c.g.s. system. It is equal to 10^{-8} weber (Wb). The unit is named for the British physicist James Clerk Maxwell (1831–79).

Maxwell–Boltzmann statistics /maks-wĕl bohlts-măn, -mahn/ The statistical rules for studying systems of identical particles in classical physics. It was tacitly assumed that particles, although identical, could be distinguished in principle. It can be shown that, for low concentrations of particles, especially at high temperatures, the classical statistics gives results similar to the more exact Fermi–Dirac and Bose–Einstein statistics. It was the failure of classical statistics to predict results in agreement with experiment in certain cases which led to the development of quantum theory. The statistics are named for Maxwell and the Austrian theoretical physicist Ludwig Edward Boltzmann (1844–1906). *See* Maxwell distribution; ultraviolet catastrophe.

Maxwell corkscrew rule A rule devised by James Clerk Maxwell (1831–79) to remember the direction of magnetic field lines associated with the electric current producing the magnetic field. If a corkscrew is being driven in the direction of the current then the direction of the lines of force is the same direction as the direction that the corkscrew is being rotated.

Maxwell distribution The laws for the distribution of speeds or kinetic energies among the molecules of a gas in equilibrium. The number of molecules per unit range of speed at speed v is given by

$$N_v = Av^2 \exp(-mv^2/2kT)$$

where m is the mass of a molecule, T is the kelvin temperature, k is the Boltzmann constant, and A is a constant. The number of molecules per unit range of kinetic energy at kinetic energy E is given by

$$N_E = \sqrt{(BE)}\exp(-E/kT)$$

where B is a constant.

This distribution is very closely obeyed by gases at ordinary pressures. For large concentrations of particles more exact laws are needed, particularly for the valence electrons in a solid.

See Bose–Einstein statistics; Fermi–Dirac statistics; Maxwell–Boltzmann statistics.

Maxwell equations A set of four equations that together form the basis of classical electrodynamics. They are: (1) Coulomb's law, (2) a statement of the nonexistence of magnetic monopoles, (3) Faraday's law of electromagnetic induction, (4) a modification of Ampère's law. It has become customary to express these laws in terms of vector calculus.

Maxwell's demon A hypothetical being in a thought experiment that was used by James Clerk Maxwell in a discussion of the second law of thermodynamics. Maxwell considered a container of gas, with the gas being divided into two volumes by a partition. A small being (named a 'demon' by Lord Kelvin) is able to distinguish individual molecules in the gas and to open the partition when a molecule with a faster than average speed approaches it on one side and when a slower than average speed molecule approaches it from the other side. Eventually, one side would have a higher proportion of fast molecules while the other side would have a higher number of slow molecules. The side with the fast molecules would thus become hotter than the side with the slow molecules in contravention of the speed of law of thermodynamics. It is thought that the resolution of this paradox involves the relations between entropy and information theory, with the realization that identifying the correct molecules and operating the partition would generate entropy.

mean free path Symbol: λ **1.** The average distance traveled by the particles of a fluid between collisions. It is given by

$$\lambda = 1/\pi r^2 n$$

where r is the particle radius and n the density of particles.

2. The average distance traveled by electrons between collisions with the lattice in conduction.

mean free time For gas atoms or molecules in a container, or electrons and impurity atoms in a semiconductor, the average time between particle collisions. *See also* mean free path.

mean life Symbol: T The average time for which a radioactive nucleus exists before it disintegrates. It is the reciprocal of the decay constant.

mean solar day *See* day.

mechanical advantage *See* force ratio.

mechanical equivalent of heat *See* Joule's equivalent.

mechanics The study of forces and their effect on objects. If the forces on an object or in a system cause no change of momentum the object or system is in equilibrium. STATICS is the study of such cases. If the forces acting do change momentum, dynamics is used. The ideas of DYNAMICS relate the forces to the momentum changes produced. KINEMATICS is the study of motion without consideration of its cause.

medium (*pl.* **mediums** or **media**) The nature of the space through which something is transmitted or propagated. It may be solid, liquid, gas, or a vacuum (although some people do not class a vacuum as a medium).

medium frequency (**MF**) A radio frequency in the range between 3 MHz and 0.3 MHz (wavelength 100–1000 m).

mega- Symbol: M A prefix denoting 10^6. For example, 1 megahertz (MHz) = 10^6 hertz (Hz).

megaton /**meg**-ă-tun/ A unit used to express the explosive power of nuclear weapons. One megaton is an explosive power equivalent to that of one million tonnes of TNT. This is equivalent to 4.2 exajoules.

Meissner effect /**mÿss**-ner/ The exclusion of magnetic flux from a substance when it is in the superconducting state. It is then perfectly diamagnetic except for the penetration of flux into a thin surface layer. When an external magnetic field becomes sufficiently intense it penetrates through the specimen, destroying its SUPERCONDUCTIVITY. The effect is named for the German physicist Fritz Walther Meissner (1882–1974). *See* diamagnetism.

meitnerium /mÿt-**neer**-ee-ŭm/ A radioactive metallic element not found naturally on Earth. Only a few atoms of the element have ever been detected; it can be made by bombarding a bismuth target with iron nuclei. The isotope ^{266}Mt has a half-life of about 3.4×10^{-3}s. The element is named for the Austrian physicist Lise Meitner (1878–1968).

Symbol: Mt; p.n. 109.

melting (**fusion**) The change from the solid to the liquid state. For a pure crystalline substance this happens at a characteristic temperature – the *melting point* (or freezing point). This temperature depends on pressure, but is usually quoted at standard atmospheric pressure. The melting point of a substance is lowered by the presence of impurities.

melting point *See* melting.

membrane *See* supermembrane.

mendelevium /men-dĕ-**lee**-vee-ŭm/ A radioactive transuranic element of the actinoid series, not found naturally on Earth. Several short-lived isotopes have been synthesized. The element is named for the Russian chemist Dimitrii Mendeleyev (1834–1907).

Symbol: Md; p.n. 101; most stable isotope ^{258}Md (half-life 57 minutes).

meniscus (*pl.* **menisci** or **meniscuses**) /mĕ-**niss**-kŭs/ The curved surface of a column of liquid in a tube or other container. Its shape depends on contact angle. *See also* capillary action.

meniscus lens *See* concavo-convex.

mercury /**mer**-kyŭ-ree/ A transition metal that is liquid at room temperatures (bromine is the only other element with this property). The vapor is very poisonous. Mercury is used in thermometers, special amalgams for dentistry, scientific apparatus, and in mercury cells.

Symbol: Hg; m.p. –38.87°C; b.p. 356.58°C; r.d. 13.546 (20°C); p.n. 80; r.a.m. 200.59.

mercury barometer *See* barometer.

mercury cell A voltaic or electrolytic cell in which one or both of the electrodes consists of mercury or an amalgam. Amalgam electrodes are used in the Daniell cell and the Weston cadmium cell.

mercury thermometer A liquid-in-glass thermometer that uses mercury as its working substance. It consists of a thin capillary tube graduated in degrees, at the base of which is a bulb containing mercury. When the temperature rises, the mercury expands along the tube. The position indicates the temperature.

meridian, magnetic *See* magnetic meridian.

meson /**mee**-zon, -son, **mess**-on, **mez**-on/ A type of ELEMENTARY PARTICLE. Mesons are more massive than muons but lighter than protons. They are bosons and form a subclass of the hadrons. Yukawa predicted (1935) that such particles would be involved in the exchange forces between nucleons, and the subsequently discovered mesons appear to conform to his theory. *See also* kaon; pion.

metallurgy /**met**-ă-ler-jee/ The branch of science devoted to the study of metals. The behaviour of metals and their alloys is the concern of *physical metallurgy* while the extraction of metals from their ores and their purification are the concerns of *process metallurgy*.

metastable state A condition of a system or body in which it appears to be in stable equilibrium but, if disturbed, can settle into a lower energy state. For example, supercooled water is liquid at below 0°C (at standard pressure). When a small crystal of ice or dust (for example) is introduced rapid freezing occurs.

meter /**mee**-ter/ Symbol: m The SI base unit of length, defined as the length of the path traveled by light in a vacuum in a time of 1/299 792 458 second. The meter has not always been defined in terms of time. As originally proposed by the French Academy of Sciences in 1791, the meter was to be one ten-millionth of the distance running from the North Pole through Paris to the equator. Part of this distance was measured over six arduous years, and the balance extrapolated. However, because the Earth is not a perfect sphere, the extrapolated distance was eventually found to be inaccurate. Subsequently the meter was defined in 1889 as the length between two fine lines on an international prototype platinum–iridium alloy bar kept under specified conditions in a vault at Sèvres, in France. Due to the increasing need for accuracy in scientific measurement the meter was again redefined in 1960, this time as 1 650 763.73 wavelengths in vacuum corresponding to the transition between the levels $2p^{10}$ and $5d^5$ of the krypton –86 atom.

meter bridge A meter length of uniform resistance wire mounted above a scale marked in millimeters, with suitable terminal points added to make the device adaptable either as a WHEATSTONE BRIDGE or a POTENTIOMETER. As a Wheatstone bridge the wire acts as two arms of the bridge resistors; and as a potentiometer the potential drop along the wire is proportional to the length of wire.

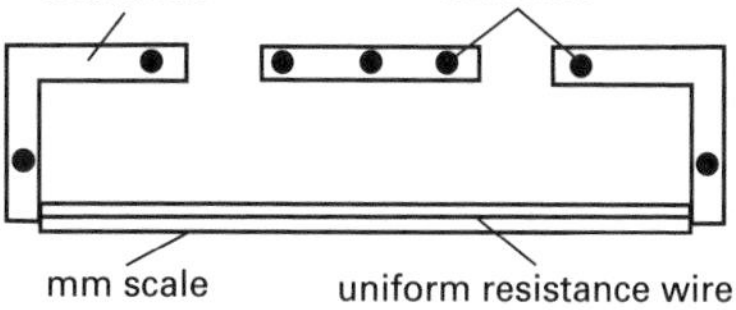

Meter bridge

metric horsepower *See* horsepower.

metric system A system of units based on the meter and the kilogram and using multiples and submultiples of 10. SI units, c.g.s. units, and m.k.s. units are all scientific metric systems of units.

metric ton *See* tonne.

MF *See* medium frequency.

MHD *See* magnetohydrodynamics.

mho /moh/ *See* siemens.

Michelson–Morley experiment /**mÿ**-kĕl-sŏn **mor**-lee/ A famous experiment carried out by the American physicists Albert Abraham Michelson (1852–1931) and Dayton Miller (1866–1941) in 1887 in an attempt to detect the ether, the medium that was supposed to be necessary for the transmission of electromagnetic waves in free space.

In the experiment, two light beams were combined to produce interference patterns after traveling for short equal distances perpendicular to each other. The apparatus was then turned through 90° and the two interference patterns were compared to see if there had been a shift of the fringes. If light has a velocity relative to the ether and there is an ether 'wind' as the Earth moves through space, then the times of travel of the two beams would change, resulting in a fringe shift. No shift was detected, not even when the experiment was repeated six months later when the ether wind would have reversed direction. *See also* relativity.

micro- Symbol: μ A prefix denoting 10^{-6}. For example, 1 micrometer (μm) = 10^{-6} meter (m).

microelectronics /mÿ-kroh-i-lek-**tron**-iks/ The branch of electronics that deals with very small electronic circuits, components, and devices, such as silicon chips and other microminiature devices.

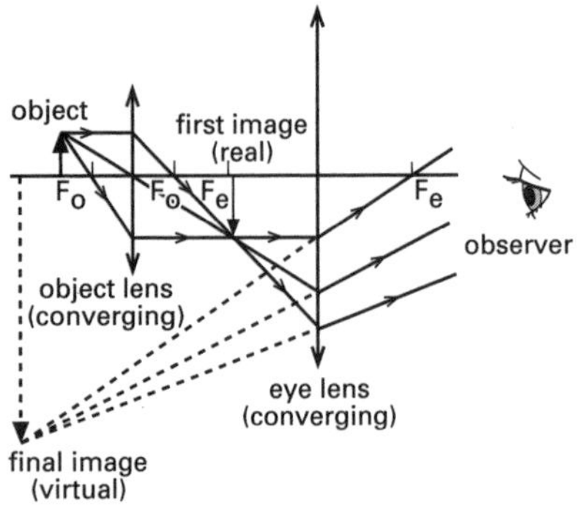

Microscope, compound

micron /**mȳ**-kron/ Symbol: μm A unit of length equal to 10^{-6} meter.

microphone A device for converting sound signals into electrical signals. The pressure variations of the sound wave are converted into variations in an electrical signal. There are various methods of doing this. A *ribbon microphone* has a thin metal ribbon held between the poles of a magnet. The sound waves vibrate this and a varying e.m.f. is induced, which can be amplified. A *moving-coil microphone* has a diaphragm connected to a small coil, which moves in a stationary magnetic field. An e.m.f. is induced in the coil. A *crystal microphone* depends for its action on the piezoelectric effect; an e.m.f is induced by stress produced by sound waves hitting a suitable crystal of quartz or Rochelle salt. A *carbon microphone* has two blocks of carbon in contact. Pressure changes produced by the sound waves cause corresponding changes in the resistance of the carbon blocks. *See also* condenser microphone.

microscope, compound A device for producing large images of close small objects with a combination of lenses. In the normal two-lens microscope each lens is converging. In practice, except in very cheap instruments, both object lens and eye lens are replaced by compound lens systems. Lighting is extremely important; light is directed onto the object by a mirror or condenser lens.

The magnifying power of the compound microscope is the product of the linear magnifications produced by objective and eyepiece.

See also blink microscope; electron microscope.

microscope, simple The magnifying glass, or simple magnifier. The small object is viewed between the lens and its focal point; an upright virtual image is obtained. In this case the image is placed at the near point of the eye – the magnifying power is then given by $(D/f) + 1$, where D is the near-point distance and f the lens focal distance. The lens can also be used to give an image at infinity; the image is then viewed with the relaxed eye, but the magnifying power is only D/f.

microwaves /**mȳ**-kroh-wayvz/ A form of electromagnetic radiation, ranging in wavelength from about 1 mm (where it merges with infrared) to about 120 mm (bordering on radio waves). Microwaves are produced by various electronic devices including the KLYSTRON; they are often carried over short distances in tubes of rectangular section called *waveguides*.

Microwave communication was until recently the most efficient form of electronic telecommunication. Since the development of cheap lasers and optical fibers, telecommunication at visible wavelengths is developing rapidly. Radar systems generally use microwaves, while their ability to

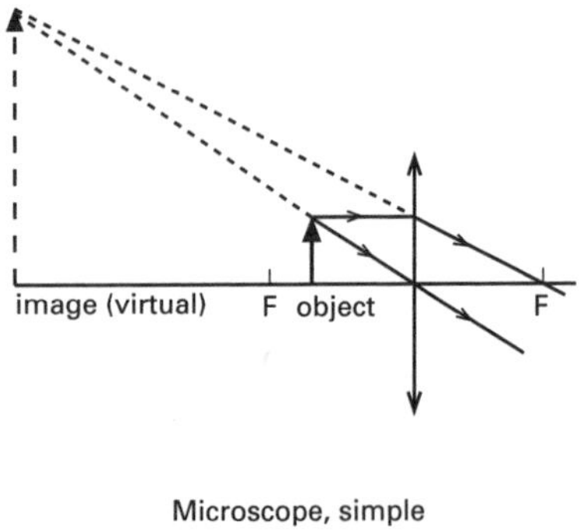

Microscope, simple

carry energy has a number of applications, including microwave cookers. *See also* electromagnetic spectrum; magnetron.

mil 1. A unit of length in f.p.s. systems, equal to 0.001 inch or 0.02554 millimetre in SI units.
2. A unit of area in f.p.s. systems, also known as the *circular mil*, used especially for wire sizes. It is based on the area of a circle having a diameter of 0.001 inch, and is equal in SI units to 506.7075 square micrometers.
3. A colloquial term for the milliliter.

mile A unit of length equal to 1760 yards (equal to 1.60934 km).

milli- Symbol: m A prefix denoting 10^{-3}. For example, 1 millimeter (mm) = 10^{-3} meter (m).

milliammeter /mil-ee-**am**-mee-ter/ An instrument for measuring electric currents of a few milliamperes. *See also* ammeter; galvanometer.

millibar *See* bar.

millimeter of mercury *See* mmHg.

minority carrier *See* semiconductor.

minor planet *See* asteroid.

minute 1. A unit of time equal to 60 seconds.
2. Symbol: ′ A unit of angle; 1/60 of a degree.

mirage An optical illusion caused by total internal reflection in the air. It is common when the ground is strongly heated by the Sun so that the air in contact with the ground is much warmer than the air above. Light from an object above the ground (including the sky) is refracted and totally internally reflected by the warm air to produce an image. Mirages are also sometimes observed next to very warm vertical surfaces.

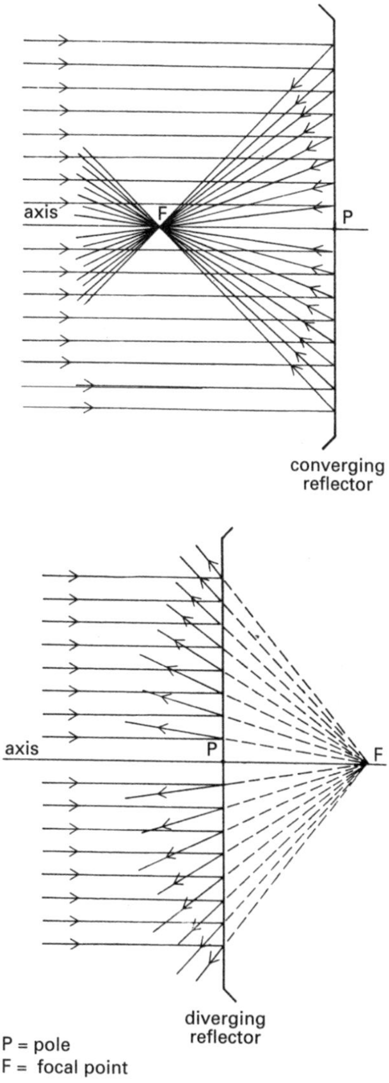

Paths of light reflected from converging and diverging mirrors

mirror (**reflector**) An optical component that reflects incident rays. A mirror is best described by its effect on a parallel beam. Plane mirrors reflect it parallel; convering mirrors reflect it convergent; diverging mirrors make it diverge. Focal distance is the usual measure of the power of a mirror

OBJECT AND IMAGE POSITIONS FOR CONVERGING MIRROR		
Object position	*Image nature*	*Image position*
infinity	real	F
beyond C	real	between F and C
C	real	C
between C and F	real	beyond C
F	virtual	infinity
between F and P	virtual	between infinity and P

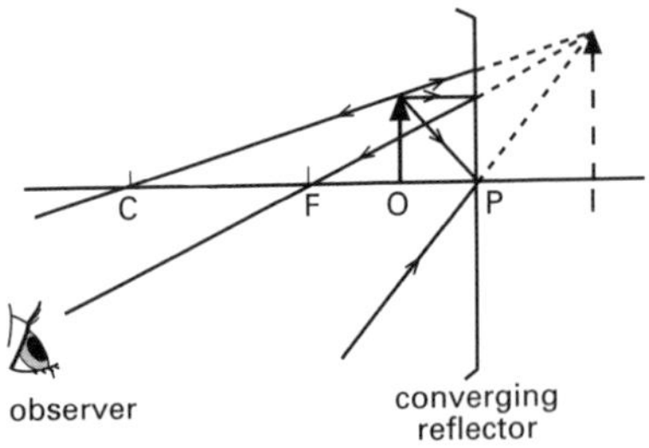

C = center of curvature
object O is between F and P
image I is virtual, upright, magnified, and behind the mirror

Formation of a virtual image by a converging mirror

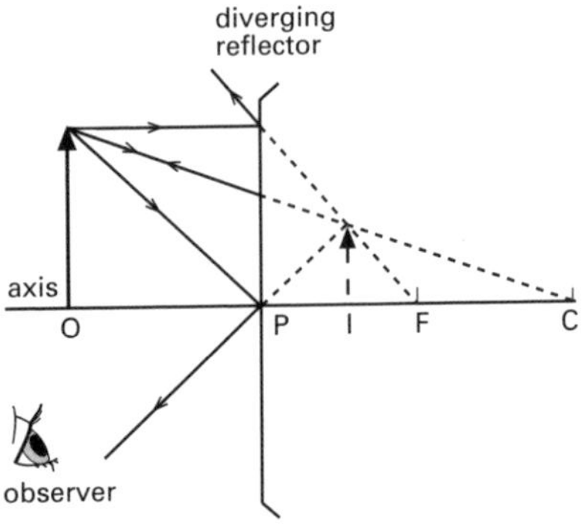

C = center of curvature
object O is anywhere
image I is between F and P, virtual, upright, and diminished

Formation of a virtual image by a real object at a diverging mirror

converging mirror (concave) plane mirror (flat) diverging mirror (convex)

Types of mirror

to converge a parallel beam. Parallel rays are reflected by converging and diverging mirrors as shown in the diagram. Strictly such focusing occurs only with rays parallel and close to the principal axis. With a parabolic mirror, wide parallel beams focus at the focal point. Rays from an object will be reflected so that they appear to come from somewhere else – the image. Two examples are given here to show how ray diagrams are drawn.

For a diverging reflector, a virtual image appears at F when the incident rays come from infinity; radiation from an object anywhere between infinity and P will form a virtual image between F and P.

Object distance (u) and image distance (v) relate to the focal distance (f) of the reflector as follows:

$$1/f = 1/v + 1/u$$

(real is positive, virtual is negative sign convention).

The formula applies to real or virtual focal points, objects, and images. All the above can be adapted for use with reflectors of non-visible radiations such as infrared and ultraviolet.

mirror image A shape that is identical to another except that its structure is reversed as if viewed in a mirror, so that the two cannot be superimposed. For example, a left hand is the mirror image of a right hand.

m.k.s.a. system A coherent metric system of units for mechanics and electromagnetics based on the meter, the kilogram, the second, and the ampere. It superseded the m.k.s. system and was in turn superseded by the SI in 1960.

m.k.s. system A system of units based on the meter, the kilogram, and the second. It superseded the c.g.s. system and was in turn superseded by the m.k.s.a. system in 1946.

mmHg (**millimeter of mercury**) A former unit of pressure defined as the pressure that will support a column of mercury one millimeter high under specified conditions. It is equal to 133.322 4 Pa. It is equivalent to the torr.

moderator /**mod**-ĕ-ray-ter/ Material used in the cores of some nuclear reactors to slow down fast-moving neutrons so that they will be more easily captured by fissile nuclei. The most effective speeds are about two thousand meters per second. Such neutrons are called *thermal neutrons* because the distribution of their speeds is such that they are in thermal equilibrium with (i.e. at the same temperature as) the surrounding materials. To achieve this slowing down in as few collisions as possible, moderators are substances of low relative atomic mass. Carbon, in the form of graphite, was used in early experimental reactors and is still used in Magnox reactors. Water – in some reactor designs, heavy water – is used and may also serve as the coolant. Paraffin and beryllium are also suitable moderators.

modulation /moj-ŭ-**lay**-shŏn/ The superimposition of a (video or audio) signal onto a CARRIER WAVE so that the information contained in the signal can be transmitted with the carrier wave. *See* amplitude modulation; frequency modulation; phase modulation. *See also* demodulation.

modulus (*pl.* **moduli**) /**moj**-ŭ-lŭs/ **1.** The magnitude of a vector quantity.
2. The absolute value of a number or quantity. For example, the modulus of –4 is 4.
3. A quantity that defines some property of a material or of some other system. *See* elastic modulus.

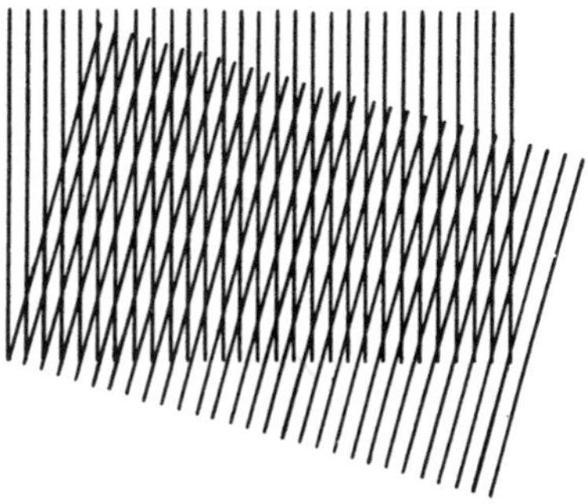

moiré fringes

moiré fringes /mwah-**ray**, **mwa**-ray/ The pattern obtained when two regular sets of lines or points overlap. The effect can be seen through folds in netting drapes. Moiré patterns can be used as models of interference patterns Another application is in comparing two diffraction gratings by superimposing them and observing the moiré pattern produced.

molar /**moh**-ler/ **1.** Denoting a physical quantity divided by the amount of substance. In almost all cases the amount of substance will be in moles. For example, volume (V) divided by the number of moles (n) is molar volume:

$$V_m = V/n.$$

2. A *molar solution* contains one mole of solute per cubic decimeter of solvent.

molar gas constant *See* gas constant.

molar thermal capacity Symbol: C_m The thermal capacity per unit amount of substance; i.e. the energy required to raise unit amount of substance (1 mol) by unit temperature (1 K). It is measured in J mol^{-1} K^{-1}. For a gas, it is common to specify two *principal* molar thermal capacities: one measured at constant pressure and the other measured at constant volume. The relationship between them are as for specific thermal capacities.

mole Symbol: mol The SI base unit of amount of substance, defined as the amount of substance that contains as many elementary entities as there are atoms in 0.012 kilogram of carbon-12. The elementary entities may be atoms, molecules, ions, electrons, photons, etc., and they must be specified. One mole contains 6.022 142 × 10^{23} entities One mole of an element with relative atomic mass A has a mass of A grams (this was formerly called one *gram-atom*). One mole of a compound with relative molecular mass M has a mass of M grams (this was formerly called one *gram-molecule*).

molecular orbital *See* orbital.

molecular sieve A method used for separating similar substances by adsorbing the molecules of one substance within the structural cavities of another, such as a natural or synthetic zeolite.

molecular spectrum The absorption or emission spectrum that is characteristic of a molecule. Molecular spectra are usually band spectra.

molecular weight *See* relative molecular mass.

molecule /**mol**-ĕ-kyool/ A single atom or a group of atoms joined by chemical bonds. It is the smallest unit of a chemical compound that can have an independent existence. Water, for instance has molecules made up of two hydrogen atoms and one oxygen atom (H_2O). Ionic compounds, such as sodium chloride, do not have distinct molecules. Sodium chloride is usually written NaCl, but crystals of sodium chloride are, in fact, a regular arrangement of sodium ions (Na^+) and chloride ions (Cl^-). Similarly, metals and certain covalently bonded solids do not have discrete molecules. In boron nitride, for instance, the covalent bonds hold the atoms together in a giant molecule.

molybdenum /mŏ-**lib**-dĕ-nŭm/ A transition element used in alloy steels, lamp bulbs and catalysts.

Symbol: Mo; m.p. 2620°C; b.p. 4610°C; r.d. 10.22 (20°C); p.n. 42; r.a.m. 95.94.

moment The moment of a force about a point is the product of the force and the perpendicular distance from the point to the line of action of the force. It can be expressed as the vector product of the force times the displacement of this point from the point where the force acts. The resultant moment of the forces acting on a body is found by vector addition. If the forces are coplanar, clockwise and anticlockwise moments are given opposite signs. For a body in equilibrium, the sum of the clockwise moments equals the sum of the anticlockwise ones (the *law of moments*). If the resultant moment is non-zero the body has an angular acceleration. *See also* couple; torque.

moment, magnetic *See* magnetic moment.

moment of inertia Symbol: I The rotational analog of mass. The moment of inertia of an object rotating about an axis is given by $I = \Sigma mr^2$, where m is the mass of an element distant r from the axis. Some important cases are listed in the table. *See also* radius of gyration; theorem of parallel axes.

momentum, conservation of /mŏ-**men**-tŭm/ *See* constant (linear) momentum (law of).

MOMENTS OF INERTIA

The value given is that for a perpendicular axis through the center of mass.

Object	*Dimensions*	*Moment of inertia*
thin rod	length l	$ml^2/12$
thin disk	radius r	$mr^2/2$
thin ring	radius r	mr^2
solid sphere	radius r	$2mr^2/5$

momentum, linear (*pl.* **momenta** or **momentums**) Symbol: p The product of an object's mass and velocity: $p = mv$. The object's momentum cannot change unless a net outside force acts. This relates to Newton's laws and to the definition of force. It also relates to the principle of constant momentum. *See also* angular momentum.

monatomic /mon-ă-**tom**-ik/ Describing a molecule that consists of a single atom. The rare gases are monatomic gases.

monochord /**mon**-ŏ-kord/ *See* sonometer.

monochromatic radiation /mon-ŏ-kroh-**mat**-ik/ Electromagnetic radiation of an extremely narrow range of wavelengths. (The word means 'of one color'.) It is impossible to produce completely monochromatic radiation, although the output of some lasers is not far off. The 'lines' in line spectra produced even in the most ideal circumstances have some width in wavelength terms. The *half-width* is the measure used. It is the range of wavelengths defined in the figure, and contains almost 90% of the energy emitted. The half-width of a sharp line in an optical spectrum is typically 10^{-6} to 10^{-7} of the wavelength (λ). Using lasers, half-widths of the order 10^{-12} λ can be obtained. Low-quantum-energy gamma rays emitted by atoms bound in crystals may have values of the order 10^{-13} λ.

Simple quantum theory leads one to expect perfectly sharp lines in a line spectrum – the energies of the levels concerned appear to be exactly defined, so that $\lambda = hc/\Delta E$. However, because of the uncertainty principle, no energy level or transition *can* be defined exactly; this means that any line is naturally broadened rather than being sharp. A second broadening influence is the Doppler effect, which is relevant as the radiating particles are always in motion. Thirdly, collisions between emitting particles will broaden the emitted line.

Compare polychromatic radiation.

monochromator /mon-ŏ-**kroh**-mă-ter/ A device for providing monochromatic radiation from a polychromatic source. In the case of light, for example, a prism or grating disperses the light and slits are used to select a small wavelength range.

monolithic IC /mon-ŏ-**lith**-ik/ *See* integrated circuit.

monomode fiber A type of optical fiber that has a narrow core of glass with a diameter of about 5 μm surrounded by a large coating (*cladding*) of glass with a smaller refractive index than the core. *See also* multimode fibre.

monostable circuit /mon-ŏ-**stay**-băl/ An electronic circuit, usually a MULTIVIBRATOR, that has one stable state but can change to

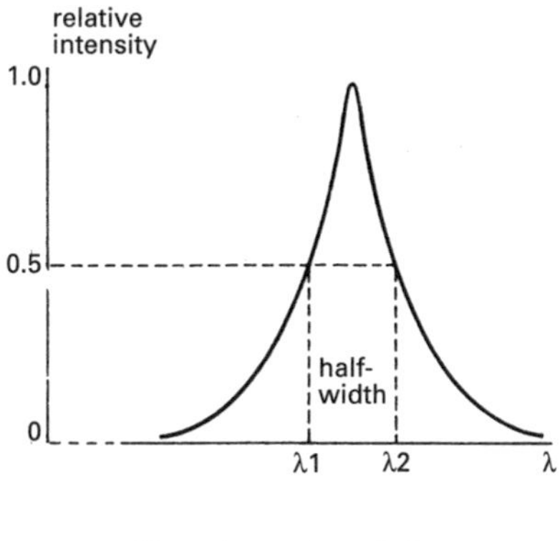

Monochromatic radiation

another state for a time after the application of a trigger pulse. Monostable circuits can be used for generating single pulses of a fixed duration, for shortening or lengthening pulses, or to delay pulses in computer logic circuits. In a monostable multivibrator the input pulse is fed to the base terminal of one transistor (*TR1*) and through a resistor *R* to the collector terminal of the other (*TR2*). The base of *TR2* and the collect of *TR1* are connected through a capacitor *C*. The output voltage at the collection of *TR2* is a pulse with a duration that depends on the values of *R* and *C*.

Moseley's law Lines in the x-ray spectra of elements have frequencies that depend on the proton number of the element. For a set of elements, a graph of the square root of the frequency of x-ray emission against proton number is a straight line (for spectral lines corresponding to the same transition). The law is named for the British physicist Henry Gwyn Jeffreys Moseley (1887–1915).

Mössbauer effect /mohs-bow-er, **moss-/** After an atomic nucleus has emitted a gamma-ray photon (during radioactive decay), the absorption of the momentum of the atom by the whole of its crystal lattice because it is so firmly bound that it cannot recoil. The same effect occurs with absorption of gamma rays. The effect is named for the German physicist Rudolph Ludwig Mössbauer (1929–).

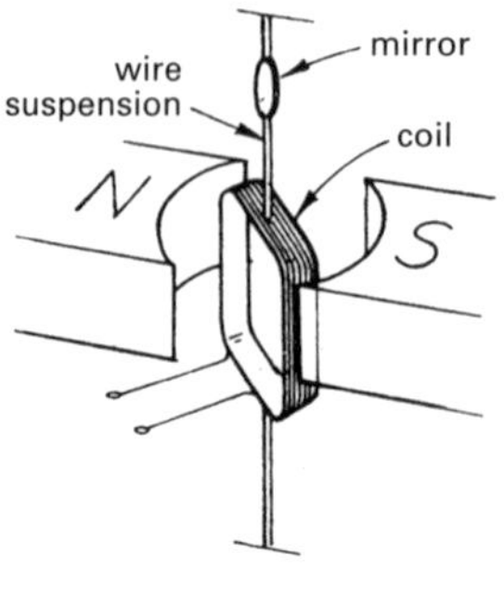

Moving-coil instrument

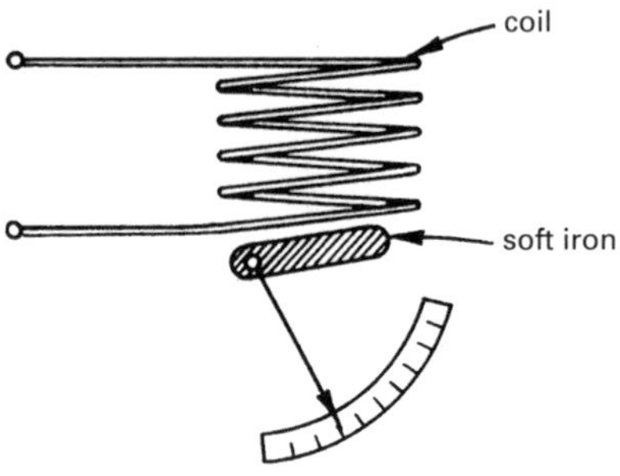

Moving-iron instrument

moving-coil instrument An electric measuring instrument that depends on the couple on a small vertical rectangular coil carrying direct current in a magnetic field. A cylindrical soft-iron core is fitted between the concave pole pieces of a strong permanent magnet, giving a radial magnetic field. The coil is supported either by a fine wire suspension or by needle points on bearings. When the current flows, there is a couple that moves the coil until torsion in the suspension or the resistance of a hair spring brings it to rest. The equilibrium angle, which is proportional to the current, is measured by a pointer or (for sensitive instruments) by reflecting a beam of light from a small mirror attached to the coil.

The torque on the coil is equal to $BIAN$, where B is magnetic flux density, I current, A coil area, and N the number of turns. The current depends on the angle of twist (θ) according to $I = k\theta/BAN$, where k is a constant of the spring or torsion wire. Typically, currents of the order of milliamps can be measured, and some instruments can measure microamps.

moving-coil microphone *See* microphone.

moving-iron instrument An electrical measuring instrument in which the current is passed through a fixed coil. The magnetic field produced attracts a piece of soft iron on a pivot. The iron is restrained by a spring and the movement detected by a pointer. Moving-iron instruments are less sensitive than moving-coil instruments, but can be used for alternating-currents without a rectifier (the attraction does not depend on the direction of current flow). The movement of the iron is roughly proportional to the square of the current.

MRI *See* magnetic resonance imaging.

M theory *See* superstring theory.

multimeter /mul-**tim**-ĕ-ter/ An electrical measuring instrument designed to measure voltage and current over a number of ranges. It is a moving-coil instrument with a switch enabling various resistors to be connected in series with the coil (for different voltage ranges) or various shunt resistances to be connected in parallel (for current ranges). Usually, an internal dry cell is incorporated so that direct measurements can also be made of resistances.

multimode fiber A type of optical fiber that has a relatively large core of glass with a diameter of about 50 μm surrounded by a relatively small amount of cladding. *See also* monomode fibre.

multiple capacitor A capacitor consisting of a series of parallel plates with a dielectric between each plate. The capacitance of such a device is *N* times the capacitance between two consecutive plates of the series, where *N* is the number of successive pairs of parallel plates.

multiplication factor Symbol: k In a nuclear chain reaction, the ratio of the rate of production of neutrons to the rate of 'loss' of neutrons (by absorption and leakage).
For subcritical reactions $k < 1$
For supercritical reactions $k > 1$
For critical reactions $k = 1$

multiplier One of a set of resistors that are used in series with a VOLTMETER to allow it to measure a wide range of voltages. *See also* photomultiplier; shunt.

multistable circuit /mul-ti-**stay**-băl/ An electronic circuit that has more than one stable state. *See* multivibrator.

multivibrator /mul-ti-**vy̆**-bray-ter/ A basic electronic circuit consisting of two transistors and other components, such as resistors and capacitors. Its main property is that it can switch its output instantaneously from one voltage level to another. Multivibrators are used to generate continous streams of voltage pulses (square waves), to store information in the form of binary digits (0 and 1) in a computer, and to form part of a logic circuit.

The transistors are connected in the common-emitter mode with the collector of each joined through other components to the base of the other. If one transistor is in the cut-off state, that is with almost zero base current and collector current, the other is saturated, that is the collector current will not increase with collector-emitter voltage. A sufficiently high voltage pulse at the base of one transistor will change its state from cut-off to saturation and vice versa. The second transistor will then change state in the other direction. A *bistable multivibrator* circuit has two overall stable conditions and has the two transistors connected to each other through resistors only. In the *astable multivibrator* the coupling is capacitive and the circuit switches continuously from one state to another. The *monostable multivibrator* has a coupling that is both resistive and capacitive and has only one stable state. *See also* astable circuit; bistable circuit; monostable circuit.

mu-meson /myoo/ *See* muon.

muon /**myoo**-on/ A type of lepton with a mass 207 times that of the electron. There are two types, having positive or negative charge. Originally the muon was classified as a meson and called a *mu-meson*. *See also* elementary particles.

mutual inductance *See* mutual induction.

mutual induction The production of e.m.f. in one circuit by change in the current in, and therefore the magnetic field around, a second nearby circuit. For example, in a transformer, an e.m.f. is induced in one coil by a changing current in the other. Mutual induction is greatly increased by the presence of an iron core that links the two coils. The mutual inductance (M) between two conductors is defined (in free space) by the equation $E = M(\mathrm{d}I/\mathrm{d}t)$, where E is the induced e.m.f. and $\mathrm{d}I/\mathrm{d}t$ is the rate of change of current. The SI unit is the henry (H). *Compare* self-induction.

myopia /mÿ-**oh**-pee-ă/ Short sight, normally caused if the distance between the lens and the back of the eyeball is too long. Light from distant objects is focused at a point in front of the retina, and thus distant objects cannot be accommodated. Only close objects can be seen clearly. The problem may be corrected by use of diverging spectacle or contact lenses.

N

NAND gate *See* logic gate.

nano- Symbol: n A prefix denoting 10^{-9}. For example, 1 nanometer (nm) = 10^{-9} meter (m).

nanotechnology /nan-oh-tek-**nol**-ŏ-jee/ The branch of technology that is concerned with the production and behavior of very small components, especially electronic devices, that have a size of a few nanometers (1 nm = 10^{-9}m). Electronic devices of this type may use the movement of only a few electrons in their action. The fabrication of nanocomponents may be effected by mechanical means or by carefully controlled chemical reactions, particularly various types of etching process.

natural abundance *See* abundance.

natural convection *See* convection.

natural frequency The frequency at which an object or system will vibrate freely. A free vibration occurs when there is no external periodic force and little resistance. The amplitude of free vibrations must not be too great. For instance, a pendulum swinging with small swings under its own weight moves at its natural frequency. Normally, an object's natural frequency is its fundamental frequency.

near point The closest point to the eye at which an object can clearly be seen without excessive strain. The eye is then fully accommodated, and has maximum power. In the normal eye, the distance to the near point increases with age, as the eye's ability to accommodate decreases. For many purposes it is convenient to take a *standard near-point distance*, D, to be 250 mm.

THE VARIATION OF NEAR-POINT DISTANCE WITH AGE

Age (years)	*Distance (mm)*
10	70
20	100
30	140
40	220
50	400
60	2000

near sight Short sight. *See* myopia.

Néel temperature /**nay**-ĕl/ *See* antiferromagnetism.

negative *See* charge; electric.

negative feedback *See* feedback.

negative lens A lens with a negative power. *See* diverging lens.

negative mirror A mirror with a negative power. *See* diverging mirror.

negative pole A south pole of a magnet. *See* pole.

neodymium /nee-ŏ-**dim**-ee-ŭm/ A toxic silvery element belonging to the lanthanoid series of metals. It occurs in association with other lanthanoids. Neodymium is used in various alloys, as a catalyst, in compound form in carbon-arc searchlights, etc., and in the glass industry.

Symbol: Nd; m.p. 1021°C; b.p. 3068°C; r.d. 7.0 (20°C); p.n. 60; r.a.m. 144.24.

neon /**nee**-on/ An inert colorless odorless monatomic element of the rare-gas group.

Neon forms no compounds. It occurs in minute quantities (0.0018% by volume) in air and is obtained from liquid air. It is used in neon signs and lights, electrical equipment, and gas lasers.

Symbol: Ne; m.p. –248.67°C; b.p. –246.05°C; d. 0.9 kg m^{-3} (0°C); p.n. 10; r.a.m. 20.18.

neper /**nay**-per, **nee**-/ Symbol: Np A unit of relative difference in power or field levels expressed logarithmically, equal to 8.686 decibels (dB). Two powers P_1 and P_2 differ by n neper, where

$$n = \tfrac{1}{2}\ln(P_2/P_1)$$

The neper is probably most commonly used as the unit of radiation attenuation; here the radiation powers P_1 and P_2 are usually given as intensities I_1 and I_2. It is also used in acoustics, electricity, and magnetism.

neptunium /nep-**tew**-nee-ŭm/ A toxic radioactive silvery element of the actinoid series of metals that was the first transuranic element to be synthesized (1940). Found on Earth only in minute quantities in uranium ores, it is obtained as a by-product from uranium fuel elements.

Symbol: Np; m.p. 640°C; b.p. 3902°C; r.d. 20.25 (20°C); p.n. 93; most stable isotope ^{237}Np (half-life 2.14×10^6 years).

Nernst calorimeter /nernst/ An apparatus for determining the specific thermal capacity of a metal. A piece of the metal (mass m) is wound with a coil of insulated platinum wire and suspended in an evacuated container. Aluminum foil is wrapped around the coil to reduce radiation loss. A current (I) is passed through the coil at measured voltage V for a time (t). The coil is also used as a resistance thermometer (with the heating current off) to measure the temperature rise (θ) in the metal. Then

$$IVt = (mc + C)\theta$$

where c is the specific thermal capacity of the specimen and C the thermal capacity of the coil and foil (this is found by a separate measurement). The calorimeter is named for the German chemist Walther Hermann Nernst (1864–1941).

Nernst effect When a magnetic field is applied at right angles to a conductor through which heat is flowing, an electric potential develops transversely across the conductor. The direction of heat flow, magnetic field, and potential gradient are all mutually perpendicular. *Compare* Hall effect.

Neumann's law /**noi**-mănz, **noi**-mahnz/ *See* electromagnetic induction.

neutral Having neither negative nor positive net charge. This will be when the body is at earth potential.

neutral equilibrium EQUILIBRIUM such that if the system is disturbed a little, there is no tendency for it to move further nor to return. *See* stability.

neutral point A point where two fields in a region are equal and opposite, so that there is no resultant force. The situation is most often met in the case of magnetic fields. Thus magnetic neutral points are found near a permanent magnet in the Earth's field.

neutrino /new-**tree**-noh/ An ELEMENTARY PARTICLE with zero charge and very small rest mass having only weak INTERACTIONS. Neutrinos are classified as LEPTONS and are FERMIONS with spin ½. There are three types of neutrino (each with a corresponding antiparticle); one type is emitted in beta decay, another in the decay of muons, and a third is associated with the tau particle.

The existence of neutrinos was postulated by Pauli (1930) and was shown to account for the continuous energy spectrum of beta radiation. Later it was found that the conservation of momentum and angular momentum, as well as energy conservation in beta decay, required the neutrino hypothesis. The development of the theory of beta radioactivity led to the concept that the particle emitted with an electron in beta decay is an antineutrino while that emitted with positrons is the neutrino.

In 1956 the antineutrinos produced by short-lived beta emitters in a nuclear reactor were detected by the reaction

$$^{1}_{1}H + \bar{\nu}; \rightarrow ^{1}_{0}n + e^{+}$$

Because of the lack of electromagnetic and strong interactions these particles have an extremely low probability of reacting with matter. For hydrogen at the density of the Sun the mean free path for the nuclear reactor antineutrinos is millions of times the solar diameter.

Thermonuclear fusion in the Sun is expected to cause the emission of neutrinos. Attempts have been made to detect these by the reaction

$$^{37}_{17}Cl + \bar{\nu} \rightarrow ^{37}_{18}Ar + e^{-}$$

These experiments have failed to detect neutrinos with the intensity predicted by theories based on the assumption that the Sun is in a steady state.

See also neutrino oscillations; solar neutrino problem.

neutrino oscillation An oscillation between the different species of neutrinos associated with electrons, muons, and tau leptons; i.e. a reversible change between one species and another. Neutrino oscillations can only occur if neutrinos have nonzero masses. There are several independent sources of evidence showing that neutrino oscillations actually occur, and they are predicted to happen in most GRAND UNIFIED THEORIES.

neutron /new-tron/ An ELEMENTARY PARTICLE with zero charge and a rest mass of $1.674\ 92 \times 10^{-27}$ kg. Neutrons are nucleons, found in all nuclides except ^{1}H. Isolated neutrons are unstable and decay with a half-life of about 12 minutes. Neutrons are fermions with spin ½. They are strongly interacting particles. As they are uncharged, they interact with nuclei more easily than do protons and other charged particles, especially at low energies. Neutrons are made up of more fundamental particles called quarks.

neutron capture The capture by an atomic nucleus of a neutron, often accompanied by the emission of gamma radiation. *See also* electron capture.

neutron diffraction The diffraction of neutrons by matter. The technique of neutron diffraction to investigate crystals depends on the fact that neutrons have an associated wavelength and, at a suitable energy, can be diffracted just as x-rays are diffracted. The technique is more difficult than x-ray diffraction because a sufficiently intense neutron source has to be available. However, neutron diffraction has certain advantages over x-ray diffraction. In particular, x-rays are scattered by electrons in atoms and consequently are more useful in investigating heavier atoms with many electrons. They do not detect the position of the hydrogen atoms present in all atomic molecules. Neutrons are scattered by atomic nuclei and are particularly useful for investigating the positions of hydrogen atoms in molecules. Also, because of their magnetic moment, neutrons are useful in studying various types of magnetic crystal.

neutron number Symbol: *N* The number of neutrons in the nucleus of an atom; i.e. the nucleon number (*A*) minus the proton number (*Z*).

neutron–proton ratio The ratio of the number of neutrons to the number of protons in an atomic nuclide. This ratio is important in the consideration of which nuclei are stable. For light nuclei, the most stable are those in which the ratio has the value of one. For nuclei heavier than those of calcium the neutron–proton ratio increases, finishing at a value of about 1.6 for uranium.

neutron star A very dense star that is protected from gravitational collapse to a BLACK HOLE by the DEGENERACY PRESSURE of the neutrons of which it is composed. Neutron stars are thought to have been formed by the collapse of stars several times more massive than the Sun in supernova explosions. *See also* pulsar.

New Cartesian convention *See* sign convention.

newton Symbol: N The SI unit of force, equal to the force needed to accelerate one kilogram by one meter second^{-2}.

$$1\ \text{N} = 1\ \text{kg m s}^{-2}$$

The unit is named for the English scientist Sir Isaac Newton (1642–1727).

Newtonian fluid A fluid in which the stress at any point is proportional to the rate of change of strain with respect to time at that point, with the constant of proportionality being called the *coefficient of viscosity*. Many liquids are Newtonian fluids. Examples of *non-Newtonian fluids* include non-drip paints, lubricating oils, and polymer solutions and melts. *See also* thixtropy.

Newtonian mechanics /new-**toh**-nee-ăn/ The system of mechanics that applies NEWTON'S LAWS OF MOTION, as they relate to stationary objects and those that are moving much slower than the speed of light (relative to the observer). *See also* relativity (theory of).

Newtonian telescope The earliest type of reflecting telescope; the system is still often used. Light from the converging mirror is reflected back onto an angled plane mirror, and from there into the eyepiece. *See also* reflector.

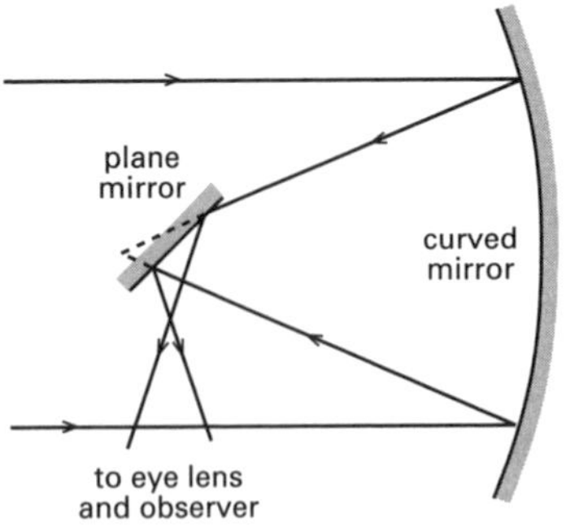

Newtonian telescope

Newton's formula For a simple lens, the distance between two CONJUGATE POINTS (p and q) and their respective FOCAL POINTS (f) are related by the equation:

$$pq = f^2$$

Newton's law of cooling When a hot body is cooling in air, the rate of transfer of heat is proportional to the temperature difference between the body and its surroundings. This is strictly true only for forced convection. *See* convection.

Newton's law of universal gravitation The force of gravitational attraction between two point masses (m_1 and m_2) is proportional to the product of the masses divided by the square of the distance (r) between them. The law is often given in the form $F = Gm_1m_2/r^2$, where G is a constant of proportionality called the *gravitational constant*. The law can also be applied to extended bodies by integrating the infinitesimal forces acting on each element. In the case of spherically symmetrical bodies, the force is the same as if the whole mass were concentrated at the center. To a first approximation, bodies such as the Earth are regarded as spherically symmetrical. It was unchallenged for over 200 years until Albert Einstein put forward his theory of relativity. *See also* Kepler's laws; relativity.

Newton's laws of motion Three laws of mechanics formulated by Sir Isaac Newton in 1687. They can be stated as:

(1) An object continues in a state of rest or constant velocity unless acted on by an external force.

(2) The resultant force acting on an object is proportional to the rate of change of momentum of the object, the change of momentum being in the same direction as the force.

(3) If one object exerts a force on another then there is a simultaneous equal and opposite force on the first object exerted by the second. The first law was discovered by Galileo, and is both a description of inertia and a definition of zero force. The second law provides a definition of force based on the inertial property of mass. The second and third laws combined give the law of conservation of momentum.

See also reaction.

Newton's rings INTERFERENCE patterns produced by placing a lens (with very large radius of curvature) on a reflecting surface and illuminating it with monochromatic light from above. Usually a plano-convex lens is used with the curved surface in contact. If the arrangement is observed from above with a microscope, light and dark rings are seen concentric with the contact point. A dark spot occurs in the center. The radius of a bright ring is given by:

$$r^2 = a(m + \tfrac{1}{2})\lambda$$

where a is the radius of curvature of the convex surface and $m = 0, 1, 2$, etc. For the first bright ring $m = 0$, for the second $m = 1$, etc. Newton's original experiments (1704) were with white light so colored rings were produced.

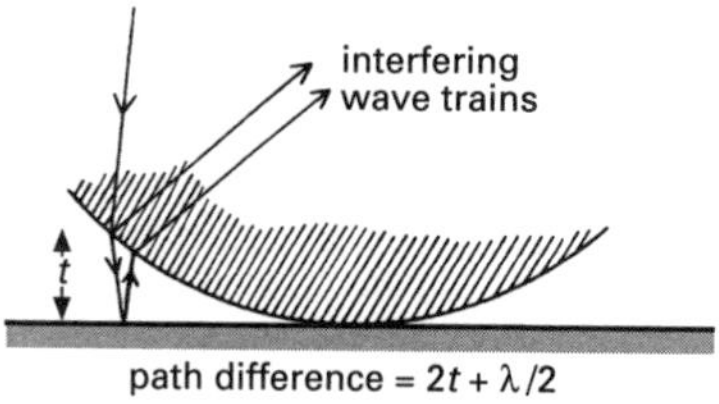

Newton's rings

Nichrome /nȳ-krohm/ (*Trademark*) An alloy of about 62% nickel, 15% chromium, and 23% iron, used for making electric heating elements and resistors. It has a high resistivity and a high melting point.

nickel /**nik**-ăl/ A transition metal that occurs naturally as the sulfide and silicate. It is used as a protective coating (on other metals) and in the manufacture of various alloys, such as Nichrome and stainless steel.

Symbol: Ni; m.p. 1453°C; b.p. 2732°C; r.d. 8.902 (25°C); p.n. 28; r.a.m. 58.6934.

nickel–iron accumulator (**Edison cell; Nife or NIFE cell**) A type of secondary cell in which the electrodes are formed by steel grids. The positive electrode is impregnated with a nickel-nickel hydride mixture. The negative electrode is impregnated with iron oxide. Potassium hydroxide solution forms the electrolyte. The nickel-iron cell is lighter and more durable than the lead accumulator, and it can work with higher currents. Its e.m.f. is approximately 1.3 volts. Modern forms of this cell use cadmium or cadmium-iron alloy for the negative electrode. It is also named the Edison cell for the American scientist and inventor Thomas A. Edison (1847–1931).

Nicol prism /**nik**-ŏl/ A device for producing plane polarized light, consisting of a crystal of calcite cut with a 68° angle, cleaved along the optic axis, and stuck together with a thin layer of Canada balsam. The Canada balsam, which is not birefringent, has the same refractive constant for both ordinary and extraordinary rays ($n = 1.66$). The latter ray passes through the prism (in calcite $n = 1.66$). However, the ordinary ray in calcite has a lower refractive constant ($n = 1.48$) than in the Canada balsam, and suffers total internal reflection at the interface. Nicol prisms are more transparent than Polaroid. It is named for the Scottish physicist William Nicol (1768–1851).

NIFE cell *See* nickel-iron accumulator.

niobium /nȳ-**oh**-bee-ŭm/ A soft silvery transition element used in welding, special steels, and nuclear reactor work.

Symbol: Nb; m.p. 2468°C; b.p. 4742°C; r.d. 8.570 (20°C); p.n. 41; r.a.m. 92.90638.

Nipkow disk /**nip**-koff/ A primitive mechanical scanning device consisting of a rotating disk with radial notches or a spiral of small holes. It was employed in the early development of television. It is named for the German physicist Paul Nipkow (1860–1940).

nit Symbol: nt A unit of luminance, equal to the luminance produced by one candela per square meter (cd m^{-2}).

nitrogen /**nȳ**-trŏ-jĕn/ A nonmetallic element; nitrogen accounts for about 78% of the atmosphere (by volume) and it also occurs as sodium nitrate in various mineral deposits. It is separated for industrial use by the fractionation of liquid air. Nitrogen has two isotopes; ^{14}N, the common isotope, and ^{15}N (natural abundance 0.366%), which is used as a marker in mass spectrometric studies.

Symbol: N; m.p. –209.86°C; b.p. –195.8°C; d. 1.2506 kg m^{-3} (0°C); p.n. 7; r.a.m. 14.

NMR *See* nuclear magnetic resonance.

nobelium /noh-**bel**-ee-ŭm/ A radioactive transuranic element of the actinoid series, not found naturally on Earth. Several very short-lived isotopes have been synthesized. The element is named for the Swedish chemist Alfred Nobel (1833–96).

Symbol: No; p.n. 102; most stable isotope ^{259}No (half-life 58 minutes).

node A place in a stationary wave where one of the two kinds of disturbance has zero (or minimum) value. Nodes occur at distances apart equal to half the wavelength of the corresponding progressive WAVE.

The nodes for one disturbance are normally antinodes (i.e. places of maximum amplitude) for the other. Thus, in stationary electromagnetic waves, electric nodes are magnetic antinodes and vice versa. In stationary sound waves the pressure nodes are antinodes for particle velocity and vice versa. Similar rules apply to other types of wave. Generally energy oscillates longitudinally between the nodes for one variable and the adjacent antinodes. *See also* interference.

noise **1.** Sound composed of a random mix of different frequencies. *White noise* is a completely random mix over a wide frequency range; it has a confusing effect on the listener. *Pink noise* – random frequencies in a selected range – is often used as a background to mask other sounds.
2. A general term describing any signal that impairs the efficient working of an electronic device. There are two principal types: white noise and impulse noise. The unwanted energy of *white noise* is distributed across a wide frequency band, in the same way that the energy of white light is distributed across a band according to color. *Impulse noise* is a consequence of one or more momentary electrical impulses. There are numerous types of white noise: thermal noise is caused by the random thermal motion of electrons superimposed on a steady current flow; random noise can also be caused by an irregular momentary disturbance originating in an atmospheric electrical disturbance or a similar disturbance in the Sun, etc.

Noise results in undesirable sounds in a loudspeaker, which mask the desired effect, and, in television, results in 'snow', an unwanted pattern on a TV screen similar to falling snow.

non-equilibrium statistical mechanics The branch of statistical mechanics concerned with systems that are not in thermodynamic equilibrium. Non-equilibrium statistical mechanics is used to calculate transport coefficients such as electrical and thermal conductivity. For systems that are far from thermodynamic equilibrium it can describe such behavior as chaos, self-organization, and turbulence.

non-equilibrium thermodynamics The branch of thermodynamics concerned with systems which are not in equilibrium. Non-equilibrium thermodynamics is used to describe such phenomena as self-organization. It is not known whether general laws exist, analogous to the well-established three laws of thermodynamics, for systems that are far from thermodynamic equilibrium.

non-inductive Describing a circuit component that has a low inductance. Non-inductive coils are made by doubling back the wire on itself before making the coil. The current then flows in both senses through the coil and a negligible magnetic field is produced.

non-ohmic Describing a substance or circuit component that does not obey Ohm's law; i.e. the current passed is not directly proportional to the potential difference across the circuit.

no parallax *See* parallax.

NOR gate *See* logic gate.

normal The perpendicular to a reflecting or refracting surface at the point of incidence of the ray concerned. Angles of incidence, reflection, and refraction are measured between the normal and the incident ray, reflected ray, and refracted ray respectively. A *normal ray* is one incident perpendicularly on a surface – the angle of incidence is zero.

normal adjustment An image formed by an optical system is in normal adjustment if it is in a similar viewing position to the object. The term really refers to the adjustment of the system. Thus, normal adjustment of a telescope gives the image at infinity; normal adjustment of a microscope puts the image at the viewer's near point. Other adjustments are quite possible, and may be preferred in some cases.

normal ray *See* normal.

normal temperature and pressure (NTP) *See* STP.

north pole *See* poles; magnetic.

note (**tone**) A sound that has a single fundamental frequency. A *pure note* (or *pure tone*) has a simple harmonic waveform (approximately). Such a note can be produced by a tuning fork. Musical instruments produce complex notes, which can be analyzed into a single fundamental frequency mixed with higher frequency overtones. *See* quality of sound.

NOT gate (**inverter gate**) *See* logic gate.

n-p-n transistor *See* transistor.

N-pole *See* poles; magnetic.

NTP Normal temperature and pressure. *See* STP.

n-type conductivity *See* semiconductor; transistor.

nuclear energy /**new**-kee-er/ Energy derived from nuclear reactions either by the fission of heavy nuclei into lighter ones or by fusion of light nuclei into heavier ones.

Nuclear energy is exploited both for weapons construction and for civil power supplies. Both fission and fusion processes have been used in weaponry but, so far, it has not proved possible to exploit fusion processes for power production. *See also* fusion; nuclear reactor; nuclear weapon.

nuclear fission The disintegration of an atomic nucleus into two or more large fragments. Fission may be spontaneous or induced. Spontaneous fission is a rare form of radioactivity shown by a few heavy nuclides. Induced fission is the instantaneous disintegration of a heavy nucleus on irradiation, usually by neutrons. In fission, nuclei usually disintegrate to form two fragments of roughly comparable mass number along with a few neutrons. For example, the neutron-induced fission of ^{235}uranium could yield ^{90}krypton and ^{143}barium as follows: $^{235}_{92}U + ^{1}_{0}n \rightarrow ^{90}_{36}Kr + ^{143}_{56}Ba + 3^{1}_{0}n$

The energy made available will be about 30 petajoules (pJ) per fission. Nuclear fission is exploited for the production of nuclear power and for nuclear weapons.

nuclear force A very strong short-range attractive force acting between nucleons. Nuclear forces act over a range up to about 2×10^{-15} m and are much stronger than electromagnetic forces, so are able to overcome the electrostatic repulsion between protons in the nucleus. They are thus responsible for holding the nucleus together. *See also* exchange force; interaction.

nuclear fusion *See* fusion.

nuclear isomers A pair of isotopes of the same proton and neutron numbers, but in different quantum states. They thus have differing stability, and behave as different nuclides having different half-lives. For example, the beta decay of ^{234}Th yields two isomers of ^{234}Pa with half-lives of about 70 seconds and about 6.7 hours respectively, which may both undergo beta decay to form $^{234}_{90}$U. About 1 in 700 nuclei of the less stable isomer (that with the shorter half-life) may emit a gamma photon and make a transition to the more stable isomer.

nuclear magnetic resonance (**NMR**) An effect that results when waves of radio frequency are absorbed by the nuclei of matter in a strong external magnetic field. It has various applications in chemistry (*NMR spectroscopy,*), medicine, and physics.

nuclear physics The branch of physics that is concerned with nuclear structure, properties, and reactions, and their applications (e.g. in producing nuclear power or using radioisotopes).

nuclear power The use of nuclear reactions for power generation (usually by generation of electricity but nuclear powered ships use the heat generated to raise steam to power the turbines directly without conversion to electricity). The reactions concerned are usually restricted to fission and fusion reactions; some people, however, include the use of energy from radioisotope decay.

nuclear reaction A reaction involving nuclei. Such reactions include the spontaneous disintegration of a single nucleus (radioactive decay or fission); the formation of a new nucleus from two smaller nuclei (fusion); or absorption of a photon, neutron, proton, or other particle. An equation for a nuclear reaction shows the nuclides and particles involved by their symbols. For example:

$$^{210}_{82}\text{Pb} \rightarrow {}^{210}_{83}\text{Bi} + {}^{0}_{-1}\text{e}$$

represents the beta decay of lead-210 to form bismuth-210.

nuclear reactor A device in which nuclear reactions take place on a large scale. A reactor may be built for the purpose of providing useful energy, producing new nuclides, or research. Present designs are *fission reactors* that depend on the controlled fission of uranium or plutonium. In general, they are constructed of a core containing the fuel (uranium); control rods to absorb surplus neutrons; a reflector to bounce straying neutrons back into the core; coolant to carry the energy from the core; containment – a barrier to prevent the escape of radioactive material; and shielding to protect workers from the intense radiation generated.

In *thermal reactors* a moderator is used to slow the neutrons so that a chain reaction can be sustained. *Fast reactors* do not have a moderator, and use fast neutrons. In this case the fuel has to be enriched (so that it contains more ^{235}U isotope). This type of reactor uses a blanket of natural uranium, which is converted to plutonium by neutrons. Some fast reactors produce more fissile ^{239}Pu from the nonfissile ^{238}U than the amount of fissile material used up. These are called BREEDER REACTORS if the original fuel is mainly plutonium and *converter reactors* if it is mainly ^{235}U. Thermal breeder reactors are possible using ^{233}U as fuel and generating more of this fissile material by capturing neutrons in nonfissile ^{232}Th. So far, this type has been little developed.

Fusion reactors are reactors for producing energy by nuclear fusion. They are potentially important because the fuel (hydrogen isotopes) is available in unlimited quantities. However, there are immense practical problems in sustaining a controlled fusion reaction, and these reactors are still in the development stage. *See also* boiling-water reactor; gas-cooled reactor; pressurized water reactor; Tokamak.

nuclear waste The unwanted radioactive by-products of the nuclear industry, particularly from the mining, extraction, and reprocessing of nuclear fuels and from decommissioned nuclear reactors. Their disposal presents an environmental (and political) problem. Methods of treatment include storage of highly radioactive waste until its activity is reduced enough to process it, containment in concrete of intermediate-level waste and burial underground or below the seabed, and storage of low-level waste in steel drums at special sites.

nuclear weapon An explosive device in which the energy release results either from the fission of heavy nuclei (such as ^{235}U or ^{239}Pu) in a rapidly built up chain reaction or from the fusion of light nuclei such as deuterium and tritium to form helium and neutrons. This latter type of device is known as a 'hydrogen bomb'. *See also* critical mass; mass defect.

nucleon /**new**-klee-on/ A particle found in the nucleus of atoms; i.e. a proton or a neutron. Nucleons are classified as baryons. *See also* elementary particles.

nucleon number (**mass number**) Symbol: *A* The number of nucleons (protons plus neutrons) in an atomic nucleus.

nucleosynthesis /new-klee-oh-**sin**-th'ĕ-siss/ The synthesis of the nuclei of the chemical elements in nuclear reactions. The term is often used in the context of the creation of chemical elements in the Universe. Nucleosynthesis can occur in several different ways. In *primordial nucleosynthesis*, which occurred very soon after the BIG BANG when the Universe was very hot, the light elements such as helium were made. Nearly all other elements have been made in *stellar nucleosynthesis*, i.e. nuclear reactions occurring inside stars. In particular, most of the elements heavier than iron are made in SUPERNOVA explosions. Some light elements, particularly beryllium and boron, were produced by *spallation*, i.e. nuclear reactions in interstellar space occurring when cosmic rays hit the nuclei of interstellar matter.

nucleus, atomic (*pl.* **nuclei** or **nucleuses**) The compact, comparatively massive, positively charged center of an atom made up of one or more nucleons (protons and neutrons) around which is a cloud of electrons. The density of nuclei is about 10^{15} kg m^{-3}. The number of protons in the nucleus defines the element, being its atomic number (or proton number). The nucleon number, or atomic mass number, is the sum of the protons and neutrons. The simplest nucleus is that of a hydrogen atom, ^{1}H, being simply one proton (mass 1.67×10^{-27} kg). The most massive naturally occurring nucleus is ^{238}U of 92 protons and 146 neutrons (mass 4×10^{-25} kg, radius 9.54×10^{-15} m). Only certain combinations of protons and neutrons form stable nuclei. Others undergo spontaneous decay.

A nucleus is depicted by a symbol indicating nucleon number (mass number), proton number (atomic number), and element name. For example, $^{23}_{11}Na$ represents a nucleus of sodium having 11 protons and mass 23, hence there are (23 – 11) = 12 neutrons.

nuclide /**new**-klÿd/ A nuclear species with a given number of protons and neutrons; for example, ^{23}Na, ^{24}Na, and ^{24}Mg are all different nuclides. Thus:

$^{23}_{11}Na$ has 11 protons and 12 neutrons
$^{24}_{11}Na$ has 11 protons and 13 neutrons
$^{24}_{12}Mg$ has 12 protons and 12 neutrons

The term is applied to the nucleus and often also to the atom.

nutation /new-**tay**-shŏn/ A periodic change in the inclination of the axis in the motion of a spinning body with respect to some reference axis. For example, the main nutation in the precessional motion of the Earth occurs because of periodic changes in the orbit of the Moon. This gives a nutation with a period of 19 years.

O

object A real or apparent source of rays in an optical system, perhaps incident on a lens or a reflector. After refraction or reflection, the rays appear to come from some other place – the image.

An object need not be real. The diagrams show how the real image I_1, produced by the lens, becomes a *virtual object* O_2 when the mirror is introduced. This now gives a real image I_2. Just as with real objects, virtual objects can appear as real or virtual images.

objective The lens or combination of lenses nearest the object in an optical instrument. The nearest lens to the object in a compound objective is often called the *object lens*. The large converging mirror in a reflecting telescope can also be described as the objective.

object plane The plane that is perpendicular to the axis and centered on the object in an optical system.

octave /ok-tiv/ The interval between a waveform and another of twice the frequency. An octave in sound corresponds to eight notes on the diatonic musical scale.

octet /ok-**tet**/ (**electron octet**) For an atom, an outer shell containing eight electrons, which confers stability. For example, the very unreactive rare gases have an octet of electrons in their outer shell. *See also* electron shell.

ocular /ok-yŭ-ler/ *See* eyepiece.

oersted Symbol: Oe A unit of magnetic field strength in the c.g.s. system. It is equal to 79.578 A m^{-1}. The unit is named for the Danish physicist Hans Christian Oersted (1777–1851).

ohm /ohm/ Symbol: Ω The SI unit of electrical resistance, equal to a resistance that passes a current of one ampere when there is an electric potential difference of one

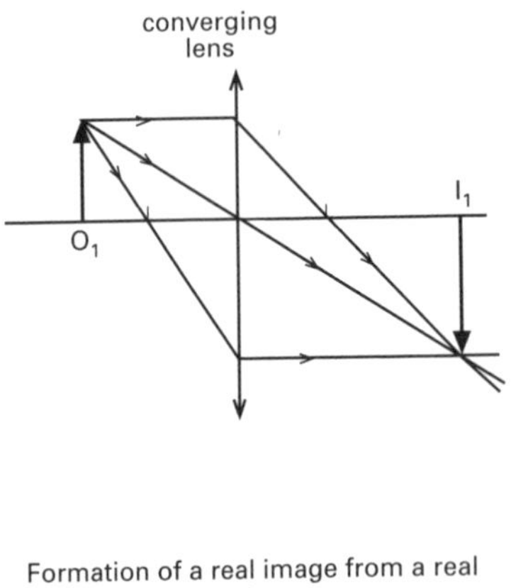

Formation of a real image from a real object

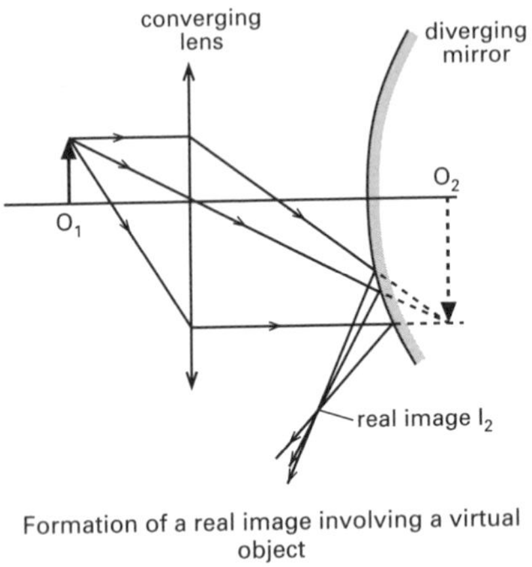

Formation of a real image involving a virtual object

volt across it. 1 Ω = 1 V A^{-1}. The unit is named for the German physicist Georg Simon Ohm (1787–1854).

ohmic /**oh**-mik/ Describing a substance or circuit component that obeys Ohm's law.

Ohm's law The current (I) in a conductor is proportional to the potential difference (V) between its ends. This leads to $V = IR$, where R is the conductor's resistance. R is constant provided the physical conditions in the conductor are unaltered. If V is in volts and I in amperes, R is in ohms. Components that conduct in accordance with Ohm's law (e.g. metal resistors) are said to be *ohmic*.

oil-drop experiment An experiment carried out by Millikan in 1913 to determine the charge on an electron. He deduced the charge by measuring the rate of descent of a charged oil drop in an electric field. *See* electron charge.

oil-immersion objective *See* immersion objective.

omega particle /oh-**meg**-ă, -**mee**-gă, -**may**-gă/ *See* elementary particles.

opaque /oh-**payk**/ Not able to pass radiant energy. A substance that is opaque to one type of electromagnetic radiation may be transparent to another. Thus glass passes visible radiation very well, but is opaque to most thermal (infrared) radiation. *Compare* translucent.

opaque projector (**episcope**) An optical instrument for projecting an image of an opaque object (e.g. a diagram or picture) onto a screen. High-intensity illumination is used and the image is projected by a combination of mirrors and lenses.

open circuit *See* circuit; electrical.

open pipe A pipe that is open at both ends. When stationary waves are set up in an open pipe the ends of the pipe have to be antinodes.

OPTICAL GLASS

Glass	*Density (kg m^{-3})*	*Refractive constant*
crown	2500–2600	1.51–1.61
flint	2700–4800	1.53–1.75

opera glasses *See* binoculars; Galilean telescope.

optical activity /**op**-tă-kăl/ The ability of certain crystals or compounds in solution to rotate the plane of polarization of plane polarized light. Compounds that are optically active have an asymmetric molecular structure such that their molecules can exist in left- and right-handed forms (the forms are mirror images of each other; the property of having such forms is called *chirality*). Particular forms of the compound are classified as *dextrorotatory* (right turning) or *levorotatory* (left turning). Levorotatory compounds rotate the plane of polarization to the left; i.e. anticlockwise as viewed facing the oncoming light. Dextrorotatory compounds rotate the plane to the right (i.e. in the opposite sense). Many naturally occurring substances (e.g. sugars) are optically active.

optical axis The principal axis of an optical system, i.e. the path of rays passing along the principal axes of the lenses or mirrors.

optical center (**optic center**) *See* pole.

optical density A measure of the extent to which a transluscent material prevents the passage of light through it. It is defined as $\log_{10}I_0/I$, where I_0 is the intensity of the incident light and I is the transmitted intensity.

optical fiber *See* fiber optics.

optical glass A type of glass with properties suitable for making lenses, prisms, etc. There are two major groups: the *crown glasses* and the *flint glasses*, differing in their density and refractive constants. Optical glasses are fairly hard yet easily polished and highly transparent to light.

optical lever A device for measuring angular displacement. A small mirror is attached to the rotating body and a narrow fixed beam of light directed onto it. The reflected beam is directed onto a screen, producing a spot of light. Movement of this spot shows small angular movements of the mirror. The angle turned through by the reflected beam is twice that turned by the mirror. The device is commonly used in torsion balances, as in the mirror galvanometer.

optical path Symbol: *d* The product of distance (*l*) traveled in a medium and the refractive constant (*n*) of the medium; i.e. $d = nl$. Phase difference ($\Delta\phi$) relates to path difference (Δd) thus:

$$\Delta\phi = 2\pi\Delta d/\lambda$$

optical pyrometer (**disappearing-filament pyrometer**) A type of pyrometer used to measure the temperature of incandescent sources. Light from the source is focused onto a tungsten filament, and the filament and the image of the source are viewed through a red filter and an eyepiece. The current in the filament, measured by an ammeter, is varied until this has the same brightness as the body (i.e. it cannot be seen against the background of the source). The ammeter is calibrated directly in degrees Celsius, with the assumption that the source emits black-body radiation. A correction can be made for the spectral emissivity of the source. *See also* total-radiation pyrometer.

optical rotary dispersion (**ORD**) The phenomenon in which the amount of rotation of plane-polarized light of an optically active substance depends on the wavelength. Optical rotary dispersion is an important technique in chemistry. Plots of rotation against wavelength can be used to give information on the molecular structure of optically active compounds.

optical telescope A TELESCOPE using visible light from a distant object to produce a magnified image.

optic axis The direction in a BIREFRINGENT CRYSTAL along which the ordinary and extraordinary rays travel at the same speed. Uniaxial crystals have one such axis; biaxial crystals have two.

optics /**op**-tiks/ The study of the nature and behavior of light and other radiations. The reflection of ultraviolet radiation and the refraction of sound (pressure) waves also follow the laws of optics. Electron diffraction and the electron microscope are branches of electron optics.

Where the wave nature of the radiation need not be considered, situations can be discussed in terms of rays. That study is traditionally called *geometrical optics*. *Physical optics* is the field of optics in which wave properties are important. Thus the use of lenses is part of geometrical optics, while the diffraction grating comes into physical optics.

optoelectronics /op-toh-i-lek-**tron**-iks/ A branch of electronics based on light beams rather than electric currents. The main advantage is that light frequencies are far higher than radio frequencies so that much more information can be carried in a suitably treated light beam than in an electronic (radio) signal. Optical fibers, of very cheap very transparent glass, carry the light beams, produced by semiconductor lasers and detected by photocells.

orbit The curved, usually closed, path, or trajectory along which a moving object travels. Examples include the elliptical orbit of a planet in the solar system, the circular orbit of an electron in a magnetic field, and the parabolic trajectory of a projectile under gravity.

orbital 1. Pertaining to an orbit. For example, an orbital electron is one that is bonded around the nucleus of an atom.
2. According to quantum theory, the electrons in an atom do not have fixed orbits around the nucleus. Instead, there is a finite probability of finding the electron in a given volume at any distance from the nucleus. The region in space in which there is a high probability of finding an electron is

an *atomic orbital*. Orbitals are usually thought of as three-dimensional states centered on the nucleus. For a hydrogen atom in its ground state the electron occupies an orbital that has a spherical distribution around the nucleus. This is known as an *s orbital*. Other types of orbital have different shapes: for example, *p orbitals* have two lobes extending out from the nucleus; *d orbitals* have four lobes. Similarly, in molecules, electrons move in *molecular orbitals* around the nuclei of the atoms.

order 1. A specific pattern in a system. For example, there is order in a perfect crystal because of the lattice structure. In a ferromagnet there is order in the form of all the magnetic moments of the atoms being aligned in the same direction.
2. An integer (m) associated with a given interference fringe or diffraction pattern. In interference a bright fringe occurs for a path difference $m\lambda$; a dark fringe is produced if the path difference is $(m + \frac{1}{2})\lambda$. A bright fringe is first order if it arises through a path difference of one wavelength ($m = 1$). Similarly, second order corresponds to $m = 2$, etc.

ordinary ray *See* birefringent crystal.

OR gate *See* logic gate.

oscillation /os-ă-**lay**-shŏn/ A regularly repeated motion or change. *See* vibration.

oscillator /**os**-ă-lay-ter/ A device that generates an alternating current signal of known frequency from a direct current input. The output is usually in the form of a sine wave or a square wave. A sinusoidal oscillation can be produced by combining an inductor and a capacitor in a resonant circuit. The frequency is dependent on the values of the inductance and the capacitance. A square wave is produced in a MULTIVIBRATOR by the alternate charge and discharge of a capacitor, the frequency being dependent on the resistance and capacitance. *See also* crystal oscillator.

oscilloscope /ŏ-**sil**-ŏ-skohp/ *See* cathode-ray oscilloscope.

osmium /**oz**-mee-ŭm/ A transition metal that is found associated with platinum. Osmium is the most dense of all metals. The metal is used in catalysts and in alloys for pen nibs, pivots, and electrical contacts.

Symbol: Os; m.p. 3054°C; b.p. 5027°C; r.d. 22.59 (20°C); p.n. 76; r.a.m. 190.23.

Otto cycle The idealized reversible cycle of four operations occurring in a perfect four-stroke (spark ignition) gasoline engine. The cylinder is filled with atmospheric air containing gasoline vapor (intake stroke); with valves closed, the air-vapor mixture is compressed adiabatically (compression stroke); the spark ignites the mixture, which burns rapidly, increasing the temperature at nearly constant volume; the hot mixture of air and combustion products expands adiabatically doing work (power stroke); a valve opens and the piston expels the gas (exhaust stroke). The cycle is then repeated.

Assuming ideal reversible processes and perfect thermal insulation, the ideal cycle gives the maximum conceivable EFFICIENCY for this type of engine. In practice, it will be considerably less efficient. The cycle is named for the German engineer Nikolaus August Otto (1832–91). *See also* Carnot cycle.

overdamping *See* damping.

overhead projector A type of PROJECTOR able to project, on a screen, large bright images of slides or transparent objects placed on a horizontal table.

overtone A component of a note that has higher frequency and (usually) lower intensity than the fundamental. The overtones have frequencies that are simple multiples of the fundamental frequencies. The word is a musical term best avoided in physics. *See* partials; quality of sound.

oxygen /**oks**-ă-jĕn/ A colorless odorless diatomic gas. It has the electronic configuration $[He]2s^22p^4$. Oxygen is the most plentiful element in the Earth's crust accounting for over 40% by weight. It is present in the atmosphere (20%) and is a constituent of the majority of minerals and rocks as well as the major constituent of the sea.

Oxygen is an essential element for almost all living things. Elemental oxygen has two forms: the diatomic molecule O_2 and the less stable molecule trioxygen (ozone), O_3, which is formed by passing an electric discharge through oxygen gas. Industrially oxygen is obtained by the fractionation of liquid air. The element occurs in three natural isotopic forms, ^{16}O (99.76%), ^{17}O (0.0374%), ^{18}O (0.2039%); the rarer isotopes are used in detailed studies of the behavior of oxygen-containing groups during chemical reactions (tracer studies).

Symbol: O; m.p. –218.4°C; b.p. –182.962°C; d. 1.429 kg m^{-3} (0°C); p.n. 8; r.a.m. 15.99.

P

packing fraction The difference between the actual mass of a nuclide and the nearest whole number (i.e., the mass defect) divided by the mass number. *See also* binding energy.

pain threshold The intensity of sound at which sound is painful to the ear. It corresponds to about 120 decibels.

pair production The production of an electron and its antiparticle (a positron) from a photon (according to the equation $E = mc^2$). The process can occur in the field of an atomic nucleus. Since the mass of an electron or positron is equivalent to 0.511 MeV, the minimum energy of a photon that can promote pair production is 1.022 MeV. Any surplus energy becomes kinetic energy of the products.

palladium /pă-**lay**-dee-ŭm/ A silvery white ductile transition metal occurring in platinum ores. It is used in electrical relays and as a catalyst in hydrogenation processes. Hydrogen will diffuse through a hot palladium barrier.

Symbol: Pd; m.p. 1552°C; b.p. 3140°C; r.d. 12.02 (20°C); p.n. 46; r.a.m. 106.42.

parabola /pă-**rab**-ŏ-lă/ A conic with an eccentricity of 1. The curve is symmetrical about an axis through the focus at right angles to the directrix. This axis intercepts the parabola at the *vertex*. A chord through the focus perpendicular to the axis is the *latus rectum* of the parabola.

In Cartesian coordinates a parabola can be represented by an equation:

$$y^2 = 4ax$$

In this form the vertex is at the origin and the x-axis is the axis of symmetry. The focus is at the point $(0,a)$ and the directrix

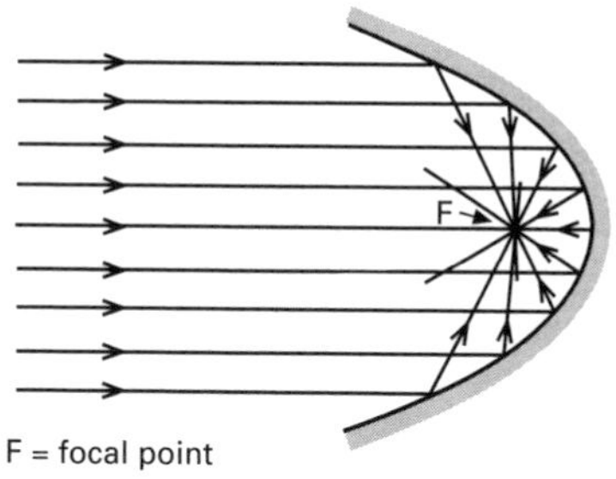

Parabolic mirror

is the line $x = -a$ (parallel to the y-axis). The *latus rectum* is $4a$.

If a point is taken on a parabola and two lines drawn from it – one parallel to the axis and the other from the point to the focus – then these lines make equal angles with the tangent at that point. This is known as the *reflection property* of the parabola, and is utilized in parabolic reflectors and antennas.

The parabola is the curve traced out by a projectile falling freely under gravity. For example, a tennis ball projected horizontally with a velocity v has, after time t, traveled a distance $d = vt$ horizontally, and has also fallen vertically by $h = gt^2/2$ because of the acceleration of free fall g. These two equations are *parametric equations* of the parabola. Their standard form, corresponding to $y^2 = 4ax$, is

$$x = at^2$$
$$y = 2at$$

where x represents h, the constant a is $g/2$, and y represents d. *See also* conic.

parabolic mirror /pa-ră-**bol**-ik/ A reflector with a parabolic section. The converging type can converge wide parallel beams accurately into its focal point (and is thus used for reflection telescopes and solar power applications). On the other hand, radiation from a source at the focal point will be reflected into a parallel beam (as used in various lighting systems).

parallax /**pa**-ră-laks/ The apparent difference in an object's position relative to that of another when viewed from two different places. If there is *no parallax* between two objects, they must be at effectively the same place. Often in experiments with light, no-parallax methods are used to locate images. When there is no parallax between the image and the object used to find it, the two are in the same position.

parallel Elements in an electrical circuit are *in parallel* if connected so that the current divides between them and rejoins at the other side. The word 'shunt' is sometimes used.

For resistors in parallel, the resulting resistance R is given by:

$$1/R = 1/R_1 + 1/R_2 + 1/R_3 + \ldots$$

For capacitors in parallel, the capacitance of the combination is given by:

$$C = C_1 + C_2 + C_3 + \ldots$$

Compare series.

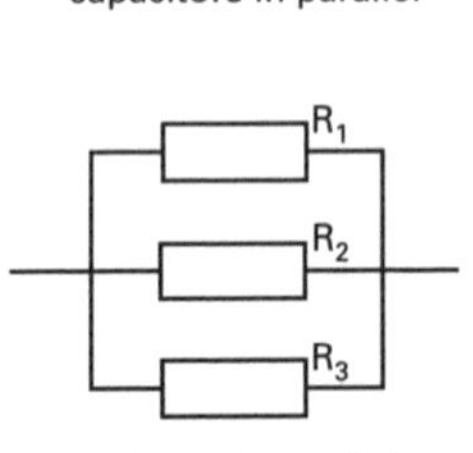

Parallel circuits

parallel axes, theorem of *See* theorem of parallel axes.

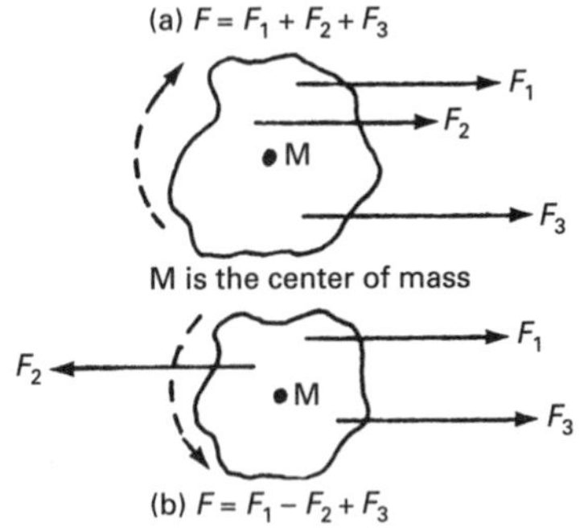

Parallel forces

parallel forces When the forces on an object pass through one point, their resultant can be found by using the parallelogram of vectors. If the forces are parallel the resultant is found by addition, taking sign into account. There may also be a turning effect in such cases, which can be found by the principle of moments.

parallelogram (law) of forces *See* parallelogram of vectors.

parallelogram (law) of velocities *See* parallelogram of vectors.

parallelogram of vectors A method for finding the *resultant* of two vectors acting at a point. The two vectors are shown as two sides of a parallelogram: the resultant is the diagonal through the starting point. The technique can be used either with careful scale drawing or with trigonometry. The trigonometrical relations give:

$$F = \sqrt{(F_1^2 + F_2^2 + 2F_1F_2\cos\theta)}$$
$$\alpha = \sin^{-1}[(F_1/F)\sin\theta]$$

paramagnetism /pa-ră-**mag**-nĕ-tiz-ăm/ The magnetic behavior of substances with a small susceptibility; this often varies with temperature according to Curie's law. The relative permeability of a paramagnetic substance is slightly greater than one. Paramagnetism results from unpaired electron

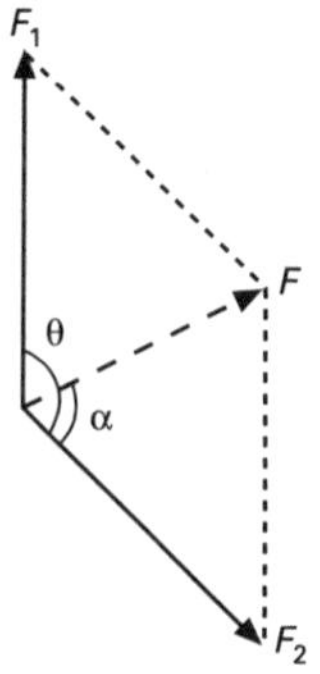

Parallelogram of vectors

spins. In an external field, a piece of paramagnetic material will slightly concentrate, and thus increase, the field; this is because the magnetic moments of the particles align themselves in the same direction. As a result, the sample itself will align parallel to the field. Above its Curie temperature, a ferromagnetic material will have no domains and will show only paramagnetic behavior. *See also* magnetism.

paraxial /pă-**raks**-ee-ăl/ Describing rays incident on a surface close and parallel to the axis. Only paraxial rays pass or appear to pass through the focal point of a spherical reflecting or refracting surface. *See also* mirror; parabolic mirror.

parent A nuclide that undergoes radioactive decay to another specified nuclide (the *daughter*).

parity (*pl.* **parities**) /**pa**-ră-tee/ A property of the *wave functions* that describe physical systems in quantum mechanics. If all three space coordinates are reversed the magnitude of the wave function does not change. If the sign is unchanged the function is said to have even parity (or +1); if the sign is reversed it has odd parity (or –1).

Each quantum state of a body has even or odd parity and the various interactions are subject to rules related to changes of parity. For example, the more usual processes of emission or absorption of electromagnetic radiation occur only between states of opposite parity. The parity of a system is the product of the parities of the parts. The parity of a system of interacting bodies is conserved in strong and electromagnetic interactions, but not in weak interactions such as beta decay.

parking orbit (synchronous orbit) An orbit of a satellite round the Earth in which the period of the orbiting satellite is 24 hours, i.e. the same as the period of the rotation of the Earth about its axis of rotation. A satellite in a parking orbit is always positioned above the same point on the Earth. A parking orbit is about 36 000 km above the surface of the Earth. Such orbits are used for communication satellites.

parsec /**par**-sek/ Symbol: pc A unit of distance used in astronomy and defined as that at which the angle subtended by one ASTRONOMICAL UNIT equals one second of one degree of arc, or 1/3600 of a degree of arc. The parsec is used with SI units, in which it is equal to approximately 30.857 petameters.

partial eclipse *See* eclipse.

partial pressure *See* Dalton's law.

partials Components of a sound wave with frequencies above the fundamental frequency. The partials include the overtones together with non-harmonic components.

particle An abstract simplification of a real object – the mass is concentrated at the object's center of mass; its volume is zero. Thus relational aspects can be ignored.

particle, elementary *See* elementary particle.

particle accelerator *See* accelerator.

particle physics The branch of physics that deals with elementary particles, their properties and interactions.

partition coefficient The ratio of the equilibrium concentrations of a solute dissolved in two specified immiscible solvents. It is independent of the actual concentrations.

pascal /**pas**-kăl/ Symbol: Pa The SI unit of pressure and stress, equal to a pressure of one newton per square meter (1 Pa = 1 N m^{-2}). The unit is named for the French mathematician, physicist, and religious philosopher Blaise Pascal (1623–62).

Paschen series /**pash**-ĕn/ A series of lines in the infrared spectrum emitted by excited hydrogen atoms. The lines correspond to the atomic electrons falling into the third lowest energy level and emitting energy as radiation. The wavelength (λ) of the radiation in the Paschen series is given by

$$1/\lambda = R(1/3^2 - 1/n^2)$$

where n is an integer and R is the Rydberg constant. The series is named for the German physicist Louis Paschen (1865–1947). *See also* spectral series.

Pauli exclusion principle /**pow**-lee/ *See* exclusion principle.

p.d. *See* potential difference.

peak value The maximum value attained by an alternating quantity (e.g. an alternating current).

Peltier effect /pel-**tyay**/ The change in temperature produced at a junction between two different metals or semiconductors when electric charge is passed through it. If the current direction is reversed then a heating effect becomes a cooling one or vice versa. The temperature change is directly proportional to the current. The effect is named for the French physicist Jean Charles Athanase Peltier (1785–1845). *Compare* Seebeck effect.

pencil A narrow beam of rays from a single point. *See* beam.

pendulum A *simple pendulum* consists of a small mass oscillating to and fro at the end of a very light string. If the amplitude of oscillation is small (less than about 10°), it moves with simple harmonic motion. The period does not depend on amplitude or mass; there is a continuous interchange of potential and kinetic energy. The period is given by

$$T = 2\pi\sqrt{(l/g)}$$

Here l is the length of the pendulum (from support to center of the mass) and g is the acceleration of free fall.

A *compound pendulum* is a rigid body swinging about a point. The period depends on the radius of gyration. For small oscillations it is given by the same relationship as that of the simple pendulum with l replaced by $[\sqrt{(k^2 + h^2)}]/h$. Here, k is the radius of gyration about an axis through the center of mass and h is the distance from the pivot to the center of mass.

pentode /**pen**-tohd/ A thermionic tube with three GRIDS, one grid more than the TETRODE. The extra grid, called the *suppressor grid*, is placed between the screen grid and the anode. It is held at cathode potential, which suppresses by repulsion the loss of secondary electrons ejected by the anode. Thus the disadvantage of the tetrode is overcome. *See also* thermionic tube.

penumbra (*pl.* **penumbrae** or **penumbras**) /pi-**num**-bră/ Partial shadow. *See* shadow.

perfect gas *See* gas laws; kinetic theory.

period Symbol: T The time for one complete cycle of an oscillation, wave motion, or other regularly repeated process. It is the reciprocal of the frequency, and is related to pulsatance, or angular frequency, (ω) by $T = 2\pi/\omega$.

periodic motion Any kind of regularly repeated motion, such as the swinging of a pendulum, the orbiting of a satellite, the vibration of a source of sound, or an electromagnetic wave.

If the motion can be represented as a pure sine wave, it is a simple harmonic motion. Harmonic motions in general are given by the sum of two or more pure sine waves. *See* simple harmonic motion.

periodic table A table of the chemical elements arranged in order of increasing atomic number. The table is usually organized in rows and columns and shows the relation between chemical properties of elements and their electron configurations.

periscope /**pe**-ră-skohp/ An optical instrument enabling the user to see over or round an obstacle. The simplest version is the mirror periscope, which has two mirrors, each at 45° to the viewed direction. Rather more effective types use internally reflecting prisms. In submarines, telescopic lenses are fitted.

permanent gas A gas that cannot be liquefied by increasing its pressure at room temperature, but must also be cooled. Such a gas has a critical temperature below room temperature. *See* critical state.

permanent magnet A sample of a substance that retains its magnetism when the external magnetic field is removed. Permanent magnets are often made from steel, other alloys, or ferrites; they have many uses. Permanent magnetism is mainly a property of ferromagnetic materials that have high coercivity. Materials like this are often said to be magnetically 'hard'. *Compare* temporary magnetism

permeability /per-mee-ă-**bil**-ă-tee/ Symbol: μ A measure of the tendency of a substance to develop magnetic lines of flux when exposed to an external source of magnetomotive force. *Absolute permeability* is the ratio of the magnetic flux density (B) in a substance to the external magnetic field strength (H); i.e.

$$\mu = B/H$$

The unit is the henry per meter (H m^{-1}). The permeability of free space (i.e. a vacuum) has a value of $4\pi \times 10^{-7}$ H m^{-1}, and is given the symbol μ_0. The *relative permeability* (μ_r) of a substance is the ratio of its absolute permeability to the permeability of free space; i.e.

$$\mu_r = \mu/\mu_0$$

Note that relative permeability, being a ratio, has no units. It is related to susceptibility (χ) by:

$$\mu_r = 1 + \chi$$

permittivity /per-mă-**tiv**-ă-tee/ Symbol: ε The mutual force between two charges Q_1 and Q_2 a distance r apart is given by the equation:

$$F = Q_1Q_2/4\pi\varepsilon r^2$$

The constant ε is the permittivity of the medium. The unit is the coulomb^2 newton^{-1} metre^{-2} (C^2 N^{-1} m^{-2}), or farad metre^{-1} (F m^{-1}). The permittivity of free space, ε_0, (also called the *electric constant*) has the value 8.854×10^{-12} F m^{-1}. The absolute permittivity of a material is equal to $\varepsilon_0\varepsilon_r$, where ε_r is the relative permittivity. *See also* Coulomb's law.

perpetual motion Constant motion occurring without any external energy supply. Perpetual-motion machines are impossible to construct because of frictional and other forces dissipating the original energy.

perversion (**lateral inversion**) The effect produced by a mirror in reversing images apparently left to right. Thus writing appears backwards when reflected; the image of a person raising the right hand appears to raise the left hand.

Perversion is not of unusual significance. In the case of a plane mirror it follows directly from the fact that each point of the object produces an image point that is directly opposite it behind the mirror.

peta- Symbol: P A prefix denoting 10^{15}. For example, 1 petameter (Pm) = 10^{15} meter (m).

phase /fayz/ **1.** A homogeneous part of a mixture distinguished from other parts by boundaries. A mixture of ice and water has two phases. A mixture of ice crystals and salt crystals also has two phases. A solution of salt in water is a single phase.
2. The stage in a cycle that a WAVE (or other periodic system) has reached at a particular time (taken from some reference point). Two waves are *in phase* if their maxima and minima coincide.

For a simple wave represented by the equation

$$y = a\sin 2\pi(ft - x/\lambda)$$

the phase of the wave is the expression $2\pi(ft - x/\lambda)$. The *phase difference* between two points distances x_1 and x_2 from the origin is $2\pi(x_1 - x_2)/\lambda$.

A more general equation for a progressive wave is

$$y = a\sin 2\pi(ft - x/\lambda - \phi)$$

Here, ϕ is the *phase constant* – the phase when t and x are zero. Two waves that are out of phase have different phase constants (they 'start' at different stages at the origin). The phase difference is $|\phi_1 - \phi_2|$. It is equal to $2\pi x/\lambda$, where x is the distance between corresponding points on the two waves. It is the *phase angle* between the two waves; the angle between two rotating vectors (phasors) representing the waves.

See also beats; interference.

phase angle *See* phase.

phase constant *See* phase.

phase difference The difference in the phases of two coherent radiations traveling in the same direction in a superposition region. If the phase difference ($\Delta\phi$) is 0, 2π, 4π, etc., the two are in phase and will interfere constructively. If $\Delta\phi$ is π, 3π, 5π, etc., the two are in anti-phase and will interfere destructively. *See* interference.

phase modulation A type of modulation in which a CARRIER WAVE is made to carry the information in a signal (audio or visual) by fluctuations in the phase of the carrier. The difference in phases between the modulated and the unmodulated carrier is proportional to the amplitude of the signal. The carrier remains at the same frequency. The carrier amplitude is constant. *See also* modulation.

phase rule (**Gibbs phase rule**) For any equilibrium system, the expression

$$P + F = C + 2$$

which relates the number of DEGREES OF FREEDOM, F, to the number of phases, P, and the number of components, C. *See* phase.

phase space A multi-dimensional space that can be used to define the state of a system. Phase space has coordinates (q_1, q_2,… p_1, p_2,…), where q_1, q_2, etc., are degrees of freedom of the system and p_1, p_2, are the momenta corresponding to these degrees of freedom. For example, a single particle has three degrees of freedom (corresponding to the three coordinates defining its position). It also has three components of momentum corresponding to these degrees of freedom. This means that the state of the particle can be defined by six numbers (q_1, q_2, q_3, p_1, p_2, p_3) and it is thus defined by a point in six-dimensional phase space. If the system changes with time (i.e. the particle changes its position and momentum), then the point in phase space traces out a path (known as the *trajectory*). The system may consist of more than one particle. Thus, if there are N particles in the system then the state of the system is specified by a point in a phase space of $6N$ dimensions. The idea of phase space is useful in CHAOS THEORY. *See also* attractor.

phase transition A change in a system from one phase to another phase. Examples of phase transitions including melting, freezing, boiling, sublimation, going from a ferromagnetic to a paramagnetic state and going from a superconducting to a non-superconducting state. A phase transition involves a change in the order in a system. Phase transitions can be described theoretically using techniques in statistical mechanics.

phase velocity The velocity with which the phase in a traveling wave is propagated. It is equal to λ/T, where T is the period. *Compare* group velocity.

phasor /**fay**-zer/ A rotating vector used to represent a sinusoidally varying quantity (e.g. an alternating current). The projection of the vector on a fixed axis represents the amplitude variation with time. A phase angle between two quantities (e.g. current and voltage) is represented by the angle between their phasors.

phi particle /fÿ/ *See* elementary particles.

phon /fon/ Symbol: p A unit for measuring the loudness of sounds on a sound-level scale. Noises of the same intensity sound louder or softer, depending on the frequency. The phon is defined using a standard reference source of 1000 hertz, with which other sounds are compared. A loudness of *n* phons is the same as that of a standard source with an intensity of *n* decibels above the threshold of hearing, which for the average person is 2×10^{-5} pascal or 0 decibels at 1000 hertz..

phonon /**foh**-non/ A quantized lattice vibration in a crystal. The concept of phonons is essential in the quantum theory of solids for analyzing thermal conductivity, electrical conductivity, and superconductivity. In electrical conductivity and superconductivity it is necessary to take electron–phonon interactions into account.

phosphor /**fos**-fer/ A substance that shows luminescence or phosphorescence.

phosphorescence /fos-fŏ-**ress**-ĕns/ **1.** The absorption of energy by molecules followed by emission of electromagnetic radiation. Phosphorescence is a type of LUMINESCENCE, and is distinguished from fluorescence by the fact that the emitted radiation continues for some time after the source of excitation has been removed. In phosphorescence the excited molecules have relatively long lifetimes before they make transitions to lower energy states. However, there is no defined time distinguishing phosphorescence from fluorescence. The process by which molecules return to the ground state in phosphorescence is a different process from fluroescence and usually takes longer.
2. In general usage the term is applied to the emission of 'cold light' – light produced without a high temperature. The name comes from the fact that white phosphorus glows slightly in the dark as a result of a chemical reaction with oxygen. The light comes from excited atoms produced directly in the reaction – not from the heat produced. It is thus an example of *chemiluminescence*. There are also a number of biochemical examples (bioluminescence); for example, phosphorescence is sometimes seen in the sea from marine organisms, or on rotting wood from certain fungi (known as 'fox fire').

phosphorus /**fos**-fŏ-rŭs/ A reactive solid non-metallic element. There are three common allotropes of phosphorus and several other modifications of these, some of which have indefinite structures.

Symbol: P; m.p. 44.1°C (white) 410°C (red under pressure); b.p. 280.5°C; r.d. 1.82 (white) 2.2 (red) 2.69 (black) (all at 20°C); p.n. 15; r.a.m. 30.973762.

phot /foht, fot/ A unit of illumination in the c.g.s. system, equal to an illumination of one lumen per square centimeter. It is equal to 10^4 lux.

photocathode /foh-toh-**kath**-ohd/ A cathode that emits electrons by the photoelectric effect.

photocell /**foh**-tŏ-sell/ Any device for producing an electric signal from electromagnetic radiation. Originally, photocells were photoelectric cells – i.e. devices in which a current was produced between two electrodes by the photoelectric effect. The present common type depends on photoconductivity. A piece of semiconductor material is held between two contacts and a voltage applied. In the absence of radiation the current is very small; when radiation falls on the sample, its resistance is reduced and the current increases. Photoconductive cells, unlike photoelectric cells, can be used in the infrared region. Other types of photocell are photodiodes or depend on the photovoltaic effect. *See also* photoelectric cell.

photoconductivity /foh-toh-kon-duk-**tiv**-ă-tee/ The increase of electrical conductivity in certain materials, produced by incident electromagnetic radiation. The photons absorbed by the solid give up energy to electrons, freeing them and thus increasing the number of electrons in the conduction band. *See* energy bands.

photodiode /foh-toh-**dÿ**-ohd/ A semiconductor diode that is sensitive to electromagnetic radiation. The p-n junction is reverse biased. When radiation falls on it, electron-hole pairs are created. Some of these additional charge carriers are immediately swept across the junction before they recombine, constituting a current that depends on the illumination. *See also* phototransistor.

photodisintegration /foh-toh-dis-in-tĕ-**gray**-shŏn/ The ejection of particles from atomic nuclei on absorption of electromagnetic radiation. Most commonly a neutron is ejected.

photoelasticity /foh-toh-ĕ-las-**tis**-ă-tee/ The double refraction observed in certain materials only when stressed. A number of synthetic materials are photoelastic, examples being Perspex and Cellophane. Polarized white light passed through a stressed sample shows colored fringes relating to the stress pattern. This is a very useful engineering test technique. Glass also shows interference fringes and the method is used for locating strains in glass laboratory apparatus.

photoelectric cell /foh-toh-i-**lek**-trik / A device in which an electric current is produced by electromagnetic radiation (visible or ultraviolet) by the photoelectric effect. A light-sensitive cathode (photocathode) and an anode are placed in an evacuated glass envelope. The photosurface contains material with low work function (e.g. potassium or cesium) and emits electrons when irradiated. A positive potential on the anode causes a current to flow between the electrodes and in an external circuit. The device used to be called a PHOTOCELL. In most applications it has been superseded by the photoconductive cell.

photoelectric effect The ejection of electrons from the surface of substances by ultraviolet radiation (or occasionally light or even infrared), or from atoms within a substance by x- or gamma radiation. The number of electrons emitted is proportional to the intensity of radiation at a given frequency. There is normally no emission below a certain threshold, and above this the amount depends in a complicated way upon frequency. The maximum kinetic energy (W) of the ejected electrons is given by Einstein's equation: $W = h\nu - \phi$ where h is the Planck constant and ϕ is the minimum energy required to remove an electron. In the case of the emission of valence electrons from solids and liquids by optical radiations ϕ is called the *work function*. For the removal of electrons from individual atoms by high quantum energy radiations ϕ is the binding energy of the electron. The threshold frequency ν_0 is given by: $h\nu_0 = \phi$.

For ultraviolet and visible light, only a very small proportion of the incident radiation causes photoelectric emission. For x-rays the probability of the effect is proportional to the fourth power of the proton number, hence the use of lead (Z = 82) in shielding.

Einstein (1905) proposed his equation as an example of the concept that electromagnetic radiation interacts with matter as quanta (photons) with energy $h\nu$. At normal intensities one photon ejects an electron. It is now possible to eject electrons by radiation with quantum energy far below ν_0 using the great intensities obtainable by focusing laser radiation. In this case one may suppose that many photons interact with one electron. *See* duality; photoionization; quantum theory.

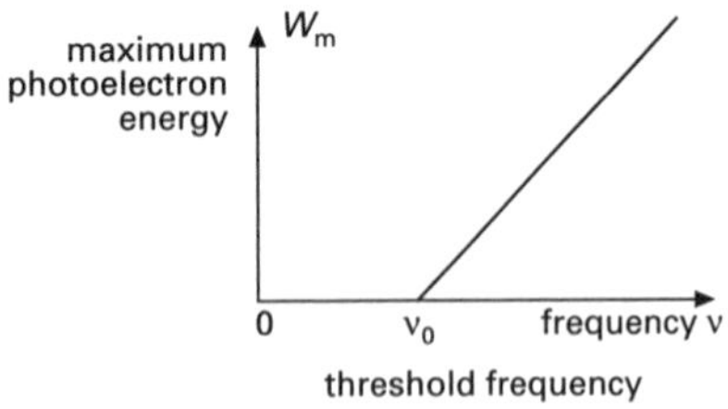

Photoelectric effect

photoelectricity /foh-toh-i-lek-**tris**-ă-tee/ A group of phenomena in which electric effects are produced by electromagnetic radiation. *See* photoconductivity; photoelectric effect; photoionization; photovoltaic effect.

photoelectron /foh-toh-i-**lek**-tron/ An electron ejected from a solid, liquid, or gas by the photoelectric effect or by photoionization.

photoemission /foh-toh-i-**mish**-ŏn/ The emission of photoelectrons by the photoelectric effect or by photoionization.

photography The production of a permanent record of an image on suitable paper or film. Normally the image is produced by the optical system of some type of camera on the light-sensitive surface of an emulsion. This is usually a layer of gelatine containing grains of silver bromide coated on a sheet of glass, paper, or plastic. Exposure to radiation releases minute quantities of silver in some of the grains forming a latent image. The material is placed in a developer, that is a solution containing a weak reducing agent. The silver of the latent image acts as a catalyst, promoting the reduction of the grains to particles of silver. The remaining silver salt is then dissolved by a fixing solution and the material is washed and dried giving the permanent photograph.

Photographic emulsions are 'sensitive' to electromagnetic radiations of wavelengths less than around 1.5 micrometers. They are also sensitive to other energetic ionizing radiations, and are important in x-ray investigations and in nuclear physics. Color photography involves mixing specified dyes into the silver halide–gelatine emulsion.

photoionization /foh-toh-ÿ-ŏ-ni-**zay**-shŏn/ The ionization of atoms or molecules by electromagnetic radiation. Photons absorbed by an atom may have sufficient photon energy to free an electron from its attraction by the nucleus. The process is $M + h\nu \rightarrow M^+ + e^-$. As in the photoelectric effect, the radiation must have a certain minimum threshold frequency. The energy of the photoelectrons ejected is given by $W = h\nu - I$, where I is the ionization potential of the atom or molecule. *See* photoelectric effect.

photoluminescence /foh-toh-loo-mă-**ness**-ĕns/ *See* luminescence.

photometer /foh-**tom**-ĕ-ter/ An instrument for determining illumination or luminous intensity by comparisons made with the eye. There are various types. The general principle is to compare a source with a standard source. The two sources each illuminate a screen, and the positions of the sources are adjusted until both give equal illumination. The illumination (E) is related to luminous intensity by $E = (I \cos i)/d^2$, where I is the luminous intensity, d the distance from the source to surface, and i the angle of incidence.

photometry /foh-**tom**-ĕ-tree/ The branch of physics concerned with measuring intensity, illumination, etc., for visual radiation.

There are two basic types of measurement. 'Luminous' quantities, such as luminous flux, luminous intensity, and illumination, are based on measurements made with the eye (e.g. by comparing the illumination of two surfaces). They thus depend on the sensitivity of the eye to different wavelengths. 'Radiant' quantities (e.g. radiant flux, radiance, and irradiance) depend on absolute measures of energy flow made by photocells. *See also* Lambert's law.

photomultiplier /foh-toh-**mul**-tă-plÿ-er/ A device in which electrons originally emitted from a photocathode initiate a cascade of electrons by secondary emission in an electron multiplier. It is much more sensitive as a radiation detector than a single photoelectric cell. *See also* photoelectric cell.

photon /**foh**-ton/ *See* duality.

photonics /foh-**tonn**-niks/ The study and development of devices that use light rather than electrons in circuits. An example is the use of light transmission through optical fibers in telecommunications.

photopic vision /foh-**top**-ik/ Vision at normal levels of light intensity (as in normal daylight). Under such conditions the cones in the retina are the main receptors. *Compare* scotopic vision.

photosphere /**foh**-tŏ-sfeer/ The visible surface layer of a star. In the Sun this layer consists of gas at an average temperature of about 5750 K. It extends this layer for a few hundred kilometres from the surface and radiates a continuous spectrum.

phototransistor /foh-toh-tran-**zis**-ter/ A transistor that is sensitive to electromagnetic radiation. When radiation falls on the emitter part of the device, more charge carriers are produced in the base region and the collector current increases. A phototransistor acts like a photodiode with an amplifier. *See also* photodiode.

photovoltaic effect /foh-toh-vol-**tay**-ik/ The effect in which irradiation of a p-n junction or the junction of a metal and a SEMICONDUCTOR by electromagnetic radiation (ultraviolet to infrared) generates an e.m.f., which can be used to deliver power to an external circuit. A solar cell consists of such a p-n junction.

physical change A reversible change in a substance's properties (such as melting), as opposed to a chemical change, which results in a change in composition that is difficult to reverse.

physical constants *See* fundamental constants.

physical optics *See* optics.

physisorption /fiz-ă-**sorp**-shŏn, **-zorp-**/ *See* adsorption.

pico- /**pÿ**-koh/ Symbol: p A prefix denoting 10^{-12}. For example, 1 picofarad (pF) = 10^{-12} farad (F).

piezoelectric effect /pee-ay-zoh-i-**lek**-trik, pÿ-ee-zoh-/ The generation of a small potential difference across certain materials when they are subjected to a stress. The effect is used in devices for pressure and vibration measurement, in gas lighters, in crystal microphones, in strain gauges, and in oscillators. *Compare* magnetostriction.

pi-meson *See* pion.

pin-hole camera A box into which light can enter only through a very small hole. Because light rays travel in straight lines in a given medium (air in this case), an inverted image is formed on the wall of the box opposite the pin hole. The wall may carry a photographic plate, or be made of a diffusing material such as tissue paper or ground glass.

Because the hole is small, the image is very faint – a long exposure is needed to make a photograph. However the image is in sharp focus, whatever the distance of the object. This is because the rays pass through the hole without deviation. Unlike the lens of a lens camera, the pin hole gives no image aberrations. If the hole is made larger, the image becomes brighter – but more blurred. A large hole can be thought of as a set of pin-holes – each pin hole gives a faint image, but the images are in slightly different positions.

pink noise *See* noise.

pion /**pÿ**-on/ (pi-meson) A type of MESON. There are three types of pion, having positive, negative, or zero charge. The charged pions have rest energy 139.6 MeV and mean life 2.6×10^{-8} s. Each decays to give a muon and a neutrino. The neutral pion has a rest energy 135 MeV and has mean life 8×10^{-17} s. It decays by emitting electromagnetic radiation. Pions are strongly interacting bosons with zero spin. *See also* elementary particle.

pitch The sensation that a sound produces in a listener as a result of its frequency (though other factors are involved). High-pitched notes are high-frequency vibrations and low-pitched notes are low-frequency vibrations.

Pitot tube /**pee**-toh/ A double tube placed with one end in a stream of fluid, parallel to the flow, to measure fluid pressure. It has two openings. One is usually at the side, in the outer section of a double-walled tube, and measures static pressure, p. The other, in the inner section, faces the fluid flow and registers the total pressure (the sum of p and dynamic pressure). The other ends are connected to pressure-measuring devices. Pitot tubes can be used to measure fluid velocity. The tube is named for the French physicist Henri Pitot (1695–1771). *See* Bernoulli effect.

Planck constant /plank/ Symbol: h A fundamental constant; the ratio of the energy (W) carried by a photon to its frequency (ν). A basic relationship in the quantum theory of radiation is $W = h\nu$. The value of h is $6.626\,069\,3 \times 10^{-34}$ joule second. The Planck constant appears in many relationships in which some observable measurement is quantized (i.e. can take only specific separate values rather than any of a range of values). The constant is named for the German physicist Max Planck (1858–1947).

Planck's formula *See* Planck's radiation law.

Planck's radiation law The energy radiated per unit area per unit time per unit wavelength range at wavelength λ from a black body at kelvin temperature T is given by

$$E_\lambda = 2\pi hc^2\lambda^{-5}/[\exp(hc/\lambda kT) - 1]$$

where h is the Planck constant, k the Boltzmann constant and c the speed of light in a vacuum.

This law was first proposed by Planck (1900). The problem of the energy distribution in black-body radiation had been studied for many years. The physical system considered is of extreme simplicity, and the failure of the existing statistical system of Maxwell and Boltzmann to solve the problem was rightly considered to be very significant (*see* ultraviolet catastrophe). Planck deduced his equation by assuming that electromagnetic radiation was not emitted and absorbed continuously as previously supposed but in quanta of energy hc/λ.

Planck's formula was found to be in good agreement with experiment and it permitted the first reasonably accurate measurement of the Boltzmann constant, and hence of related quantities such as the charge of the electron and the masses of atoms. Little progress however was made with developing the ideas until Einstein (1905) argued that the quantum hypothesis would explain the photoelectric effect and Stokes' law of fluorescence, and later proposed other applications. This work led eventually to the development of quantum mechanics.

See Bose–Einstein statistics.

plane of polarization For historical reasons, this is defined as the plane containing incident and reflected rays in cases of polarization by reflection. It is therefore the plane containing the magnetic vector B, rather than the electric vector E.

plane polarization A type of POLARIZATION of electromagnetic radiation, normally described with reference to a ray. In plane-polarized radiation, the magnetic oscillations are all parallel and occur in one plane containing the ray. The electric oscillations are all parallel and occur in a plane containing the ray and at right angles to the plane containing the magnetic oscillations. The electric and magnetic fields are both perpendicular to the direction of propagation. Plane polarization can be produced by reflection or by transmission through a Nicol prism or Polaroid. *See also* circular polarization.

planetoid *See* minor planet.

plano-concave lens /play-noh-**kon**-kayv/ A diverging LENS with one plane face and once concave face.

plano-convex lens /play-noh-**kon**-veks/ A converging LENS with one plane face and one convex face.

plasma /**plaz**-mă/ A mixture of free electrons and ions or atomic nuclei. Plasmas occur in thermonuclear reactions, as in the Sun. The glowing region of ions and electrons in a discharge tube is also a plasma. Sometimes plasmas are referred to as a fourth state of matter.

plasticity /plas-**tis**-ă-tee/ The tendency of a material to suffer a permanent deformation; i.e. not to return to its original dimensions after a deforming stress has been removed. An elastic material becomes plastic above its yield point. *See* elasticity.

platinum /**plat**-ă-nŭm/ A silvery-white malleable ductile transition metal. It occurs naturally either free or in association with other platinum metals. Platinum is used as a catalyst for ammonia oxidation (to make nitric acid) and in catalytic converters. It is also used in jewelry.

Symbol: Pt; m.p. 1772°C; b.p. 3830 ± 100°C; r.d. 21.45 (20°C); p.n. 78; r.a.m. 195.08.

platinum black A finely divided black form of platinum produced, as a coating, by evaporating platinum onto a surface in an inert atmosphere. Platinum-black coatings are used as pure absorbent electrode coatings in experiments on electric cells. They are also used, like carbon-black coatings, to improve the ability of a surface to absorb radiation.

plutonium /ploo-**toh**-nee-ŭm/ A radioactive silvery element of the actinoid series of metals. It is a transuranic element found on Earth only in minute quantities in uranium ores but readily obtained, as ^{239}Pu, by neutron bombardment of natural uranium. The readily fissionable ^{239}Pu is a major nuclear fuel and nuclear explosive. Plutonium is highly toxic because of its radioactivity; in the body it accumulates in bone.

Symbol: Pu; m.p. 641°C; b.p. 3232°C; r.d. 19.84 (25°C); p.n. 94; most stable isotope ^{244}Pu (half-life 8.2×10^7 years).

p-n-p transistor *See* transistor.

point defect *See* defect.

point source A source of exactly spherical wave fronts. All sources can be considered to be point-like if viewed from a large enought distance. The stars provide an obvious example.

In a number of practical situations, sources that are effectively point sources are required. Normally a small hole in an otherwise opaque illuminated surface is used. Holes can be made as small as 100 μm; the main problem is to arrange adequate illumination of the hole by the real source.

poise Symbol: P A c.g.s. unit of dynamic viscosity. It is equal to 0.1 pascal second. It is named for the French physician and physicist Jean Léonard Marie Poiseuille (1797–1869).

Poiseuille's formula /**poi**-zew-eelz, pwa-ză-**eelz**/ (**Hagen's formula**) An equation that describes the laminar flow of liquid through a pipe. The volume flow per second in a circular pipe of radius r and length l is given by:

$$\pi r^4 \Delta p / 8\eta l$$

where Δp is the fluid pressure difference and η is its viscosity. It is named for Poiseuille (*see* poise) and for the German physicist Gotthilf Heinrich Ludwig Hagen (1797–1884).

Poisson equation /**pwass**-awn/ The fundamental equation of POTENTIAL THEORY. In the case of electrostatics it states that $\nabla^2 V = \rho/\varepsilon$, where ∇^2 is the Laplacian operator, V is the electric potential, ρ is the electric charge density, and ε is the electric permittivity. There is an analogous equation in the case of the gravitational potential. When the right hand of the Poisson equation is zero the Poisson equation becomes the LAPLACE EQUATION. The equation was first put forward by the French mathematician and physicist Siméon-Denis Poisson in 1813.

Poisson ratio Symbol: υ A measure of how a material changes shape when it is stretched. It is equal to minus the fractional change in cross-sectional area ($\Delta a/a$) divided by the fractional change in length ($\Delta l/l$). For many solids, including many metals, the value is about 0.3. It cannot exceed 0.5 for any material. The ratio is named for the French mathematician and physicist Siméon-Denis Poisson (1781–1840).

Poisson's spot A bright spot observed in the middle of the shadow of a circular opaque obstacle. This spot was shown to be a consequence of the wave theory of light by Siméon-Denis Poisson in 1818. It was discovered experimentally by his colleague Dominique François Jean Arago (and is sometimes known as *Arago's spot*).

polar coordinates A method of defining the position of a point by its distance and direction from a fixed reference point (pole). The direction is given as the angle between the line from the origin to the point, and a fixed line (axis). On a flat surface only one angle, θ, and the radius, r, are needed to specify each point. For example, if the axis is horizontal, the point $(r,\theta) = (1,\pi/2)$ is the point one unit length away from the origin in the perpendicular direction. Conventionally, angles are taken as positive in the anticlockwise sense.

In a rectangular Cartesian coordinate system with the same origin and the x-axis at $\theta = 0$, the x- and y-coordinates of the point (r,θ) are

$$x = r\cos\theta$$
$$y = r\sin\theta$$

Conversely

$$r = \sqrt{(x^2 + y^2)}$$

and

$$\theta = \tan^{-1}(y/x)$$

In three dimensions, two forms of polar coordinate systems can be used. *See* cylindrical polar coordinates; spherical polar coordinates. *See also* Cartesian coordinates.

polarimeter /poh-lă-**rim**-ĕ-ter/ (**saccharimeter**) A device for measuring the angle of rotation of a plane-polarized beam caused by an optically active sample. Typically, light from a source is passed through a Nicol prism (the polarizer) to plane-polarize it, then through the sample. The amount of rotation is measured by a second prism (the analyzer), which can be turned on an angular scale. The position of minimum transmission is observed. As this angle depends, among other things, on the concentration of an active solute, polarimeters are used to measure this. *See also* optical activity.

polarization /poh-lă-ri-**zay**-shŏn/ Restriction of the vibrations in a transverse wave. Normally in a transverse wave the vibrations can have any direction in the plane perpendicular to the direction of propagation. If the directions of the vibrations are restricted in any way the radiation is said to be polarized. The simplest case is that of PLANE POLARIZATION. In a plane-polarized transverse acoustic wave in a solid, all the vibrations are parallel to each other. In plane-polarized electromagnetic radiation, all the electric oscillations are parallel to each other and at right angles to the magnetic oscillations.

Light can be plane-polarized by reflection or on passing through certain substances (e.g. Polaroid). Light polarized in a certain plane by passing through one polarizer is totally absorbed by a second polarizer set to polarize at right angles. Electromagnetic waves such as light and those of shorter wavelength generally interact with matter by means of their electric fields (although energy is transferred equally by both). Thus, for example, the directions in which photoelectrons are ejected may depend upon the plane of polarization.

Radio waves are normally emitted plane-polarized. Thus if the electric vibrations are vertical a rod-shaped antenna, which responds to these vibrations, must be set vertically to detect them. An antenna consisting of a coil with ferrite core will have to be set so that the magnetic vibrations are perpendicular to its plane.

See also circular polarization; elliptical polarization; birefringent crystal; Nicol prism; optical activity; plane of polarization.

polarization, angle of *See* Brewster angle.

polarization, electric The separation of the charges in the molecules of an insulator as an effect of an electric field. One face of an insulator in a field gains a net positive charge with the other becoming negative.

polarization, electrolytic Changes in the characteristics of a voltaic cell when it produces large currents. Generally the e.m.f. is reduced while the internal resistance may increase (or in certain cases, decrease). In extreme cases a layer of bubbles of hydrogen may form on the cathode. Many cells contain a depolarizer such as manganese(IV) oxide to reduce this effect. Polarization usually disappears on resting the cell.

polarized light *See* polarization.

polarizer /**poh**-lă-rÿ-zer/ A device or arrangement producing plane-polarized radiation from incident unpolarized radiation. A Nicol prism or Polaroid sheets are normally used.

polarizing angle *See* Brewster angle.

Polaroid (*Trademark*) A synthetic doubly refracting substance, that strongly absorbs polarized light in one plane, while easily passing polarized light in another plane at right angles. Unpolarized light passed through a sheet of Polaroid is plane-polarized. Spectacle lenses made of this material normally absorb light vibrating horizontally – as produced by reflection from horizontal surfaces. They thus reduce reflected glare.

The first type was announced by Land in 1934. The modern form is made from a plastic sheet, highly strained to align the molecules and make it birefringent, then stained with iodine to make it dichroic.

pole (**optical center; optic center**) The geometric center of a surface or lens. In ray diagrams rays can be drawn undeviated through the pole of a LENS. At the pole of a MIRROR the incident and reflected rays make equal angles with the principal axis. All distances from a refracting surface, reflector, or lens – object and image distances, focal distances, radii of curvature, etc. – are measured from the pole.

P= pole, or optic center

Pole of a lens

poles, magnetic The regions of a magnetic field where the forces appear strongest. Thus, in a bar magnet, the lines of force appear to diverge from and converge to two small regions near the ends, these being the poles of the magnet. The concept of magnetic 'pole', however, has no more reality than that of 'regions in the field', as in the first sentence above.

A north-seeking pole (*north pole* or *N-pole*) is attracted approximately in the direction of geographical North; a south-seeking pole (*south pole* or *S-pole*) tends to move toward geographical South. Magnetic poles can be found only in opposite (unlike) pairs. The Earth's magnetic poles – regions where the Earth's field is strongest – are close to the geographical poles.

pole strength The force exerted by a given magnetic pole when it is one meter from a unit pole in a vacuum. A unit pole is defined as that pole which, when placed one meter from a similar pole in a vacuum, exerts a force of one newton. It is equal to 125.663 7 microweber (μWb). *See also* poles, magnetic.

polonium /pŏ-**loh**-nee-ŭm/ A radioactive metallic element belonging to group 16 of the periodic table. It occurs in very minute quantities in uranium ores. Over 30 radioisotopes are known, nearly all alpha-particle emitters. Polonium is a volatile metal and evaporates with time. It is also strongly radioactive; a quantity of polonium quickly reaches a tempertaure of a few hundred degrees C because of the alpha emission. For this reason it has been used as a lightweight heat supply in space satellites.

Symbol: Po; m.p. 254°C; b.p. 962°C; r.d. 9.32 (20°C); p.n. 84; stablest isotope ^{209}Po (half-life 102 years).

polyatomic /pol-ee-ă-**tom**-ik/ Describing a molecule that consists of several atoms (three or more). Examples are benzene (C_6H_6) and methane (CH_4).

polychromatic radiation /pol-ee-krŏ-**mat**-ik/ Electromagnetic radiation that has a mixture of different wavelengths. *Compare* monochromatic radiation.

polygon of vectors A closed polygon that represents a set of vectors in equilibrium. Each vector is represented, both in size and direction, by an arrow, with the arrows being arranged head to tail. If the arrows did not close to form a complete polygon then there would be a net force and the forces would not be in equilibrium. The TRIANGLE OF VECTORS is a special case.

polymer A compound consisting of large molecules in which basic molecular units are repeated. A common example is the synthetic plastic polythene, which is a polymer of the compound ethylene ($CH_2 = CH_2$). Many similar synthetic polymers have been produced, but natural polymers are of fundamental importance in biochemistry. In physics, the physical properties of polymers can be analyzed theoretically using the methods of statistical mechanics.

population inversion *See* laser.

positive *See* charge.

positive column A luminous area in an electrical discharge in a gas, occurring near the positive electrode.

positive feedback *See* feedback.

positive lens A lens with a positive power. *See* converging lens.

positive mirror A curved mirror with a positive power. *See* converging mirror.

positive pole A magnetic N-pole. *See* pole.

positive rays Streams of positive ions in an electrical discharge.

positron /**poz**-ă-tron/ The antiparticle of the electron; a particle with the same mass as the electron but with a positive charge. Positrons are found in cosmic-ray showers, where they result from pair production. They are also produced by a type of beta decay. They are annihilated when they encounter an electron, so have a short separate existence. Positrons can be detected by their tracks in cloud or bubble chambers, where they show the opposite deflection to electrons in a magnetic field. *See* elementary particles.

positronium /poz-ă-**troh**-nee-ŭm/ A fleeting combination of an electron and a positron to form an analog to a hydrogen atom. When the two particles have their spins parallel the half-life is about 1.5×10^{-7} s; when they are antiparallel the half-life is shortened to 10^{-10} s. A positronium 'atom' decays to form two photons by annihilation. A combination of two electrons and two positrons also appears to exist, and is known as a positronium 'molecule', analogous to a hydrogen molecule.

Post Office box A box containing resistances that can be switched into the circuit, suitable for use as a WHEATSTONE BRIDGE or potentiometer.

potassium /pŏ-**tass**-ee-ŭm/ A soft reactive metal. The atom has the argon electronic configuration plus an outer $4s^1$ electron.

Symbol: K; m.p. 63.65°C; b.p. 774°C; r.d. 0.862 (20°C); p.n. 19; r.a.m. 39.0983.

potassium–argon dating A method of radioactive dating used for estimating the age of certain rocks. It is based on the decay of ^{40}K to ^{40}Ar (half-life 1.28×10^9 years). The technique is useful for time periods from 10^5 years ago back to the age of the Earth (4.6×10^9 years).

potential difference (**p.d.**) Symbol: *V* The difference in electric potential between two points; the work done per unit charge when a charge is displaced between the points, by any path. The SI unit is the volt. In a vacuum, the total energy of the charge is constant; the change in potential energy is balanced by change of kinetic energy. In a material the average kinetic energy of the displaced charges remains constant, because of resistance, and energy is dissipated. Potential difference can also be expressed as power divided by current. *See also* electromotive force.

potential divider *See* voltage divider.

potential energy Symbol: *V* The work an object can do because of its position or state. There are many examples. The work an object at height can do in falling is its gravitational potential energy. The ENERGY 'stored' in elastic or a spring under tension or compression is elastic potential energy. Potential difference in electricity is a similar concept, and so on. In practice the potential energy of a system is the energy involved in bringing it to its current state from some reference state, or vice versa.

potential theory The theory of physical fields, such as electric, gravitational, and magnetic fields, in terms of the potentials of these fields. In the cases of electric and gravitational fields, the potentials are described by the POISSON EQUATION and the LAPLACE EQUATION.

potentiometer /pŏ-ten-shee-**om**-ĕ-ter/ An instrument for measuring electrical potential difference by balancing two opposing potentials, so that no current flows through a galvanometer. In a simple form a battery maintains a steady current in a uniform resistance wire. A sliding contact connected with a standard cell (as shown in the diagram) is moved until the galvanometer is undeflected, the contact being at distance L_s from the left-hand end. The unknown p.d. *V* now replaces the standard cell, and the new distance L_v for zero current is found. Then $V = ELv/L_s$; where *E* is the e.m.f. of the standard cell.

In more accurate instruments the resistance wire is replaced by a series of high-grade resistors. The instrument is then capable of extremely high precision. One additional advantage is that it draws very little current from the system under test. (If a galvanometer is not sensitive enough an electrometer is used.)

Current can also be measured by finding the p.d. across a standard resistance. Also resistances can be compared by measuring the potential differences across them when connected in series.

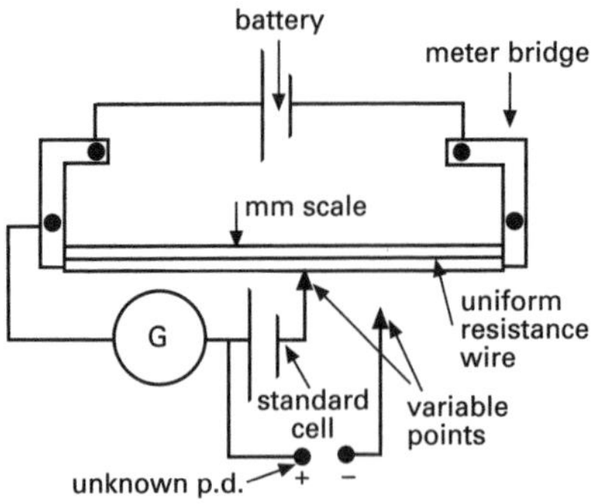

Potentiometer

pound **1.** An f.p.s. unit of weight or mass in the avoirdupois system, subdivided into 16 ounces and equal to 0.453 592 37 kg in SI units.
2. An f.p.s. unit of weight or mass in the troy system, subdivided into 12 ounces and equal to 0.373 241 72 kg in SI units.

poundal Symbol: pdl The unit of force in the f.p.s. system. It is equal to 0.138 255 N.

pound force Symbol: lbf An f.p.s. unit of gravitational force, equal to the force exerted on a one-pound mass by gravity at sea level. It is equal to 9.806 65 meters per second squared or, in SI derived units, to 4.448 222 newtons.

power (**focal power**) Symbol: *P* A measure of the ability of a lens or mirror to converge a parallel beam, given by the reciprocal of the focal distance *f*.

$$P = 1/f$$

Generally, f is in meters in which case P is in DIOPTERS (D).

Strictly, power measures the ability of a reflecting or refracting surface, or of a lens, to change the curvature of incident wave fronts..

power Symbol: P The rate of energy transfer (or work done) by or to a system. The unit of power is the watt – the energy transfer in joules per second. In an electrical system power is given by VI, where V is the potential difference across a conductor and I the current through it. If V and I are not in phase the power absorbed is $IV\cos\phi$, where ϕ, the phase angle, is known as the 'power factor'.

Poynting vector Symbol: S The vector product of the electric field vector $\boldsymbol{E}$ and the magnetic field vector $\boldsymbol{H}$ (= $\boldsymbol{B}_0/\mu_0$) in an electromagnetic wave. The Poynting vector gives, in magnitude and direction, the power radiated through unit area at any instant. The unit is the watt per square meter. In a simple harmonic wave, the average value is $\frac{1}{2}\boldsymbol{E}_0\boldsymbol{H}_0$, where $\boldsymbol{E}_0$ and $\boldsymbol{H}_0$ are the amplitudes. The vector is named for the British physicist John Henry Poynting (1852–1914).

praseodymium /pray-zee-oh-**dim**-ee-ŭm/ A soft ductile malleable silvery element of the lanthanoid series of metals. It occurs in association with other lanthanoids. Praseodymium is used in several alloys, as a catalyst, and in enamel and yellow glass for eye protection.

Symbol: Pr; m.p. 931°C; b.p. 3512°C; r.d. 6.773 (20°C); p.n. 59; r.a.m. 140.91.

precession /pri-**sesh**-ŏn/ If a body is spinning on an axis, the axis of rotation can itself move around another axis at an angle to it. This occurs when a couple is applied about an axis which is always perpendicular to the axis of rotation. For gyroscopes and tops this couple is caused by gravity and the support force. The gravitational fields of the Moon and Sun acting on the equatorial bulge of the Earth causes the *precession of the equinoxes*, that is the axis of rotation of the Earth sweeps out a cone with semi-angle 23½° in a period of 26 000 years. Similarly the planes of the orbits of planets and satellites precess. A normal to the plane of the orbit of the Moon sweeps out a cone of semi-angle 5° and period 18.6 years, causing the sequence of eclipses.

presbyopia /prez-bee-**oh**-pee-ă/ The normal development with age of far sightedness. The distance to the eye's near point increases as the eye's ability to accommodate is reduced (the lens becomes more rigid with age). *See* near point.

pressure Symbol: p A quantity that expresses the condition of a fluid such that it exerts a perpendicular force on any surface in contact with it: pressure = force/area. Gases can only exert repulsive forces so their pressure is always positive. Liquids also normally have positive pressure, but in some circumstances liquids free from dissolved gas can sustain a tension, that is a negative pressure. The SI unit of pressure is the pascal.

Uniform isotropic solids within fluids are compressed uniformly (without distortion) so the concept of pressure can be extended to them. More usually, especially when solids interact with each other, the bodies are distorted by uneven systems of forces. There is then a STRESS, which is distinct from fluid pressure. When the *pressure* on a body is increased the distances between the molecules all decrease in the same proportion, whereas for stresses in solids, different separations change differently. For example, when a heel thrusts downward upon a floor the molecules move closer together in the vertical direction and farther apart horizontally. Around the edge of the heel the floor material is in tension; it is here that damage is usually done. Similarly, the human body is insensitive to changes in fluid pressure yet detects slight local stresses caused by gentle contact with a solid.

On going downward in a fluid the pressure increases by an amount equal to: increase in depth × mean density × g. The density of a liquid can usually be assumed to be constant, but for gases it is necessary to integrate. At uniform temperature this gives an exponential variation of pressure with depth.

See also gas laws; pressure of the atmosphere; upthrust.

pressure broadening The increase in the width of a spectral line resulting from collisions between atoms or molecules. It relates to pressure, as there are more collisions at high pressure. *See* monochromatic radiation.

pressure gauge An instrument for measuring fluid pressure, such as a BAROMETER or MANOMETER. *See* Bourdon gauge.

pressure of the atmosphere The pressure at a point near the Earth's surface due to the weight of air above that point. Its value varies around about 100 kPa (100 000 newtons per square meter). Barometers are used to measure atmospheric pressure, which is important because the small changes relate to imminent weather changes.

Because the pressure at depth in a fluid depends on depth, barometers can be used as altimeters; they can be marked to indicate distance above or below sea level.

pressurized-water reactor (**PWR**) A nuclear reactor in which water, used as moderator, coolant, and reflector, is kept under pressure to prevent it from boiling at the operating temperature. The hot water is used to raise steam in a heat exchanger to operate a steam turbine. This is by far the most common form of reactor used for electricity generation and for nuclear submarines. The fuel is enriched, typically containing a few percent of uranium–235 in the form of UO_2. *Compare* boiling-water reactor.

Prévost's theory of exchanges /pray-vosts/ The theory that all bodies emit and absorb radiation continuously. The rate of emission depends on the body's surface and its temperature. The rate of absorption depends on the surface and the temperature of the surroundings. If there is no other process of energy transfer and the temperature of the surroundings is uniform and constant, the body will eventually attain the same temperature as that of the surroundings. The rates of emission and absorption of radiation must then be equal. This shows that the absorptance of a surface equals its emissivity. The theory is named for the Swiss physicist Pierre Prévost (1751–1839). *See also* black body; Kirchhoff's law of radiation.

primary cell A voltaic cell in which the chemical reaction that produces the e.m.f. is not reversible. This is not the case with a secondary cell (accumulator).

primary colors A set of three colored lights (hues) that when mixed together in the right proportions, give the sensation of white. Normally the hues are chosen from the red, green, and blue regions.

An infinite number of sets of primary hues exists. The standard choice is one of convenience. According to the trichromatic theory of color, any set of three primary colors has only one requirement – no one of them can be matched by mixing the other two. *See also* color.

primary winding *See* transformer (voltage).

principal axis (**axis**) The line joining the centers of curvature of the faces of a LENS, or the line normal to a reflector at the pole. *See also* mirror.

principal focus *See* focal point.

principal plane The plane perpendicular to the principal axis of a lens centered on the pole (optical center). The pole is equidistant between the principal focal points. This is strictly true only for a thin lens. A thick lens has two principal planes; each is centered on a principal point. The focal distance of a thick lens is the distance between each focal point and the principal plane on that side of the lens.

principal points Two points of a thick lens corresponding to the single pole (opti-

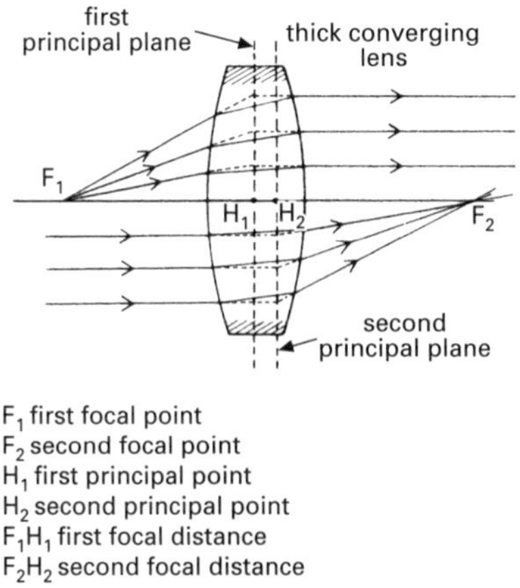

Principal plane

cal center) of a thin lens. *See* principal plane.

principle of least action A VARIATIONAL PRINCIPLE regarded as the fundamental principle underlying Newtonian mechanics, stating that the integral over time of the kinetic energy minus the potential energy (the *action*) for a system is always less for the actual motion of the system than for any other possible motion.

principle of moments When an object or system is in equilibrium, the sum of the moments in any direction equals the sum of the moments in the opposite direction. Because there is no resultant turning force, the moments of the forces can be measured relative to any point in the system or outside it.

principle of superposition When two (or more) waves of the same type pass through the same region, the amplitude of vibration at any point is the algebraic sum of the individual amplitudes. Sign (crest or trough, i.e. positive or negative) must be taken into account. The waves emerge from the region of superposition unaffected. *See also* interference.

printed circuit A circuit that is constructed as an entity on an insulating board

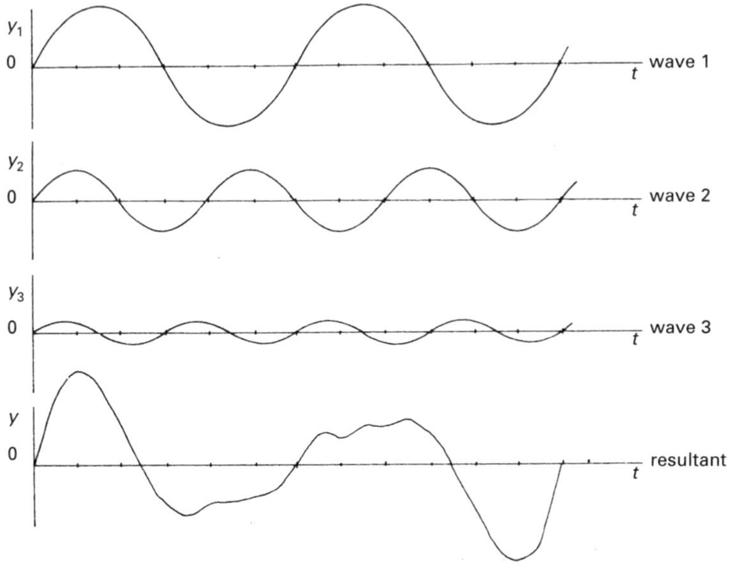

Principle of superposition

without the need for connecting wires, the connections being incorporated on the surface of the board as a conducting film. The board is first completely coated with a conductor film, usually copper. The circuit is then 'printed' on the board by a photographic process that superimposes another film, this time of protective material in a predetermined pattern. The exposed conducting film is then removed by etching, and the remainder acts as a network of conductors between components or ICs built into the circuit.

Boards are often printed on both sides, and some have two or more conducting layers, insulated from each other and separately etched. *See also* integrated circuit.

prism, optical An important component of many optical systems: a block (usually of glass for work with visible radiation) with two nonparallel sides. Prisms are often triangular in section. Prism action depends on two successive refractions, each with DEVIATION (bending) in the same direction. This is effective, for instance, in spreading different wavelengths into a spectrum. Many prisms have been designed for special purposes for use in optical instruments. *See* binoculars; erecting prism.

prism binoculars *See* binoculars.

progressive wave *See* wave.

projectile An object falling freely in a gravitational field, having been projected at a speed v and at an angle of elevation θ to the horizontal. In the special case that $\theta = 90°$, the motion is linear in the vertical direction. It may then be treated using the equations of motion. In all other cases the vertical and horizontal components of velocity must be treated separately. In the absence of friction, the horizontal component is constant and the vertical motion may be treated using the equations of motion. The path of the projectile is an arc of a parabola. Useful relations are:

time to reach maximum height

$$t = v\sin\theta/g$$

maximum height

$$h = v^2\sin^2\theta/2g$$

horizontal range

$$R = v^2\sin 2\theta/g$$

Note that similar treatment can be applied to the case of an electric charge projected into an electric field. *See also* orbit.

projectors Optical arrangements able to produce a large bright real image on a screen.

There are two main types of projector – those in which light from the source is passed through the object, and those in which the light is reflected by the object. The *opaque projector* (episcope) is the main example of the latter. The former type, the diascope, includes the overhead projector and slide and film projectors.

Light from the source, concentrated by a converging mirror, passes through the slide (or is reflected by the opaque object). A condenser (condensing lens) produces an image; the converging projection lens forms an image of this on the screen.

promethium /prŏ-**mee**-th'ee-ŭm/ A radioactive element of the lanthanoid series of metals. It is not found naturally on Earth but can be produced artificially by the fission of uranium. It is used in some miniature batteries.

Symbol: Pm; m.p. 1168°C; b.p. 2730°C (approx.); r.d. 7.22 (20°C); p.n. 61; stablest isotope ^{145}Pm (half-life 18 years).

proof plane A small shaped piece of foil with an insulating handle, used (often with an electroscope) to investigate the distribution of charge on the surface of an object.

propagation /prop-ă-**gay**-shŏn/ The transmission of energy in the form of waves, such as light, radio waves, and other forms of electromagnetic radiation, and sound.

proportional counter A counter for ionizing radiation in which the potential difference applied is high enough for multiplication of ions, so that the height of the pulse is proportional to the number of ions produced by the particle, and thus to its energy loss. A counter used in this way

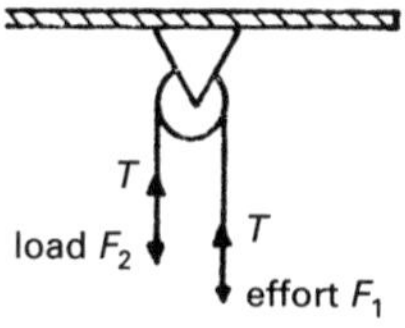

single fixed pulley

force ratio = $F_2/F_1 = T/T = 1$
distance ratio =1

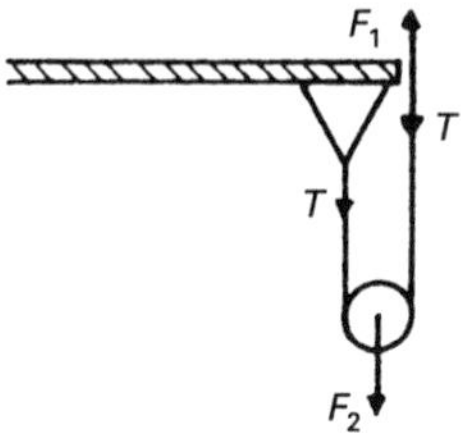

single moving pulley

force ratio = $F_2/F_1 = 2T/T = 2$
distance ratio = 2

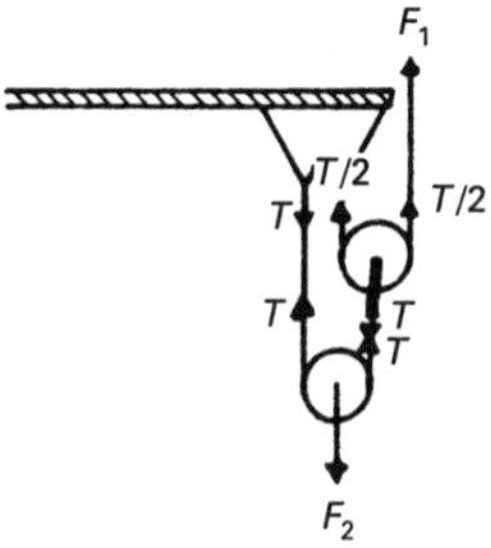

multiple moving pulleys (example)

force ratio = $F_2/F_1 = 2T/T/2 = 4$
distance ratio = 4

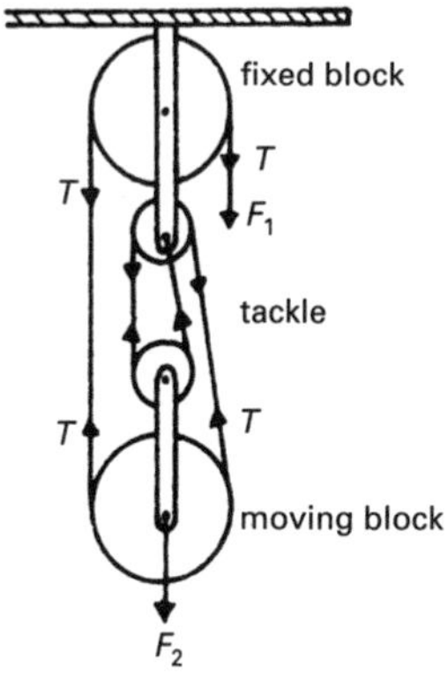

block and tackle (simplified example)

force ratio = $F_2/F_1 = 4T/T = 4$
distance ratio = 4

Pulley

its energy loss. A counter used in this way is said to be in the *proportional region.*

proportional limit *See* elasticity; Hooke's law.

proportional region *See* proportional counter.

proton An ELEMENTARY PARTICLE with a positive charge ($+1.602\,192 \times 10^{-19}$ C) and rest mass $1.672\,614 \times 10^{-27}$ kg. Protons are nucleons, found in all nuclides. They are fermions with spin ½ showing strong interactions. Protons are made up of more fundamental particles called quarks.

proton number (**atomic number**) Symbol: Z The number of protons in the nucleus of an atom. The proton number determines the chemical properties of the element because the electron structure, which determines chemical bonding, depends on the electrostatic attraction to the positively charged nucleus.

proton–proton chain reaction A series of nuclear fusion reactions by which hydrogen is converted into helium. It is thought that the reactions are responsible for energy production in the Sun and similar stars. There are in fact three possible sequences. The main one is:

$$^{1}H + {}^{1}H \rightarrow {}^{2}H + \nu + e^{+}$$
$$^{2}H + {}^{1}H \rightarrow {}^{3}He + \gamma$$
$$^{3}He + {}^{3}He \rightarrow {}^{4}He + 2{}^{1}H$$

The two other sequences lead to ^{4}He also, through ^{7}Li and ^{8}Be. *See also* carbon cycle.

psi particle /sÿ, psÿ/ *See* elementary particles.

p-type conductivity *See* semiconductor; transistor.

pulley A class of machine. In any pulley system power is transferred through the tension in a string wound over one or more wheels. The FORCE RATIO and DISTANCE RATIO depend on the relative arrangement of strings and wheels. Efficiency is not usually very high as work must be done to overcome friction in the strings and the wheel bearings, and to lift any moving wheels. There are two types of single pulley system and two types of multiple pulley system. In the diagram, T is string tension; in each case it is assumed that efficiency is 100%.

pulsar /**pul**-sar/ An astronomical body that emits electromagnetic radiation in the form of highly regular pulses. Pulsars are thought to be rotating neutron stars. The pulses are due to a 'lighthouse-beam' effect caused by synchrotron radiation from electrons in the very strong magnetic field of the neutron star.

pulsatance /**pul**-să-tăns/ *See* angular frequency.

pulse generator *See* astable circuit.

pupil /**pyoo**-păl/ The aperture of the eye, adjustable in size by the circular iris muscle. The pupil appears black as very little incident light is reflected by the retina.

pure note (**pure tone**) *See* note.

PWR *See* pressurized-water reactor.

pyrolysis /pÿ-**rol**-ă-sis, pi-/ The decomposition of chemical compounds by subjecting them to very high temperature.

pyrometer /pÿ-**rom**-ĕ-ter, pi-/ A device for measuring very high temperatures (above about 1000°C) by the radiation emitted by the body. There are two main types. *See* optical pyrometer; total-radiation pyrometer.

pyrometry /pÿ-**rom**-ĕ-tree, pi-/ The measurement of high temperatures using pyrometers.

Q

QCD *See* quantum chromodynamics.

quality of sound (**timbre**) The characteristic of a musical note that is determined by the frequencies present. It enables a listener to tell the difference between two notes of the same fundamental frequency played on different musical instruments. A pure tone has no overtones present and its waveform is a sine wave. Musical instruments produce notes that have more complex waveforms, and it can be shown that these are formed by mixing a pure tone (the fundamental) with different higher frequencies. These frequencies are simple multiples of the fundamental frequency (f); i.e. $2f$, $3f$, and so on. The fundamental together with the 'overtones' are harmonics of the note; the different contributions characterize its quality.

quantized /kwon-tÿzd/ A physical quantity is said to be quantized when it can change only in definite steps – it does not have a continuous range of values. To explain the photoelectric effect, for instance, the energy of the electromagnetic radiation must be quantized. Angular momentum is quantized in atoms and molecules.

quantum /kwon-tŭm/ (*pl.* **quanta**) A definite amount of energy released or absorbed in a process. Energy often behaves as if it were 'quantized' in this way. The quantum of electromagnetic radiation is the photon. *See also* photoelectric effect.

quantum chromodynamics (QCD) A QUANTUM FIELD THEORY that describes the strong nuclear interactions. The basic entities in QCD are QUARKS and GLUONS. QCD is mathematically much more complicated than quantum electrodynamics.

quantum computer A computer that depends on quantum mechanics in the logic of its processing. There has been a lot of interest in the idea that the indeterminacy inherent in quantum mechanics could be exploited to increase computing power. So far, these ideas are mainly theoretical.

quantum electrodynamics The use of quantum mechanics to describe how particles and electromagnetic radiation interact.

quantum entanglement *See* Bell's paradox.

quantum field theory A FIELD theory that also involves QUANTUM MECHANICS. All the non-gravitational interactions of matter can be described by quantum field theories, and these are also in accord with special relativity theory. Quantum field theory can also be used to analyze many-body problems in quantum mechanics. At present, it does not appear to be possible to have a quantum field theory that unifies all the INTERACTIONS known in Nature.

quantum Hall effect A version of the Hall effect found at low temperatures in which it is found that the coefficient of the Hall effect is quantized. In the *integer quantum Hall effect* the coefficient is quantized in integers. This can be explained in terms of non-interacting electrons. In the *fractional quantum Hall effect*, fractional number quantization for the coefficient can occur. This can be explained in terms of interactions between electrons.

quantum mechanics *See* quantum theory.

quantum number An integer or half integer that specifies the value of a quantized physical quantity (energy, angular momentum, etc.). *See* atom; Bohr theory; spin.

quantum state A state of a physical system, such as a crystal or an ATOM, specifying the condition of its constituent particles and defined by a unique set of quantum numbers. In the case of bosons any number of identical particles can occupy a given state whereas for fermions (including electrons) a state cannot contain more than one particle of a given type.

A particular quantum state under defined conditions has a certain energy, but the energy may be changed by external electric or magnetic fields. A number of states may have the same energy. For example, an atom has two quantum states with principal quantum number $n = 1$. The numbers l and m are both zero, but the spin quantum number m_s can be $+\frac{1}{2}$ or $-\frac{1}{2}$ corresponding to opposite orientations of the spin axis. A hydrogen atom is said to be in the ground state if either of these states is occupied by its one electron. There is a minute difference in energy caused by the interaction between the electron's intrinsic magnetic moment and that of the nucleus. A helium atom normally has both these states occupied. As the nucleus has no magnetic moment these states have exactly the same energy unless the atom is in a magnetic field.

When atoms are combined in a structure such as a molecule or a crystal each quantum state of each atom becomes a state of the system. Normally the energies associated with the states of the system are different from those of the separate atoms. *See* energy bands.

quantum theory A mathematical theory originally introduced by Max Planck (1900) to explain the black-body radiation from hot bodies. Quantum theory is based on the idea that energy (or certain other physical quantities) can be changed only in certain discrete amounts for a given system. Other early applications were the explanations of the photoelectric effect and the Compton effect, and the Bohr theory of the atom.

Quantum mechanics is a system of mechanics that developed from quantum theory and is used to explain the behavior of atoms, molecules, etc. In one form it is based on de Broglie's idea that particles can have wavelike properties – this branch of quantum mechanics is called *wave mechanics*.

quark A type of elementary particle that is thought to be one of the fundamental building blocks of matter. There are six *flavors* of quark, with there being two flavors in each of the three generations of elementary particles. The quarks in the first generation are called the *up quark* and the *down quark*. Those in the second generation are called the *strange quark* and the *charm quark*, while those in the third generation are called the *bottom quark* (sometimes called *beauty quark*) and the *top quark* (sometimes called *truth quark*). All quarks are spin 1/2 particles with an electric charge which is either +2/3 or −1/3. All HADRONS are made up of quarks. Quarks take part in strong interactions, as governed by QUANTUM CHROMODYNAMICS. This means that a quark is associated with a *color* charge. There are three types of color charge: *red*, *green*, and *blue*.

Isolated quarks have never been observed. This has given rise to the hypothesis of QUARK CONFINEMENT. Nevertheless, there is still a great deal of experimental evidence in favor of quarks.

In the course of weak interactions of hadrons one flavor of quark turns into another flavor of quark. For example, in the beta decay of a neutron to a proton (with the emission of an electron and an antineutrino) a neutron, which has two down quarks and up quark, is converted into a proton, which has two up quarks and one down quark. Since quarks are electrically charged they take part in electromagnetic interactions and since they have nonzero rest masses they also take part in gravitational interactions.

quarter-wave antenna An antenna that has a length of $\lambda/4$, where λ is the wavelength of the radio wave. Many aerials for medium-wavelength radio waves are quarter-wave aerials.

quarter-wave plate *See* retardation plate.

quasar /kway-zar/ Abbreviation for quasi-stellar object. Quasars are astronomical objects that have large gravitational red-shifts. It is thought that quasars are super-massive BLACK HOLES at the centers of galaxies, each having a mass of about 200 million times the mass of the Sun and a rapidly accreting mass.

quasicrystal /kway-sÿ-**kriss**-tăl, **kway**-zÿ-, **kwah**-see-, **kwah**-zee-/ A type of solid that does not have the long-range order of a true crystal but has orientational order involving shapes such as pentagons and icosahedra, which would normally be forbidden in crystallography. The two-dimensional analog of a quasicrystal is called a *Penrose pattern*. For example, an alloy of aluminum and manganese, Al Mn, can form a quasicrystal with icosahedral symmetry.

quenching The rapid cooling of a hot metal by immersion in a cold liquid. Quenching is sometimes used to improve the properties of a metal. For example, defects in the crystal, occurring at high temperature, can be frozen in by rapid cooling to harden the metal.

R

racemic mixture /ray-**see**-mik, ră-/ A mixture that contains equal amounts of the dextrorotatory and levorotatory forms of an optically active substance; the mixture is not optically active. *See* optical activity.

rad A m.k.s. unit of absorbed dose of ionizing radiation, equal to 1/100 gray.

radar /**ray**-dar/ An acronym for *ra*dio *d*irection *a*nd *r*anging. A system for the precise location of distant objects (especially ships and aircraft) by means of RADIO. A microwave radio beam is transmitted, reflected by the distant object, and received on return to the transmitter. The directional antenna points in the direction of the object, the distance of which is calculated from the time of travel of the wave at the speed of electromagnetic waves, 3×10^8 m s^{-1}. Alternatively the beam can be made to scan in continuous circles, and so locate all objects in any direction, the result being shown on a plan position indicator (PPI), a cathode-ray tube with a rotating time base.

radial field A field in which the field lines are directed either towards or away from the source of the field, as along the radii of a circle. The electric field of a single point charge is an example of a radial field.

radian Symbol: rad The SI unit of plane angle; 2π radian is one complete revolution (360°).

radiance Symbol: L_e In a given direction, the radiant intensity of a source of radiation per unit area projected at right angles to the direction. The unit is the watt per steradian per square meter (W sr^{-1} m^{-2}).

radiant exitance *See* exitance.

radiant flux Symbol: Φ_e The rate of flow of radiant energy; the total power emitted or received in the form of electromagnetic radiation. It is measured in watts.

radiant heat Heat emitted from a body in the form of electromagnetic radiation. At low temperatures it is nearly all in the infrared region but at a few thousand degrees the amount of visible light becomes important, and at the radiating temperature of the Sun (6000 K) ultraviolet is becoming significant. The fact that sunlight was part of the heat radiation of the Sun was discovered by Herschel (1800). *See also* black body.

radiant intensity Symbol: I_e The radiant flux from a point source per unit solid angle. The unit is the watt per steradian (W sr^{-1}).

radiation /ray-dee-**ay**-shŏn/ Any particles or waves emitted by a source. There are two main uses of the term: **1.** *Electromagnetic radiation* produced by any mechanism. In particular the emission of radiant heat by hot bodies, in contrast to heat transfer by convection and conduction.
2. The emission of particles (alpha or beta particles) and electromagnetic radiation (gamma rays) by radioactive substances.

radiation formula *See* Planck's radiation law.

radiation pressure Pressure exerted on a surface by electromagnetic radiation. A beam of electromagnetic radiation of energy E has momentum equal to E/c, where c is the speed of light. Hence, if radiation of intensity I is totally absorbed by a surface at perpendicular incidence, the force/area is equal to *I*/*c*. For perfect specular reflection, force/area is 2*I*/*c*. For example, solar radiation totally absorbed by a surface at the distance of the Earth will exert a pressure of 4.5×10^{-6} N m^{-2}.

There is an analogous phenomenon in acoustics.

radiator **1.** An object that emits radiation. **2.** A heat exchanger. *See* heat exchanger.

radio /**ray**-dee-oh/ The process of communication across space by the transmission and reception of an electromagnetic wave of radiofrequency without the use of connecting wires or other material link.

radioactive /ray-dee-oh-**ak**-tiv/ Describing an element or nuclide that exhibits natural radioactivity.

radioactive dating (**radiometric dating**) Any method of measuring the age of materials that depends on radioactivity. *See* carbon dating; dating; fission-track dating; potassium–argon dating; rubidium–strontium dating; thermoluminescent dating; uranium–lead dating.

radioactive series A series of radioactive nuclides, each being formed by the decay of the previous one. All such series terminate in the product of a stable nuclide. An example of a radioactive series is the *thorium series*, which starts with ^{232}Th, and which, after six alpha decays and four beta decays, terminates in ^{208}Pb. This series is known as the 4*n* series because all its members have mass numbers that are multiples of 4. The *actinium series*, passing from ^{235}U to ^{207}Pb is known as the $4n - 1$ series. The *uranium-radium series*, starting with ^{238}U and ending with ^{206}Pb, is known as the $4n + 2$ series. The *neptunium series*, terminating in ^{209}Bi, is not found in nature as it contains no long-lived member.

radioactive waste *See* nuclear waste.

radioactivity /ray-dee-oh-ak-**tiv**-ă-tee/ The spontaneous disintegration of certain unstable nuclides with emission of radiation. There are various types classified as alpha, beta, and gamma decay, and fission.

radiocarbon dating /ray-dee-oh-**kar**-bŏn/ *See* carbon dating.

radio frequency (**RF**) A frequency between 3 KHz and 300 GHz (wavelength 1 cm–100 km).

radiography /ray-dee-**og**-ră-fee/ The production of an image of an object using short-wavelength radiation. The term is frequently taken to mean the production of an image in medicine using x-rays. A photograph or other image produced in this way is called a *radiograph*.

radioisotope /ray-dee-oh-**ÿ**-sŏ-tohp/ A radioactive isotope of an element. Tritium, for instance, is a radioisotope of hydrogen. Radioisotopes are extensively used in research as sources of radiation and as tracers in studies of chemical reactions. Thus, if an atom in a compound is replaced by a radioactive nuclide of the element (a *label*) it is possible to follow the course of the chemical reaction. Radioisotopes are also used in medicine for diagnosis and treatment.

radioluminescence /ray-dee-oh-loo-mă-**ness**-ĕns/ *See* luminescence.

radiometer /ray-dee-**om**-ĕ-ter/ An instrument for detecting and measuring radiant (i.e., electromagnetic) radiation. *See* actinometer; bolometer; thermopile.

radiometric dating /ray-dee-oh-**met**-rik/ *See* radioactive dating.

radio telescope A telescope that detects radiation in the radio region of the electromagnetic spectrum from beyond the atmosphere of the Earth. There are many types of *radio source*, both inside and outside the Galaxy.

radio waves A form of electromagnetic radiation with wavelengths greater than a few millimeters. There is no upper limit –

FREQUENCY BANDS OF RADIO WAVES

Band	*Frequency range (Hz)*	*Wavelength range (m)*
extremely low frequency (ELF)	-3×10^3	10^5–
very low frequency (VLF)	3×10^3–3×10^4	10^4–10^5
low frequency (LF)	3×10^4–3×10^5	10^3–10^4
medium frequency (MF)	3×10^5–3×10^6	10^2–10^3
high frequency (HF)	3×10^6–3×10^7	10 –10^2
very high frequency (VHF)	3×10^7–3×10^8	1 –10
ultra high frequency (UHF)	3×10^8–3×10^9	0.1 –1

applications have been found for radio waves tens of kilometers long.

Artificial radio waves of any frequency can be produced when suitable electronic circuits feed alternating signals to suitable antennas. They are generated by the oscillating motion of the electrons in the conductor.

The main application of radio waves is to carry information – for 'wireless telegraphy' and television for instance. The information is used to modulate a carrier wave, suitably changing its amplitude, frequency, or phase during transmission. The receiver is able to separate the carried information from the carrier.

See also electromagnetic spectrum.

radium /**ray**-dee-ŭm/ A white radioactive luminescent metallic element of the alkaline-earth group. It has several short-lived radioisotopes and one long-lived isotope, radium-226 (half-life 1602 years). Radium is found in uranium ores, such as the oxides pitchblende and carnotite. It was formerly used in luminous paints and radiotherapy.

Symbol: Ra; m.p. 700°C; b.p. 1140°C; r.d. 5 (approx. 20°C); p.n. 88; r.a.m. 226.0254 (^{226}Ra).

radius of curvature Symbol: *r* The radius of the sphere of which a MIRROR or LENS surface is part. In the case of a reflector, the radius of curvature is twice the focal distance, *f*.

$$r = 2f$$

In the case of a lens it is of value to use the points at 2f and 2f′, twice as far from the pole as f and f′. The distance to either of these, double the focal distance, is the nominal radius of curvature of the lens. It can be related to the radii of curvature of the two lens surfaces.

radius of gyration Symbol: *k* For a body of mass *m* and moment of inertia *I* about an axis, the radius of gyration about that axis is given by $k^2 = I/m$. In other words, a point mass *m* rotating at a distance *k* from the axis would have the same moment of inertia as the body.

radon /**ray**-don/ A colorless monatomic radioactive element of the rare-gas group, now known to form unstable compounds. It has 19 short-lived radioisotopes; the most stable, radon-222, is a decay product of radium-226 and itself disintegrates into an isotope of polonium with a half-life of 3.82 days. ^{222}Rn is sometimes used in radiotherapy. Radon occurs in uranium mines and is also detectable in houses built in certain areas of the country.

Symbol: Rn; m.p. –71°C; b.p. –61.8°C; d. 9.73 kg m^{-3} (0°C); p.n. 86.

rainbow The effect of refraction in, for example, rain drops causing sunlight to separate into the colors of the spectrum. The Sun is behind the observer as he or she faces the rainbow.

The primary bow is formed by light that is partially internally reflected once within each drop, and is refracted on entering and leaving. (The appearance may be affected by diffraction, especially with very small drops.) In good viewing conditions, a faint secondary bow with the colors in reverse order may be seen outside the primary bow. This involves two partial internal reflections in each drop. Sometimes other bows (*supernumary bows*) can also be seen.

Raman effect /**rah**-măn/ The scattering of monochromatic light as it passes through a transparent medium. Interaction with the molecules of the medium cause the light to be scattered, with some at wavelengths above and below that of the incident light. The Raman effect is used extensively in the determination of molecular structure. The effect is named for the Indian physicist Sir Chandrasekhara Venkata Raman (1888–1970).

random walk A concept applied to Brownian movement and diffusion based on the idea of finding the distance a walker is from a starting point if he or she can walk forward or backward, making the choices entirely randomly. After taking N steps, the average distance from the starting point is $\sqrt{N}$.

range **1.** The length over which a force operates. For example, both strong and weak nuclear forces are short range since they do not extend outside a nucleus but electromagnetic and gravitational forces are long range since they extend over macroscopic distances.
2. The average distance that a particle causing ionization in matter travels before it comes to rest.

Rankine scale /**rank**-in/ An absolute temperature scale on which a temperature interval of one degree Rankine (°R) corresponds to a temperature interval of one degree Fahrenheit. On the Rankine scale absolute zero is 0°R (–273.15°C), the freezing point of water is 491.67°R (0°C), and the boiling point of water is 671.67°R (100°C). The Rankine scale is named for the Scottish engineer William John Macquorn Rankine (1820–72).

Raoult's law /rah-**oolz**/ At a particular temperature, the relative lowering in vapor pressure of a solution is proportional to the mole fraction of the solute dissolved in it. The law is true only for ideal solutions. It is named for the French chemist François-Marie Raoult (1830–1901).

raster /**ras**-ter/ The scanning-line pattern on a visual display such as the screen of a cathode-ray tube.

ratemeter /**rayt**-mee-ter/ An instrument that measures or indicates the rate of decay of a source of radioactivity. This rate can be manifested as a reading in an electrical instrument and/or a click in a loudspeaker.

ratio arms *See* Wheatstone bridge.

rationalized units A system of units in which the units are related to one another by only a few mathematical operations involving addition or multiplication. SI units form a rationalized system of units.

ray A very narrow beam of radiation. By considering how selected rays behave, ray diagrams can be drawn to relate object and image positions for any lens or mirror.

Theoretically a ray is infinitely narrow, like a mathematical line. Then, even if part of a convergent or divergent beam, it will not converge or diverge itself. There is no evidence that beams are really formed of rays in this way; however rays are extremely useful in discussion and in ray diagrams.

ray diagrams Accurate or approximate diagrams showing how selected rays pass through an optical system (such as a lens, prism, or microscope). The rays are selected on the basis of known behavior. Suitable ray diagrams can show details of the image formed of any object by the optical system. They are also important in system design. Many examples appear in this book.

Rayleigh criterion /**ray**-lee/ *See* resolving power.

Rayleigh–Jeans law The law that describes black-body radiation according to classical physics. This law shows the necessity of quantum physics since it leads to the *ultraviolet catastrophe* of infinity in the limit of the wavelength of the radiation going to zero. The Rayleigh–Jeans law was derived by the British scientist Lord Rayleigh (1842–1919) in 1900 and modified by the British physicist and astronomer Sir James Jeans (1877–1946) in 1905.

Rayleigh scattering The scattering of electromagnetic radiation, usually light, in which the wavelength of the electromagnetic radiation is not changed. Rayleigh scattering involves no exchange of energy between the radiation and the particles (although it does involve a change of phase and direction). This type of scattering is responsible for the blue color of the sky.

reactance /ree-**ak**-tăns/ Symbol: X A measure of the opposition of inductance or capacitance to alternating current. Reactance is the ratio of the peak voltage to the peak current. Like resistance it is measured in ohms; it causes the current to become out of phase with the voltage. Reactance depends on the frequency of the supply.

For a pure capacitance (C), the current leads the voltage by one quarter cycle. The reactance is given by $1/2\pi fC$, where f is the frequency. For a pure inductance (L) the current lags behind the voltage by one quarter cycle. The reactance is given by $2\pi fL$. *See also* alternating-current circuit; impedance; resonance.

reaction Newton's third law of motion states that whenever object A applies a force on object B, B simultaneously applies an equal force on A in the same line but the opposite direction. These forces arise from an interaction; neither should be considered the cause or effect of the other force. Newton expressed this law in a Latin phrase translated as 'to every action there is an equal and opposite reaction'. Students very often seriously misunderstand this. The words *action* and *reaction* are just names given arbitrarily to the two forces in a third-law pair. Often it is wrongly supposed that the 'reaction' is a consequence of the 'action', and comes afterwards. Sometimes 'reaction' is thought to mean any effect of a force. It is advisable to avoid the use of the word *reaction* in mechanics.

reactor *See* nuclear reactor.

real gas An actual gas, as opposed to a hypothetical ideal gas (*see* gas laws). In a real gas the molecules have non-zero sizes and there are intermolecular forces between the molecules. The equation of state for a real gas is more complicated than for an ideal gas. *See* van der Waals' equation.

real image *See* image.

real is positive, virtual is negative *See* sign convention.

Reaumur scale An obsolete temperature scale in which water freezes at 0° and boils at 80° at standard pressure. A temperature interval of 1 Reamur (Re) is equivalent to a temperature interval of 1.25 degrees Celsius. The scale is named for the French physicist René-Antoine de Réaumur (1683–1757).

rectifier /**rek**-tă-fÿ-er/ An electrical conductor that allows current to flow through it in one direction only, thus enabling the conversion of a.c. to d.c. The commonest type of rectifier is the semiconductor diode, which consists of a single p-n junction. A metal rectifier, which is also a junction rectifier, consists of a metal in contact with a semiconductor or an oxide of the metal. Typical examples are copper-selenium and copper-copper oxide rectifiers. The thermionic diode is another type of rectifier.

rectilinear propagation Passage of radiation in straight lines in a constant medium.

red shift A displacement of lines in the spectra of certain celestial objects toward longer wavelengths (i.e. toward the red end of the visible spectrum). The spectral lines appear at slightly longer wavelengths than they would under 'normal' laboratory conditions. It is caused by the Doppler effect and indicates that the observed galaxy is moving away from the Earth. Some objects show a *blue shift*, indicating movement toward the observer. These shifts are known as *Doppler shifts*.

Not all red shifts are thought to be caused by the Doppler effect. Some may result from the source being at a very low gravitational potential. These are explained by the general theory of relativity, and are known as *gravitational red shifts* or *Einstein shifts*. It is customary to express a red shift as $\Delta\lambda/\lambda$, where $\Delta\lambda$ is the change in the wavelength for an electromagnetic wave with a wavelength λ.

reed switch A type of switch that is used either as a sensor or in the fast charging and discharging of capacitors. It gets its name because the simplest type of reed switch consists of a glass capsule with three metal strips. Two of these strips, including the one called the reed can become magnetized. The third strip cannot become magnetic. In the presence of a magnetic field the reed moves from the strip which cannot become magnetic to the magnetized strip which attracts the reed.

reflectance /ri-**flek**-tăns/ (**reflection factor**) Symbol: ρ The fraction of incident radiation reflected by a surface; i.e. the ratio of the reflected flux to the incident flux. For each surface this will depend on wavelength; often too it depends on angle of incidence. There is no unit. If the material is sufficiently thick, so that the reflecting properties do not depend on thickness, it is called *reflectivity*.

reflecting telescope *See* reflector.

reflection /ri-**flek**-shŏn/ The process in which radiation meeting the boundary between two media 'bounces back', to stay in the first medium. Any kind of radiation – wave or stream of particles – can be reflected. Whenever an incident ray is reflected, the laws of reflection can be used to find the direction into which the ray is turned (deviated). Reflection from a source makes the radiation appear to come from somewhere else – the image of the source. For reflection by a plane surface:

1. The image is the same distance behind the surface as the object is in front.
2. The line joining each object point with its image is normal (perpendicular) to the surface.
3. The image is the same size as the object.
4. The image is the same way up as the object, but perverted (laterally inverted).
5. The image is virtual (no light actually passes through it).

For reflection by a curved surface, the relations between object and image depend on how far the object is from the surface compared with the focal distance (f) of the surface. According to the real is positive, virtual is negative convention, the image distance (v) and object distance (u) are related by:

$$1/f = 1/v + 1/u$$

Image size (y) relates to object size (x) thus:

$$y = x(v/u)$$

The focal distance of a curved reflector is half the radius of curvature (r).

Regular reflection (*specular reflection*) occurs from smooth surfaces. The beam is not scattered on reflection and undistorted images are formed. In *diffuse reflection* the reflecting surface is rough and the reflected beam is scattered randomly.

See also diffusion; mirror; reflection, laws of; total internal reflection.

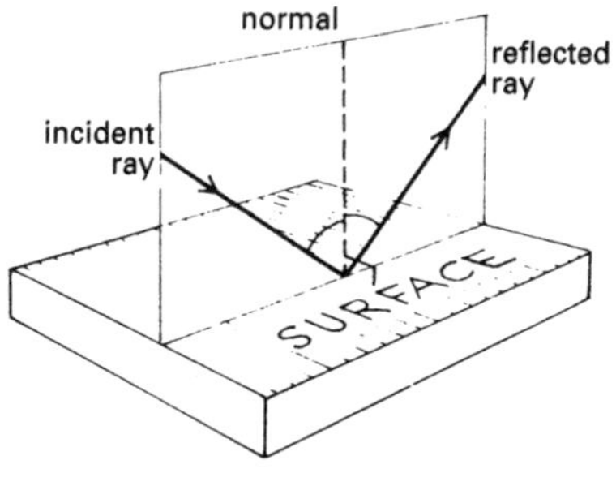

Reflection

reflection, angle of *See* angle of reflection.

reflection, laws of Two laws that apply whenever any form of radiation is reflected, whatever the circumstances. In order to state the laws, the *normal* is defined – the perpendicular to the surface at the point of incidence. All angles named are measured from the normal.
(1) The reflected ray is in the same plane as the incident ray and the normal.
(2) The angle of reflection equals the angle of incidence and lies on the other side of the normal.

reflection, total internal *See* total internal reflection.

reflection factor *See* reflectance.

reflectivity /ree-flek-**tiv**-ă-tee/ *See* reflectance.

reflector /ri-**flek**-ter/ (**reflecting telescope**) An optical telescope in which the objective (the light collector) is a converging reflector. The largest ones in use have mirrors several meters in diameter. The problem with reflecting telescopes is that the first image is formed in the incident beam. There are a number of types distinguished by different optical systems for avoiding this and taking the light to the eyepiece. *See* Cassegrainian telescope; coudé telescope; Gregorian telescope; Newtonian telescope.

refracting telescope *See* refractor.

refraction /ri-**frak**-shŏn/ The process in which radiation incident on the boundary between two media passes on into the second medium. Any kind of radiation – wave or stream of particles – can be refracted. Unless the radiation meets the surface at a right angle (down the normal) it will change direction on refraction. The behavior of refraction in any case depends on the refractive constant (index) of the pair of media. Whenever a ray is refracted, the laws of refraction can be used to find the direction into which the ray is turned (deviated).

Diffuse refraction occurs when the radiation is scattered on transmission through the medium, as on passage through translucent media such as wax or frosted glass. The opposite effect is *regular refraction*, in which the beam is not scattered and undistorted images can be formed.

Refraction also occurs within a medium in which there is a gradient of properties. The apparent position of a star, especially near the horizon, is affected by refraction in the atmosphere the density of which changes with altitude. Sound waves are refracted in the atmosphere as a result of temperature variation with height.

See also refraction, laws of; refractive index.

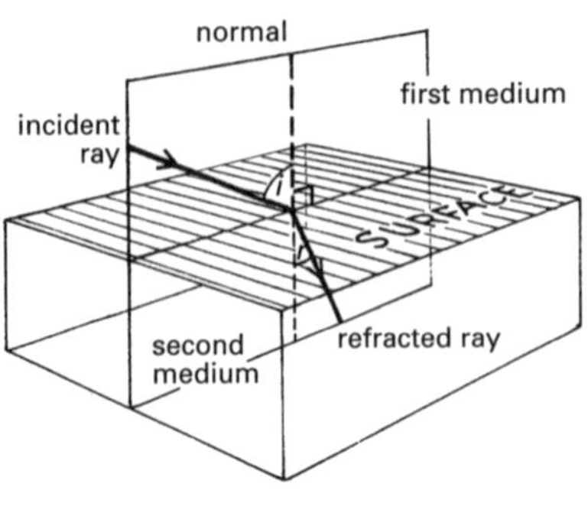

Refraction

refraction, angle of *See* angle of refraction.

refraction, laws of Two laws that apply whenever any radiation is refracted, whatever the circumstances. In order to state the laws, the normal is defined – the perpendicular to the surface at the point of refraction. All angles named are measured from the normal.
(1) The refracted ray is in the same plane as the incident ray and the normal.
(2) The sine of the angle of incidence divided by the sine of the angle of refraction is constant for a given wavelength and a given pair of media; the angles lie on opposite sides of the normal.

The second law is sometimes called *Snell's law* (named for the Dutch mathematician and physicist Willebrord van Roijen Snell (1591–1626)). It leads to a definition for the refractive constant (n)

concerned with a given wavelength and a given pair of media:

$$n = (\sin i)/(\sin r)$$

Special care needs to be taken with the laws of refraction in situations related to polarization of the radiation.

See also birefringent crystal; refractive constant.

refractive constant /ri-**frak**-tiv/ (**refractive index**) Symbol: n When monochromatic radiation passes through the interface between two media it is refracted so that

$$_1n_2 = \sin i/\sin r \text{ (\textit{Snell's law})}$$

where $_1n_2$ is defined as the *refractive constant* for radiation passing from medium 1 to medium 2. For radiation in the reverse direction, $_2n_1 = 1/_1n_2$. For successive refractions at the boundaries between media 1, 2, and 3:

$$_1n_3 = {}_1n_2 \times {}_2n_3$$

Partial reflection also occurs at the interface. For two media such that the angle of refraction is greater than the angle of incidence, transmission only occurs for angles of incidence up to a critical angle p, at which the angle of refraction is 90°. At angles above p there is total internal reflection:

$$_1n_2 = \sin p$$

The *absolute refractive constant* of a medium is that for radiation passing from a vacuum into the medium. For many purposes it is sufficiently accurate to regard air as equivalent to a vacuum.

If λ_1 and λ_2 are the wavelengths in two media and v_1 and v_2 are the phase speeds, then

$$_1n_2 = \lambda_1/\lambda_2 = v_1/v_2$$

The apparent depth of a medium is related to its real depth by $_1n_2$ = real depth/apparent depth. Refractive constants often vary with the frequency of the radiation (*see* dispersion; anomalous dispersion). For optical materials it is common to give values for a standard radiation, usually the yellow light from sodium vapor (D-lines).

For electromagnetic radiation at ordinary frequencies the absolute refractive constants of all materials are greater than unity. For x-radiation generally the values are slightly less than one, such that total internal reflection can be obtained at grazing incidence going from vacuum to the material. Thus the phase speeds of x-rays in materials are greater than the speed of light in a vacuum. The *group speed* (at which quantities such as momentum, energy, and mass are propagated) cannot in any circumstances be greater than the speed of light in a vacuum.

refractive index *See* refractive constant.

refractivity /ree-frak-**tiv**-ă-tee/ A measure of the ability of a medium to bend (deviate) radiation entering its surface.

refractometer /ree-frak-**tom**-ĕ-ter/ A device for measuring the refractive constant of a medium. There are many types, each using one or other of the relations giving the refractive constant, directly or as applied in special circumstances. In the elementary laboratory, methods based on the apparent-depth and critical-angle expressions are most effective.

refractor /ri-**frak**-ter/ (**refracting telescope**) An optical TELESCOPE in which the objective (the light collector) is a converging lens. Practical problems preclude the design of such telescopes with lenses more than about one meter in diameter. Telescopes used in astronomy produce an inverted image; KEPLERIAN TELESCOPES are invariably used. For non-astronomical use a *terrrestrial telescope* is employed – either a GALILEAN TELESCOPE or a Keplerian telescope with an added inverting lens.

refrigerant /ri-**frij**-ĕ-rănt/ The working fluid in a refrigerator.

refrigerator /ri-**frij**-ĕ-ray-ter/ (**icebox**) A refrigerator is equivalent to a heat engine run in reverse. A working substance is taken through a cycle of operations in which work is done upon it (usually by an electric motor) so that it can take heat from a cold box and discharge heat to the atmosphere. By the first law of thermodynamics the heat discharged equals the work done on the working substance plus the heat taken from the cold box. Certain refrigerators are operated by a heat source such as a gas flame. In such devices a heat engine and a refrigerator are effectively combined in one device. *See also* heat pump.

regelation /ree-jĕ-**lay**-shŏn/ The refreezing of ice melted by pressure that takes place when the pressure is removed.

Regnault's apparatus /ren-**yohz**/ An apparatus for measuring the absolute thermal expansivity (γ) of a liquid. It consists of a shaped U-tube (see illustration) with one limb at ice temperature and the other at a constant temperature θ (usually the steam temperature). The value of γ can be calculated from the difference between the liquid heights in the two limbs of the tubes. The apparatus is named for the French scientist Henri Victor Regnault (1800–78).

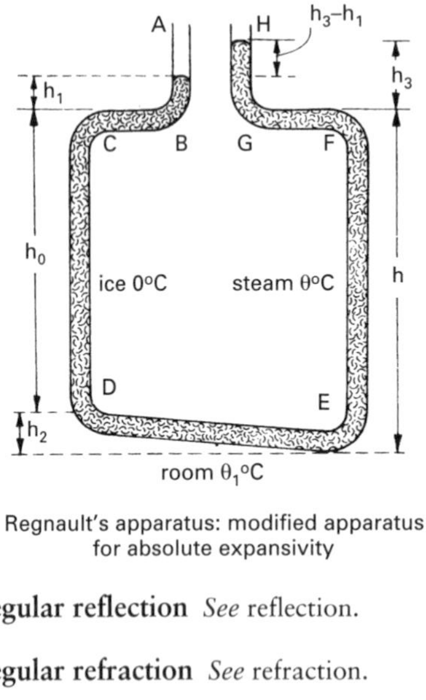

Regnault's apparatus: modified apparatus for absolute expansivity

regular reflection *See* reflection.

regular refraction *See* refraction.

rejector circuit *See* resonance.

relative atomic mass Symbol: A_r The ratio of the average mass per atom of the naturally occurring element to 1/12 of the mass of an atom of nuclide ^{12}C. It was formerly called *atomic weight*.

relative density Symbol: *d* The density of a substance divided by the density of a reference substance. For liquids the density of water at 4°C (the temperature at which its density is a maximum) is usually used. The temperature of the substance is either stated or is understood to be 20°C. Relative densities are also stated for gases, usually relative to air at STP. Relative density was formerly called *specific gravity*.

relative humidity A measure of the amount of water vapor in the air, expressed as a proportion of the maximum amount the air could hold at the temperature concerned. (The higher the temperature the more water vapor air can hold.) For a fixed amount of water and air, the relative humidity therefore decreases as temperature rises and increases as it falls. If the temperature falls to the point where the relative humidity equals one, the water vapor starts to condense out as droplets on surfaces or in the air (dew or mist). This temperature is the saturation temperature or the *dew temperature* (dew point). *Compare* absolute humidity.

relative molecular mass Symbol: M_r The ratio of the average mass per molecule of the naturally occurring form of an element or compound to 1/12 of the mass of an atom of nuclide ^{12}C. This was formerly called *molecular weight*. It does not have to be used only for compounds that have discrete molecules; for ionic compounds (e.g. NaCl) and giant-molecular structures (e.g. BN) the formula unit is used.

relative permeability *See* permeability.

relative permittivity Symbol: ε_r For a parallel-plate capacitor (say) with a given material between the plates, the relative permittivity is the ratio of the capacitance observed (C_m) to the capacitance (C_0) if the material were removed (free space between the plates); i.e. $\varepsilon_r = C_m/C_0$. Relative permittivity measures the effect of a material on an electric field relative to free space. It is equal to $\varepsilon/\varepsilon_0$, where ε is the (absolute) PERMITTIVITY of the material and ε_0 the permittivity of free space (the electric constant). An early term for relative permittivity was *dielectric constant*.

relative velocity If two objects are moving at velocities v_A and v_B in a given direction, the velocity of A relative to B is $v_A - v_B$ in that direction. In general, if two objects are moving in the same frame at nonrelativistic speeds their relative velocity is the vector difference of the two velocities.

relativistic mass /rel-ă-ti-**vis**-tik/ The mass of an object as measured by an observer at rest in a frame of reference in which the object is moving with a velocity v. It is given by $m_0 = m\sqrt{(1 - v^2/c^2)}$, where m_0 is the rest mass, c is the speed of light, and m is the relativistic mass. The equation is a consequence of the special theory of RELATIVITY, and is in excellent agreement with experiment. No object with a nonzero REST MASS can travel at the speed of light because its mass would then be infinite.

relativistic particle A particle that is traveling at a RELATIVISTIC VELOCITY.

relativistic velocity Any velocity that is sufficiently high to make the mass of an object significantly greater than its rest mass. It is usually expressed as a fraction of c, the speed of light. At a velocity of $c/2$ the RELATIVISTIC MASS of an object increases by 15% over the REST MASS.

relativity, theory of /rel-ă-**tiv**-ă-tee/ A theory put forward in two parts by Albert Einstein. The special theory (1905) referred only to non-accelerated (inertial) frames of reference. The general theory (1915) is also applicable to accelerated systems.

The *special theory* was based on two postulates:

(1) That physical laws are the same in all inertial frames of reference.

(2) That the speed of light in a vacuum is constant for all observers, regardless of the motion of the source or observer.

The second postulate seems contrary to 'common sense' ideas of motion. Einstein was led to the theory by considering the problem of the 'ether' and the relation between electric and magnetic fields in relative motion. The theory accounts for the negative result of the Michelson–Morley experiment and shows that the Lorentz–Fitzgerald contraction is only an apparent effect of motion of an object relative to an observer, not a 'real' contraction. It leads to the result that the mass of an object moving at a velocity v relative to an observer is given by:

$$m = m_0/\sqrt{(1 - v^2/c^2)}$$

where c is the speed of light and m_0 the mass of the object when at rest relative to the observer. The increase in mass is significant at high velocities. Another consequence of the theory is that an object has an energy content by virtue of its mass, and similarly that energy has inertia. Mass and energy are related by the famous equation $E = mc^2$.

The *general theory* of relativity seeks to explain the difference between accelerated and non-accelerated systems and the nature of the forces acting in both of them. For example, a passenger in a car rounding a bend at constant speed is in an accelerated frame of reference because his or her velocity is changing. He experiences a fictitious force known as a centrifugal force. This force is fictitious because to an outside observer the force is simply a result of the passenger's tendency to continue in a straight line. This analysis of forces led Einstein to the view that gravitation is a consequence of the geometry of the universe. He visualized a four-dimensional space-time continuum in which the presence of a mass affects the geometry – the space is 'curved' by the mass.

relay /**ree**-lay/ An electromagnetic switching device consisting of a coil of insulated wire with a central soft iron core acting as armature. When the coil is energized, the armature moves and operates a switch in another circuit. THYRISTORS can be used as solid-state relays. *See also* solenoid.

reluctance ('magnetic resistance') Symbol: R The ratio of magnetomotive force to total magnetic flux in a MAGNETIC CIRCUIT.

The SI unit of reluctance is the henry meter^{-1} (H m^{-1}).

reluctivity /rel-ŭk-**tiv**-ă-tee/ The reciprocal of magnetic permeability.

rem /rem/ (*r*adiation *e*quivalent *m*an) A m.k.s. unit formerly used for measuring equivalent ionizing radiation doses absorbed by living tissue. One rem is equivalent to a dosage of one roentgen of x-rays or gamma rays absorbed by an average adult male. In SI derived units it is equal to 10^{-2} sievert.

remanence /**rem**-ă-nĕns/ *See* hysteresis cycle.

resistance, electrical Symbol: *R* The ratio of the potential difference across an electrical element to the current in it. Resistance measures the opposition of the component to the flow of charge. The SI unit is the ohm. In an alternating-current circuit resistance is only one factor in the response of the component. *See* impedance. *See also* effective resistance; Ohm's law.

resistance box A device for providing variable known resistances. One type has a thick brass bar with gaps at intervals across which are connected fixed resistances. Combinations of these resistances can be chosen by inserting brass plugs in the gaps, thus shorting out selected resistors. Other types of resistance box have multi-switches to give the selection.

resistance thermometer A thermometer that measures temperature by the change in electrical resistance of a conductor. Usually a small coil of pure fine platinum wire is used, wound on a mica former. It is incorporated into one arm of a Wheatstone bridge; the other arm has a variable resistor and a pair of 'dummy' leads, to compensate for temperature effects in the leads of the measuring coil.

The resistance varies with temperature according to an approximate equation of the form:

$$R = R_0(1 + a\theta)$$

where R is the resistance at θ°C, R_0 the resistance at 0°C, and a is a constant.

More strictly:

$$R = R_0(1 + a\theta + b\theta^2)$$

where b is a second constant.

resistance wire Wire used to introduce a resistance into an electrical circuit. Resistance wire is made from alloys. Where the wire is likely to attain high temperatures, Nichrome is suitable. Constantan is used for resistors for accurate work with alternating current as it has a very low temperature coefficient of resistance, but it is less suitable for d.c. as it gives disturbing thermoelectric e.m.f.s in conjunction with copper connectors. Manganin is suitable for precise d.c. resistors.

resistivity /ree-zis-**tiv**-ă-tee/ Symbol: ρ The tendency of a material to oppose the flow of an electric current. The resistivity is a property of the material at a given temperature and does not depend on the sample. The resistance R of a sample is given by $\rho l/A$, where A is the cross section area and l the length. The unit is the ohm meter (Ωm). Resistivity was formerly called *specific resistance*. At normal temperatures the resistivities of pure metals are typically of the order 10^{-8} to 10^{-7} ohm. Alloys generally have higher values than pure metals, typically 10^{-7} to 10^{-6} ohm. The values for insulators can be very much higher – in the range 10^{6} to 10^{18} ohm. Semiconductor resistivities can vary between these extremes, their values depending greatly on the purity of the specimen.

resistor A component included in an electric circuit because of its resistance. Resistors may be variable or have a fixed value; they are made of resistance wire or carbon. Some types have a ceramic coating incorporating a color coding in the form of circular bands or stripes, which identify the value of the resistance.

resolution The ability to distinguish two close objects separately, rather than as one single object. The resolution obtained with an optical system depends on the magnification and on the RESOLVING POWER.

resolution of vectors The determination of the components of a vector in two given directions at 90°. The term is sometimes used in relation to finding any pair of components (not necessarily at 90° to each other).

resolving power A measure of the ability of an optical system to form detectably separate images of close objects or to separate close wavelengths of radiation.

The resolving power depends on the aperture (a) and the wavelength (λ) of the radiation. For separation of images it is the smallest angular separation of the images. It can be defined ideally as λ/a (the unit being the radian). In practice, diffraction effects limit the resolution. In a telescope forming images of two stars, each image will be a bright central portion surrounded by light and dark rings. The *Rayleigh criterion* for resolution (named for the English physicist John William Strutt, Lord Rayleigh (1842–1919)) is that the central portion of one image falls on the first minimum of the other. The resolving power is then $1.22\lambda/a$. The resolving power of a microscope is often given as the actual minimum distance between two points that can be separated (a small distance is a high resolving power). Again it depends on wavelength and aperture. Larger apertures give increased resolving power. The immersion objective is a method of increasing the effective aperture. Electron microscopes have higher resolving powers because the wavelengths of electrons are much smaller than those of visible radiation. The resolving power of the human eye is about 10^{-3} rad. At the standard near-point distance (250 mm) points about 100 micrometer apart can just be resolved.

resonance **1.** The large-amplitude vibration of an object or system when given impulses at its natural frequency. For instance, a pendulum swings with a natural frequency that depends on its length. If it is given a periodic 'push' at this frequency – for example, at each maximum of a complete oscillation – the amplitude is increased with little effort. Much more effort would be required to produce a swing of the same amplitude at a different frequency.
2. An electric circuit containing both capacitance and inductance can be arranged so that a very large (resonant) response occurs at a particular (resonant) frequency of alternating supply. An inductance (L) and capacitance (C) in series are known as a series resonant or *acceptor circuit*, and in it there is a large alternating current at the resonant frequency (f). Resonance occurs when $f.L = 1/f.C$. In a parallel resonant or *rejector circuit*, L and C are in parallel; then a minimum current flows at the resonant frequency. In a tuning circuit, the resonant frequency may be varied by altering the capacitance or the inductance. This is how radio receivers are tuned to the broadcast frequency. *See also* impedance; reactance.
3. One of a large number of excited states of elementary particles. Many resonances are known; all have very short lifetimes.

restitution, coefficient of /res-tă-**tew**-shŏn/ Symbol: e For the impact of two bodies, the elasticity of the collision is measured by the coefficient of restitution. It is the relative velocity after collision divided by the relative velocity before collision (velocities measured along the line of centers). For spheres A and B:

$$v_A{}^1 - v_B{}^1 = -e(v_A - v_B)$$

v indicates velocity before collision; v^1 velocity after collision. Kinetic energy is conserved only in a perfectly elastic collision. Different possibilities are shown in the table.

RESTITUTION COEFFICIENTS

Type of collision	*restitution coefficient (e)*
perfectly inelastic	0
inelastic	>0, <1
perfectly elastic	1
superelastic	>1

rest mass Symbol: m_0 The mass of an object at rest as measured by an observer at rest in the same frame of reference. *See also* relativistic mass.

resultant /ri-**zul**-tănt/ A vector with the same effect as a number of vectors. Thus, the resultant of a set of forces is a force that has the same effect; it is equal in magnitude and opposite in direction to the equilibrant. Depending on the circumstances, the resultant of a set of vectors can be found by different methods. *See* parallel forces; parallelogram of vectors; principle of moments.

retardation plate /ree-tar-**day**-shŏn/ A thin transparent plate cut from a piece of birefringent material (e.g. quartz or mica) parallel to the optic axis. For light transmitted by the plate normal to the optic axis, the ordinary and extraordinary rays travel at different speeds. The thickness is designed to induce a desired phase difference between the transmitted rays. The *half-wave plate* introduces a phase difference of π, and the ordinary and extraordinary rays are out of step by one-half wavelength. The *quarter-wave plate* produces a phase difference of π/2 and the rays are out of step by one-quarter wavelength.

retina (*pl.* **retinas** or **retinae**) /**ret**-ă-nă/ The light-sensitive inner surface of the EYE: its function is to convert incident light into nerve signals.

reverberation /ri-ver-bĕ-**ray**-shŏn/ Persistence of sound after the source has stopped as a result of repeated reflections from walls, ceilings, and other surfaces. Reverberation characteristics are important in the design of concert halls, theaters, etc.

reversal A change in the direction of the magnetic field of the Earth. Reversals occur because of changes in the motion of liquid iron inside the Earth's core. Experimental evidence for reversals is provided by the 'zebra stripe' pattern of the magnetism of rocks on the floor of the Atlantic Ocean. They seem to occur over relatively short time periods but at long intervals of tens of thousands of years.

reversible change A change in the pressure, volume, or other properties of a system, in which the system remains at equilibrium throughout the change. Such processes could be reversed; i.e. returned to the original starting position through the same series of stages. They are never realized in practice. An isothermal reversible compression of a gas, for example, would have to be carried out infinitely slowly and involve no friction, etc. Ideal energy transfer would have to take place between the gas and the surroundings to maintain a constant temperature.

In practice, all real processes are *irreversible changes* in which there is not an equilibrium throughout the change. In an irreversible change, the system can still be returned to its original state, but not through the same series of stages. For a closed system, there is an entropy increase involved in an irreversible change.

Reynolds number /**ren**-ŏldz/ Symbol: *Re* A dimensionless quantity used to describe fluid flow. It is equal to $\rho cl/\eta$, where ρ is the fluid's density, c its velocity, η its viscosity, and l is a length characteristic of the geometry. For a long pipe, for example, l would be the internal diameter. The Reynolds number is used in making scale models of fluid flow. The size (l) may be scaled down by a factor of four, for instance, if the model fluid has a value of ρ/η equal to a quarter that of the fluid being modeled. Because *Re* is unchanged the flow pattern will be similar and the CRITICAL SPEED will be the same. The number is named for the British physicist Osborne Reynolds (1842–1912).

RF *See* radio frequency.

rhenium /**ree**-nee-ŭm/ A rare silvery transition metal that usually occurs naturally with molybdenum. The metal is chemically similar to manganese and is used in alloys and catalysts.

Symbol: Re; m.p. 3180°C; b.p. 5630°C; r.d. 21.02 (20°C); p.n. 75; r.a.m. 186.207.

rheology /ree-**ol**-ŏ-jee/ The study of fluid flow.

rheostat /**ree**-ŏ-stat/ A variable resistor, usually operated by a sliding contact on a coil. *See also* potentiometer.

rhodium /**roh**-dee-ŭm/ A rare silvery hard transition metal. It is difficult to work and highly resistant to corrosion. Rhodium occurs native but most is obtained from copper and nickel ores. It is used in protective finishes, alloys, and as a catalyst.
Symbol: Rh; m.p. 1966°C; b.p. 3730°C; r.d. 12.41 (20°C); p.n. 45; r.a.m. 102.90550.

ribbon microphone *See* microphone.

Richardson–Dushman equation /**rich**-erd-sŏn **dŭsh**-măn/ An equation that describes the emission of electrons from a hot conductor. It states that $J = AT^2 \exp(-W/kT)$, where J is the current density, A is a constant, T is the absolute temperature, W the work function of the conductor, and k the Boltzmann constant. This equation was derived from the electron theory of metals by Sir Owen Richardson soon after the discovery of electrons and subsequently modified by Saul Dushman in the 1920s by taking quantum mechanics into account.

right-hand rule *See* Fleming's rules.

rigid-body dynamics The dynamics of rigid bodies, i.e. bodies in which the distance between any two particles making up the body is always constant. Rigid-body dynamics can be applied to the motion of flywheels, gyroscopes, etc. and, to a certain extent, the rotation of the Earth.

rigidity modulus *See* elastic modulus.

ring main A way of wiring electric sockets (outlets) in parallel round a building so that they form a ring or chain from and back to the supply point.

ripple tank A device for producing waves (ripples) on the surface of water and viewing these under stroboscopic light. It is used to demonstrate interference effects and other wave properties.

RMS value *See* root-mean-square value.

Rochon prism /**roh**-shon/ A type of polarizing prism. The ordinary ray passes through undeviated; the extraordinary ray leaves from the side. This result is the reverse of that given by the Nicol prism.

rods The more sensitive cells of the retina of the EYE. They are of most importance in vision in poor light, but cannot provide color information. Their action is not known in detail. It appears to be based on the photochemical breakdown of a reddish dye called rhodopsin (formerly visual purple).

roentgen /**rent**-gĕn/ Symbol: R A c.g.s. unit of radiation exposure, formerly used for x-rays and γ-rays, defined in terms of the ionizing effect of one electrostatic unit of electricity on one cubic centimeter of air. In SI units one roentgen is equal to 2.58×10^{-4} coulomb. The unit is named for the German physicist Wilhelm Conrad Röntgen (1845–1923).

roentgenium /**rent**-gĕn-ee-ŭm/ A radioactive element produced synthetically. It is named for Wilhelm Roentgen
Symbol: Rg; p.n. 111.

Roentgen rays A former name for x-rays.

rolling friction *See* friction.

root-mean-square value (**RMS value**) **1.** The square root of the average of the squares of a group of numbers or values. For example, the RMS value of {2, 4, 3, 2.5} is

$$\sqrt{[2^2 + 4^2 + 3^2 + 2.5^2)/4]} = 5.937$$

2. For continuous quantities, such as alternating electric current, the root-mean-square value is used to measure the average effect. The root-mean-square value of an alternating current (also called the *effective value*) is the value of the direct current that would produce the same energy transfer per second in a given resistor. For a sine wave, this is $I_m/\sqrt{2}$, where I_m is the peak value.

rotary converter A combination of an a.c. electric motor and a d.c. dynamo, used to convert alternating current into direct current.

rotational motion Motion of a body turning about an axis. The physical quantities and laws used to describe linear motion all have rotational analogs; the equations of rotational motion are the analogs of the equations of motion (linear).

They normally include $T = I\alpha$, the analog of $F = ma$, as well as the 'kinematic' equations themselves. Here T is the turning moment, or torque, I is the moment of inertia, and α is the angular acceleration.

The other equations relate the angular velocity, ω_1, of the object at the start of timing to its angular velocity, ω_2, at some later time, t, and thus to the angular displacement θ. The examples in the table are equations that apply to motion about a fixed axis with uniform angular acceleration. *See* equations of motion.

rotor The rotating part, or armature, of an electric motor or dynamo, or similar device. *See also* stator.

RSI value *See* r-value.

rubidium /roo-**bid**-ee-ŭm/ A soft silvery highly reactive element. Naturally occurring rubidium comprises two isotopes, one of which, ^{87}Rb, is radioactive (half-life 5×10^{10} years). The element is used in vacuum tubes, photocells, and in making special glass.

ROTATIONAL EQUATIONS OF MOTION

Variables	*Equation*
$\alpha, \varpi_1, \varpi_2, t, (\theta)$	$\varpi_2 = \varpi_1 + \alpha t$
$\varpi_1, \varpi_2, t, \theta, (\alpha)$	$\theta = (\varpi_1 + \varpi_2)t/2$
$\alpha, \varpi_1, t, \theta, (\varpi_1)$	$\theta = \varpi_1 t + \alpha t^2/2$
$\alpha, \varpi_2, t, \theta, (\varpi_1)$	$\theta = \varpi_2 t - \alpha t^2/2$
$\alpha, \varpi_1, \varpi_2, \theta, (t)$	$\varpi_2^2 = \varpi_1^2 + 2\alpha\theta$

Symbol: Rb; m.p. 39.05°C; b.p. 688°C; r.d. 1.532 (20°C); p.n. 37; r.a.m. 85.4678.

rubidium–strontium dating A method of radioactive dating used for estimating the age of certain rocks. It is based of the decay or ^{87}Rb to ^{87}Sr (half-life 5×10^{10} years). The technique is useful for very large ages, of the order 1×10^9 years to 4×10^9 years.

ruthenium /roo-**th'ee**-nee-ŭm/ A transition metal that occurs naturally with platinum. It forms alloys with platinum that are used in electrical contacts. Ruthenium is also used in jewelry alloyed with palladium.

Symbol: Ru; m.p. 2310°C; b.p. 3900°C; r.d. 12.37; p.n. 44; r.a.m. 101.07.

rutherford Symbol: rd A unit of radioactivity equal to that of a radioactive nuclide undergoing 10^6 disintegrations per second, equal to 10^6 becquerels in SI derived units. It is named for the New Zealand physicist Ernest Rutherford, First Baron Rutherford of Nelson (1871–1937).

rutherfordium /ru*th*-er-**for**-dee-ŭm/ A radioactive metal not found naturally on earth. It is the first transactinide metal. Atoms of rutherfordium are produced by bombarding ^{249}Cf with ^{12}C or by bombarding ^{248}Cm with ^{18}O.

Symbol: Rf; m.p. 2100°C (est.); b.p. 5200°C (est.); r.d. 23 (est.); most stable isotope ^{261}Rf (half-life 65s).

Rutherford–Royds experiment /roidz/ An experiment performed by Ernest Rutherford and T. Royds in 1909 that demonstrated that alpha particles were helium nuclei. The alpha particles given off by radon were collected and the resulting helium gas was characterized by its spectrum.

Rutherford scattering /**ru*th***-er-ferd/ The scattering of charged particles, such as alpha particles, by atomic nuclei. This type of scattering was analyzed by Ernest Rutherford in 1911 to explain the results of experiments, which he correctly interpreted as indicating the existence of atomic nuclei.

r-value A measure of the *thermal insulance* (resistance to heat flow) of a given thickness of a material such as insulation, with higher values indicating greater thermal insulance. The SI deried unit of thermal insulance is the square meter degree Celsius per watt ($m^2{}^\circ CW^{-1}$). The SI value of thermal insulance is known as *RSI value*

Rydberg constant /**rid**-berg/ *See* Bohr theory.

Rydberg formula An equation that gives the wave number of a line in a spectrum:

$$1/\gamma = R(1/n^2 - 1/m^2),$$

where n and m are positive whole numbers and R is the Rydberg constant (equal to $1.096\,77 \times 10^7\ m^{-1}$ for hydrogen). The formula is named for the Swedish physicist Johannes Robert Rydberg (1854–1919).

S

saccharimeter /sak-ă-**rim**-ĕ-ter/ *See* polarimeter.

salt bridge An inverted U-tube containing a concentrated solution of a salt (such as potassium chloride), used to connect two HALF CELLS.

samarium /să-**mair**-ee-ŭm/ A silvery element of the lanthanoid series of metals. It occurs in association with other lanthanoids. Samarium is used in the metallurgical, glass, and nuclear industries.

Symbol: Sm; m.p. 1077°C; b.p. 1791°C; r.d. 7.52 (20°C); p.n. 62; r.a.m. 150.36.

satellite A body that orbits a planet, with the orbiting body being much smaller than the planet. The satellite can be a *natural satellite*, such as the Moon, or an *artificial satellite*, such as the satellites used for communications and astronomy.

saturated vapor pressure *See* vapor pressure.

saturation The freedom from white of a color. A pure spectral color (single wavelength) is a hue, and is said to be *saturated*. Unsaturated colors are tints; i.e. hues mixed with white.

saturation, magnetic A magnetic material is saturated if it can be magnetized no more strongly. This can easily be explained using the domain theory; i.e. it is the state in which all the magnetic domains are aligned. *See also* hysteresis cycle.

sawtooth wave A waveform generated electronically (such as the variation of voltage with time), having a uniform increase that regularly and rapidly drops to the initial value. A sawtooth wave is used as the time base for scanning circuits in a cathode-ray tube.

scalar /**skay**-ler/ A measure in which direction is unimportant or meaningless. For instance, distance is a scalar quantity, whereas displacement is a VECTOR. Mass, temperature, and time are examples of scalars – they are each quoted as a pure number with a unit.

scalar product A multiplication of two vectors to give a scalar. The scalar product of $\boldsymbol{A}$ and $\boldsymbol{B}$ is defined by $\boldsymbol{A}.\boldsymbol{B} = AB\cos\theta$, where A and B are the magnitudes of $\boldsymbol{A}$ and $\boldsymbol{B}$ and θ is the angle between the vectors. An example is a force $\boldsymbol{F}$ displaced $\boldsymbol{s}$. Here the scalar product is energy transferred (or work done):

$$W = \boldsymbol{F}.\boldsymbol{s}$$
$$W = Fs\cos\theta$$

where θ is the angle between the line of action of the force and the displacement. A scalar product is sometimes called a *dot product*. *See also* vector product.

scaler An electronic counting device that measures the amount of energy transfer or charge that results from radiation, recording one 'count' after a certain number of input pulses. *See* counter.

scandium /**skan**-dee-ŭm/ A lightweight silvery element. It is found in minute amounts in over 800 minerals, often associated with lanthanoids. Scandium is used in high-intensity lights and in electronic devices.

Symbol: Sc; m.p. 1541°C; b.p. 2831°C; r.d. 2.989 (0°C); p.n. 21; r.a.m. 44.955910.

scanning electron microscope *See* electron microscope.

scattering The 'spreading out' of a beam of radiation as it passes through matter, reducing the energy moving in the original direction. Depending on the circumstances, scattering can follow any combination of three processes as the radiation interacts with matter particles – reflection (elastic scattering), absorption followed by re-radiation (inelastic or Compton scattering), and diffraction. Thus sunlight is scattered (or diffused) as it passes through cloud and dust in the atmosphere. However, even perfectly clear air scatters sunlight, making the sky color blue – high frequencies are scattered more than low frequencies.

schlieren technique /**shleer**-ĕn/ A method of studying turbulent flow by high-speed photography. If turbulent flow occurs in a fluid, there are localized short-lived regions of low density. These can be recorded by photography because they have a different refractive constant, and appear as streaks on the photograph.

schorl /shorl/ *See* tourmaline.

Schottky defect /**shot**-kee/ *See* defect.

Schrödinger equation /**shroh**-ding-er/ An equation for the wave function of a particle, such as an electron in an atom (i.e., it treats the electron's behavior as a three-dimensional stationary wave). The solution of the equation gives the probability that the electron will be located at a particular place. The equation is named for the Austrian physicist Erwin Schrödinger (1887–1961).

Schrödinger's cat A thought experiment first put forward by Erwin Schrödinger in 1935. In quantum mechanics, a particle is described by a wave function, which expresses the probability of finding it in a particular region of space and time. The conventional view of physicists is that the particle exists in some type of indeterminate or indefinite state until a measurement (or observation) is made, at which point it becomes a definite particle. It is said that the observation 'collapses' the wave function. Schrödinger put forward his thought experiment to show that this idea might have logical consequences that seem ridiculous.

He suggested a closed box containing a sample of radioactive material, a tube of cyanide, and a cat. The apparatus was to be constructed so that decay of a nucleus ejected a particle, which activated a mechanism to break the tube of cyanide. After a certain time, if a nucleus had decayed the cat would have died; if not the cat would still be alive.

Schrödinger pointed out that, according to our way of interpreting quantum mechanics, the nucleus had neither decayed nor not decayed until an observation was made, and it followed that the cat was neither dead nor alive until someone opened the box and observed it. The idea has caused considerable argument among people interested in the nature of QUANTUM MECHANICS. Most physicists would be happy with the idea that a particle could be in an indeterminate form until observed. Few would accept that a cat could be neither alive nor dead. The problem, which is still not fully resolved, concerns the difference in nature between events on the microscopic scale (e.g. nuclei) and events on the macroscopic scale (e.g. cats). *See also* Bell's paradox.

scintillation counter /sin-tă-**lay**-shŏn/ A detector of radiation in which radiation produces scintillations (flashes of light) in a phosphor. These are used to produce an electric pulse by means of a photomultiplier. As well as crystalline scintillators various liquids can be used.

sclerotic /skli-**rot**-ik/ The hard outer coating of the eyeball. *See* eye.

scotopic vision /skoh-**top**-ik/ Vision at low levels of light intensity (night-time vision). Under such conditions the rods in the retina are the main receptors and no colors can be distinguished. *Compare* photopic vision.

screen grid *See* tetrode.

screw A type of simple machine, related to the inclined plane, and, in practice, to the class two lever. The efficiency of screw systems is very low because of friction. Even so, the force ratio (F_2/F_1) can be very high.

The distance ratio is given by $2\pi r/p$, where r is the radius and p the pitch of the screw (the angle between the thread and a plane at right angles to the barrel of the screw).

seaborgium /see-**bor**-gee-ŭm/ A radioactive metallic element not occurring naturally on Earth. It can be made by bombarding ^{249}Cf with ^{18}O nuclei using a cyclotron. Six isotopes are known. The element is named for the American physicist Glenn Theodore Seaborg (1912–99).

Symbol: Sg; most stable isotope ^{266}Sg (half-life 27.3s).

search coil A small coil, in which induced currents are used to measure magnetic field strengths. *See* fluxmeter.

Searle's bar /serlz/ An apparatus for measuring the thermal conductivity of a good conductor. The specimen is a long narrow lagged bar heated at one end by a temperature bath or electric coil. It is cooled at the other end by a coil of copper tubing through which a steady stream of water is passed. The temperatures of the input and output water are measured (θ_3 and θ_4 respectively). In this way a temperature gradient is produced along the bar, which is found by measuring the temperatures (θ_2 and θ_3) at two intermediate points a distance l apart. These temperatures are measured by thermocouples or by thermometers placed in holes in the bar. Under steady-state conditions, thermal conductivity is calculated from the equation:

$$kA(\theta_2 - \theta_1)/l = mc(\theta_4 - \theta_3)$$

m being the mass of water passed in unit time, c the specific thermal capacity of the water, and A the bar's cross-sectional area.

second **1.** Symbol: s The SI base unit of time. Since 1967 it has been defined as the duration of 9 192 631 770 periods of a particular wavelength of radiation corresponding to a transition between two hyperfine levels in the ground state of the cesium-133 atom. Prior to 1967 the second was defined as 1/86 400 of a mean solar day or as a fraction of a tropical year. Due to irregularities in the Earth's rotation, however, these definitions proved too inaccurate for modern scientific and technical work. For less exacting purposes the second is considered to be a duration of 1/60 of a minute or 1/3600 of an hour.
2. Symbol: ″ A unit of angle: 1/3600 of a degree.

secondary cell *See* accumulator.

secondary colors The colors formed by mixing pairs of primary colors (hues). The effects can be shown by overlapping the beams from three projectors, each projecting a primary color (see diagram).

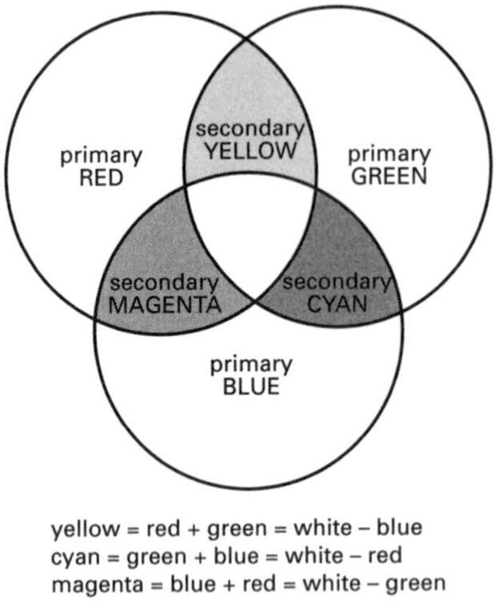

Secondary colors

secondary emission The emission of electrons from a surface as a result of the impact of high-energy electrons or ions. The incident particles must have sufficient energy to eject the 'free' electrons in the solid; i.e. to overcome the work function of the material.

secondary wavelets *See* Huygens' construction.

secondary winding *See* transformer (voltage).

Seebeck effect /**see**-bek/ The production of an e.m.f. in a circuit having two conductors, when the junctions between the two conductors are at different temperatures. The Seebeck effect is the phenomenon producing the e.m.f. in thermocouples. The effect is named for the Estonian-born German physicist Thomas Johann Seebeck (1770–1831). *Compare* Peltier effect.

seismic waves Waves that move through the Earth or on the surface of the Earth because of an earthquake. The different types of seismic wave have been used to obtain information about the interior structure of the Earth.

selenium /si-**lee**-nee-ŭm/ A metalloid element existing in several allotropic forms. The common gray metallic allotrope is very light-sensitive and is used in photocells, solar cells, some glasses, and in xerography. The red allotrope is unstable and reverts to the gray form under normal conditions.

Symbol: Se; m.p. 217°C (gray); b.p. 684.9°C (gray); r.d. 4.79 (gray); p.n. 34; r.a.m. 78.96.

selenium cell A photoconducting cell consisting of a piece of light-sensitive selenium semiconductor, which decreases its electrical resistance when light falls on it. It is powered by a battery. *See also* photocell.

self-absorption (**self-shielding**) The absorption by a large radioactive source of some of the radiation it produces, resulting in an overall decrease in external radiation.

self-inductance *See* self-induction.

self-induction The production of an e.m.f. in a conductor (e.g. a coil) caused by changes in the current in the conductor itself. The electromotive force is sometimes known as the *back e.m.f.* as its direction opposes the current change in accordance with Lenz's law. The relationship between the e.m.f. (E) and the rate of change of current (dI/dt) is given by the equation:

$$E = L(dI/dt)$$

where L is a constant for the coil known as its *self-inductance*. The SI unit of inductance is the henry (H). *Compare* mutual induction.

semiconductor /sem-ee-kŏn-**duk**-ter/ A material, such as silicon or germanium, that has a resistivity midway (on a logarithmic scale) between that of conductors and that of insulators. In a pure semiconductor material, called an *intrinsic semiconductor*, the concentrations of negative charge carriers (electrons) and positive charge carriers (holes) are the same. In practice, absolutely pure semiconductor material does not exist, but the intrinsic conductivity is used as a reference level for an impure semiconductor. The conductivity increases considerably when certain impurities are added, depending strongly on the type and concentration of the impurity. Such a material is called an *extrinsic semiconductor*. The process of adding impurity to control the conductivity is called *doping*.

Addition of, for example, phosphorus increases the number of electrons available for conduction, and the material is described as n-type, i.e. the charge carriers are negative. The impurity, or *dopant*, is called a *donor* impurity in this case.

Addition of, for example, boron has the effect of removing electrons; the dopant in this case being called an *acceptor* impurity because its atoms accept electrons for the completion of atomic bonds, leaving behind positive holes. The material is described as p-type, i.e. the charge carriers are positive; the holes are regarded as mobile.

Holes in p-type and electrons in n-type semiconductors are described as *majority carriers*; there are also electrons in p-type and holes in n-type semiconductors, but they are in the minority and are called *minority carriers*.

The electrical properties of semiconductors are related to the fact that in such materials there is only a small energy gap between the valence and conduction bands.

Semiconductors have very many uses. They distinguish solid-state devices from thermionic tubes, now mostly superseded by transistors. *See also* diode; integrated circuit; transistor.

semiconductor counter A device for detecting ionizing radiation, consisting of a semiconductor crystal with a potential difference across it. Particles striking the crystal produce electron-ion pairs, which increase the conductivity of the crystal for a brief period. Pulses of current are thus produced, and can be counted.

semiconductor diode *See* diode.

series Elements in a circuit are *in series* if connected so that each carries the same current in turn.

For resistors in series, the resulting resistance is given by:

$$R = R_1 + R_2 + R_3 + \ldots$$

For cells in series, the resulting e.m.f. is given by:

$$E = E_1 + E_2 + E_3 + \ldots$$

For capacitors in series, the resulting capacitance is given by:

$$1/C = 1/C_1 + 1/C2 + 1/C_3 + \ldots$$

Compare parallel.

series-wound motor An electric motor in which the field winding is connected in series with the armature winding. Series-wound motors are used for applications in which a high starting torque is required (e.g. in cranes). Initially the current is very high through both armature and field coil, and the torque is large. Series-wound motors, however, are less suitable than shunt-wound motors for maintaining steady speeds. *Compare* shunt-wound motor.

sextant /**sex**-tănt/ A navigational instrument used for the accurate measurement of the angle of a heavenly body above the horizon. Two mirrors are used: one sights the object; the other, part silvered, reflects an image of that object as well as showing the horizon.

shadow The result of the interception of

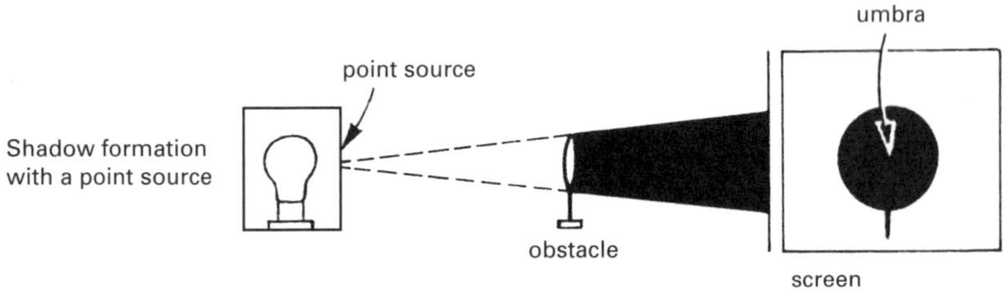

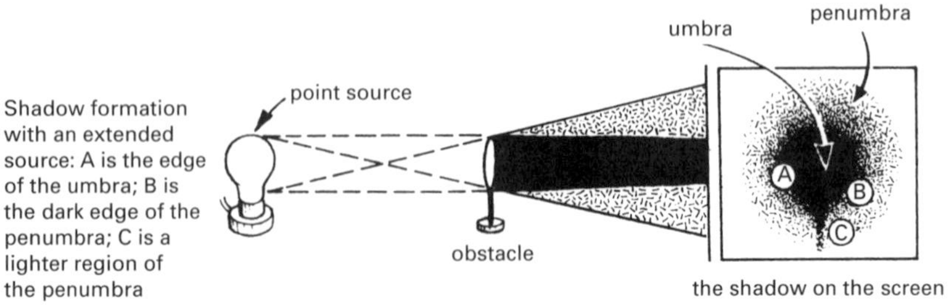

Shadow

light by an opaque obstacle. As light rays travel in straight lines, light from the source cannot be detected behind the obstacle. A point source of light will produce a very sharp dark shadow behind an obstacle. This is called an *umbra* (total shadow). A larger extended source will also produce an umbra behind an obstacle. However this will be surrounded by a *penumbra* (partial shadow). Each point in the penumbra will receive some light, so will not be in total shadow. Exactly the same ideas apply to shadows formed in other radiations.

shear modulus *See* elastic modulus.

shear strain *See* strain.

shell *See* electron shell.

SHF *See* superhigh frequency.

shielding The existence of a barrier that prevents bodies on one side of the barrier from the full effects of a field on the other side of the barrier. For example, barriers can be put round a body to protect it from electric or magnetic fields or, in the case of a nuclear reactor, to prevent dangerous radiation from escaping from the reactor. In atoms, the inner electrons can be regarded as a barrier that prevents the outer electrons from feeling the full electrostatic attraction of the positive nuclear charge.

s.h.m. *See* simple harmonic motion.

shock wave A high-intensity sound wave in air resulting from lightning or an explosion, or produced by an object traveling faster than the speed of sound in air.

short circuit The connection of two points in an electric circuit across a very low resistance, so that most of the current bypasses part of the circuit. Short circuits may occur accidentally because of wiring faults.

short sight *See* myopia.

shunt An element connected in parallel with another component and used to take part of the current in an electrical circuit. Thus, a low resistance in parallel with an ammeter, whose value can be selected in order to change the range of the meter, is called a shunt.

shunt-wound motor An electric motor in which the field winding is connected in parallel with the armature winding. Shunt-wound motors are used for maintaining a fairly steady speed (e.g. in machine tools). If the load increases the back e.m.f. falls and the current increases, tending to maintain the speed. *Compare* series-wound motor.

SI *See* SI units.

sideband A range of frequencies contained in a modulated carrier wave, either above or below the unmodulated frequency (hence upper and lower sidebands). The existence of sidebands is a consequence of the modulation process. For instance, in an amplitude modulated wave, if the carrier frequency is f_c and the modulating signal frequency is f_s, then the modulated wave has three components of frequency $f_c - f_s$, f_c, and $f_c + f_s$. *See also* carrier wave; modulation.

sidereal /sÿ-**deer**-ee-ăl/ Based on the stars, used especially of measurements.

sidereal time *See* day; time; year.

siemens /**see**-mĕnz/ (**mho**) Symbol: S The SI unit of electrical conductance, equal to a conductance of one ohm^{-1}. The unit is named for the German electrical engineer Ernst Werner von Siemens (1816–92).

sievert /**see**-vert/ Symbol: Sv. The derived SI unit of dose equivalent of ionizing radiation on living tissue, equal to one joule per kilogram. It is the absorbed dose multiplied by a factor that takes account of the relative effectiveness of different types of radiation of a given energy in causing biological damage. The statutory limits for exposure of the whole body, or specified parts of the body, of various categories of person are expressed in terms of this unit.

The unit is named for the Swedish physicist Rolf Sievert (1898–1966). *See* gray; rem; ionizing radiation.

sigma particle *See* elementary particles.

signal-to-noise ratio The ratio of the amplitude of an electrical signal to the average amplitude of the unwanted electrical disturbances (the noise).

sign convention A set of rules for determining the signs of distances related by various formulae in optics. In the *real is positive, virtual is negative* system, distances to real objects, images, and focal points are positive; distances to virtual points are negative. The other main sign convention is the *New Cartesian convention.* Intended to relate directly to the sign convention in the mathematical treatment of Cartesian coordinates, it treats distances to the right of the pole as positive; distances to the left of it are negative.

silicon /**sil**-ă-kŏn/ A hard brittle gray metalloid element. It has the electronic configuration of neon with four additional outer electrons; i.e., $[Ne]3s^23p^2$. Silicon accounts for 27.7% of the mass of the Earth's crust and occurs in a wide variety of silicates with other metals, clays, micas, and sand, which is largely SiO_2. It is used widely in semiconductor applications.

Symbol: Si; m.p. 1410°C; b.p. 2355°C; r.d. 2.329 (20°C); p.n. 14; r.a.m. 28.0855

silicon chip An INTEGRATED CIRCUIT manufactured from a thin wafer of silicon. It consists of alternate semiconductor and insulator layers into which the circuit is etched.

silver A transition metal that occurs native and as the sulfide and chloride. It is extracted as a by-product in refining copper and lead ores. Silver darkens in air due to the formation of silver sulfide. It is used in coinage alloys, tableware, and jewelry. Silver compounds are used in photography.

Symbol: Ag; m.p. 961.93°C; b.p. 2212°C; r.d. 10.5 (20°C); p.n. 47; r.a.m. 107.8682.

simple harmonic motion (**s.h.m.**) Any motion that can be drawn as a sine wave. Examples are the simple oscillation (vibration) of a pendulum or a sound source and the variation involved in a simple wave motion. Simple harmonic motion is observed when the system, moved away from the central position, experiences a restoring force proportional to the displacement from this position.

The equation of motion for such a system can be written in the form:

$$m\mathrm{d}^2x/\mathrm{d}t^2 = -\lambda x$$

λ being a constant. During the motion there is an exchange of kinetic and potential energy, the sum of the two being constant (in the absence of damping). The period (T) is given by:

$$T = 1/f$$

or

$$T = 2\pi/\omega$$

where f is frequency and ω is the pulsatance.
Other relationships are:

$$x = x_0\sin\omega t$$
$$\mathrm{d}x/\mathrm{d}t = \pm\omega\sqrt{(x_0^2 - x^2)}$$
$$\mathrm{d}^2x/\mathrm{d}t^2 = -\omega^2 x$$

Here x_0 is the maximum displacement; i.e. the amplitude of the vibration. In the case of angular motion, as for a pendulum, θ is used rather than x.

TYPICAL FORMS OF SIMPLE HARMONIC MOTION

System	*Period*	*Symbol*	*Notes*
mass on spring	$2\pi\sqrt{(m/k)}$	k = spring constant	Hooke's law region
uniform bobbing float	$2\pi\sqrt{(h/g)}$	h = immersion at rest	
liquid in U-tube	$2\pi\sqrt{(l/2g)}$	l = column length	
simple pendulum	$2\pi\sqrt{(l/g)}$	l = pendulum length	approximate s.h.m.

simple pendulum *See* pendulum.

sine wave The waveform resulting from plotting the sine of an angle against the angle. Any motion that can be plotted so as to give a sine wave is a simple harmonic motion.

singularity A point in space–time, such as the center of a BLACK HOLE or the source of the BIG BANG, for which general relativity theory predicts that certain physical quantities have an infinite value. The existence of singularities is usually interpreted as a limitation of general relativity theory. It is hoped that a quantum theory of gravity will deal with the problem of singularities in physics.

sinusoidal /sÿ-ŭ-**soid**-ăl/ Describing a quantity that has a waveform that is a sine wave.

siphon A device consisting of an inverted U-tube full of liquid used for moving a liquid from one place to another at a lower level.

siren /**sÿ**-rĕn/ A device for producing loud sounds consisting of a disk with radial louvers which is rotated at high speed. Air passes at high pressure through the louvers, generating a wailing sound. The term siren is also applied to various devices that produce loud noises electronically.

SI units (*S*ystème *I*nternational d'Unités) The modern, coherent, rationalized internationally adopted metric system of units. It has seven base units (the kilogram, second, kelvin, meter, ampere, mole, and candela) and two dimensionless units, formerly called supplementary units (the radian and steradian). Derived units are formed by multiplication and/or division of base units; a number have special names. Standard decimal prefixes are used for multiples and submultiples of SI units.

skin effect The phenomenon that causes very high-frequency electric currents to travel in only the thin outermost layer of metal in a conductor (or the innermost layer of a waveguide).

sky wave A radio wave that is transmitted between points by reflection from the ionosphere (a region of electrons and charged particles high in the Earth's atmosphere). *Compare* ground wave.

slide projector *See* projectors.

sliding friction *See* friction.

slip A way in which plastic deformation of a metal occurs. Slabs consisting of many layers of atoms within the metal grains slip over each other in one or more slip planes. *See also* creep.

slow neutron (**thermal neutron**) A neutron with low energy which can be captured by an atomic nucleus, particularly to initiate NUCLEAR FISSION. *See* fast neutron; moderator.

Snell's law *See* refraction; laws of.

sodium /**soh**-dee-ŭm/ A soft reactive metal. It has the electronic configuration of a neon structure plus an additional outer 3s electron. Electronic excitation in flames or the familiar sodium lamps gives a distinctive yellow color arising from intense emission at the so called 'sodium-D' line pair. The metal has a body-centred structure.

Symbol: Na; m.p. 97.81°C; b.p. 883°C; r.d. 0.971 (20°C); p.n. 11; r.a.m. 22.989768.

soft iron A form of iron (containing little carbon) that can easily be magnetized, but does not retain its magnetization when the external field is removed. Many such substances are now known. They are of great value as the core material of electromagnetic machines, such as transformers, generators, and electromagnets.

soft vacuum *See* vacuum.

solar cell A device for generating electricity using solar energy. *See* photocell.

BASE AND SUPPLEMENTARY SI UNITS

Physical quantity	*Name of SI unit*	*Symbol for SI unit*
length	meter	m
mass	kilogram	kg
time	second	s
electric current	ampere	A
thermodynamic temperature	kelvin	K
luminous intensity	candela	cd
amount of substance	mole	mol
*plane angle	radian	rad
*solid angle	steradian	sr

*supplementary units

DERIVED SI UNITS WITH SPECIAL NAMES

Physical quantity	*Name of SI unit*	*Symbol for SI unit*
frequency	hertz	Hz
energy	joule	J
force	newton	N
power	watt	W
pressure	pascal	Pa
electric charge	coulomb	C
electric potential difference	volt	V
electric resistance	ohm	Ω
electric conductance	siemens	S
electric capacitance	farad	F
magnetic flux	weber	Wb
inductance	henry	H
magnetic flux density	tesla	T
luminous flux	lumen	lm
illuminance (illumination)	lux	lx
absorbed dose	gray	Gy
activity	becquerel	Bq
dose equivalent	sievert	Sv

DECIMAL MULTIPLES AND SUBMULTIPLES USED WITH SI UNITS

Submultiple	*Prefix*	*Symbol*	*Multiple*	*Prefix*	*Symbol*
10^{-1}	deci-	d	10^{1}	deca-	da
10^{-2}	centi-	c	10^{2}	hecto-	h
10^{-3}	milli-	m	10^{3}	kilo-	k
10^{-6}	micro-	μ	10^{6}	mega-	M
10^{-9}	nano-	n	10^{9}	giga-	G
10^{-12}	pico-	p	10^{12}	tera-	T
10^{-15}	femto-	f	10^{15}	peta-	P
10^{-18}	atto-	a	10^{18}	exa-	E
10^{-21}	zepto-	z	10^{21}	zetta-	Z
10^{-24}	yocto-	y	10^{24}	yotta-	Y

solar constant *See* irradiance.

solar eclipse *See* eclipse.

solar energy Energy from the Sun, mainly electromagnetic radiation consisting of light (including ultraviolet) and heat. The Sun's rays are directly used to heat water in a heat exchanger (radiator) or, directed by mirrors, to produce high temperatures in a solar furnace. Solar energy can be converted into electricity using photocells. *See* heat exchanger; photocell.

solar neutrino problem The problem that, in experiments, fewer neutrinos are detected being emitted from the Sun than theories of the nuclear processes in the Sun predict. It is now thought that NEUTRINO OSCILLATIONS between the different types of neutrino can modify the theory to explain the discrepancy.

solar system The system consisting of the Sun, the planets that orbit round it, the moons of the planets, and the asteroids, comets, and other celestial objects that have their motion dictated by their gravitational interactions with the Sun. The Sun contains well over 99% of the mass of the solar system.

solar time *See* day; time; year.

solenoid /**soh**-lĕ-noid/ A tight coil of insulated wire with a diameter that is small compared with the length. When carrying a current, the solenoid's external magnetic field is similar to that of a bar magnet. It can be used as an electromagnet if a core of soft iron is used.

The flux density is given by $\mu_0 NI/l$, where N is the number of turns, I the current, and l the length.

solid *See* states of matter.

solid angle Symbol: Ω The three-dimensional analog of angle; it is subtended at a point by a surface (rather than by a line). The unit is the STERADIAN (sr), which is defined analogously to the radian (rad) – the solid angle subtending unit area at unit distance. As the area of a sphere is $4\pi r^2$, the solid angle corresponding to the revolution (2π radians) is 4π steradians.

solidification /sŏ-lid-ă-fă-**kay**-sŏn/ *See* freezing.

solid solution A solid homogeneous phase consisting of two or more substances. Many metal alloys are solid solutions.

solid state Describing electronic devices that use semiconductor circuit elements rather than thermionic tubes.

solid-state physics The branch of physics that deals with matter in the solid phase, especially SEMICONDUCTORS and superconductors. *See* superconductivity.

solubility /sol-yŭ-**bil**-ă-tee/ A measure of how much solute will dissolve in a given solvent; usually the temperature is also specified because solubility can vary widely with temperature. Solubility is generally expressed as mass per unit volume or as a percentage.

solute /**sol**-yoot, **soh**-loot/ A substance that dissolves in a solvent to form a solution.

solution A homogeneous mixture, usually liquid, of a solute and a solvent. *See also* molar; solid solution.

solvent /**sol**-vĕnt/ A substance, usually liquid, in which a solute dissolves to form a solution.

sonar /**soh**-nar/ (**asdic**) An acronym for *so*und *na*vigational *r*anging. A technique for locating objects underwater by transmitting a pulse of ultrasonic 'sound' and detecting the reflecting pulse. The time delay between transmission of the pulse and reception of the reflected pulse indicates the depth of the object. An apparatus for determining depths in this way is called an *echo sounder*.

sone /**sol**-vĕnt/ A unit of loudness, equal to 40 PHON or 40 decibels above the average threshold of hearing.

sonic boom A shock wave that is created when an object is moving faster than the speed of sound in the atmosphere of the Earth. The loud boom occurs when the wave given off by the object hits the Earth's surface.

sonometer /sŏ-**nom**-ĕ-ter/ (**monochord**) A device for investigating the vibration of a fixed string or wire. Essentially, it is a hollow box with the wire stretched across the top. One end of the wire is fixed; the other passes over a pulley and supports a load, which gives it tension. The vibrating length of wire is determined by two inverted V-shaped bridges under the wire. The wire is plucked and the note determined by comparison with tuning forks. In this way the effect of length and tension can be investigated. The frequency (f) produced is given by

$$f = (1/2l)\sqrt{(T/m)}.$$

where l is the length, T the tension, and m the mass per unit length of the string.

sorption /**sorp**-shŏn/ Absorption of gases by solids.

sorption pump A type of vacuum pump in which gas is removed from the system by absorption onto a porous solid (e.g. zeolite or charcoal) cooled in liquid nitrogen.

sound Vibrations transmitted by air, or some other material medium, in the form of alternate compressions and rarefactions of the medium. Sound is often called a pressure wave. Sound is a longitudinal waveform, and is characterized by its pitch (frequency), loudness (intensity), and quality. The speed of sound depends on the medium transmitting it. In air at 0°C it is 331.3 meter second^{-1}, in water at 25°C it is 1498 meter second^{-1}, and in glass at 20°C it is 5000 meter second^{-1}. The speed of sound in a medium is given by:

$$V = \sqrt{(E/\rho)}$$

where ρ is the density of the medium and E a modulus of elasticity. For a solid rod, E is the Young modulus; for a broad solid specimen it is the axial modulus; for liquids and gases, it is the adiabatic bulk modulus. Except at high pressures, this gives for gases:

$$V = \sqrt{(\gamma p/\rho)}$$

where p is the pressure and γ the ratio of the principal specific heat capacities. This is equivalent to

$$V = \sqrt{(\gamma RT/M)}$$

where R is the molar gas constant, T the thermodynamic temperature, and M the mass of one mole of the gas. V is independent of pressure and is proportional to the square root of the temperature.

Solids also transmit transverse and torsional vibrations at a speed given by

$$V = \sqrt{(G/\rho)}$$

where G is the shear modulus.

source *See* field-effect transistor.

south pole *See* poles; magnetic.

space charge A region in a vacuum, gas, or semiconductor in which there is a net electric charge resulting from an excess or deficiency of electrons.

space-time In Newtonian (pre-relativity) physics space and time are separate and absolute quantities; that is they are the same for all observers in any FRAME OF REFERENCE. An event seen in one frame is also seen in the same place and at the same time by another observer in a different frame.

After Einstein had proposed his theory of relativity, Minkowski suggested that since space and time could no longer be regarded as separate continua, they should be replaced by a single continuum of four dimensions, called space-time. In space-time the history of an object's motion in the course of time is represented by a line called the 'world curve'. *See also* relativity.

spark, electric *See* electric spark.

spark chamber An apparatus for detecting ionizing particles and making their tracks visible. A stack of up to 100 plates are connected alternately to a source of positive and negative high-voltage electricity and the chamber is filled with helium and neon. An incoming particle creates ion pairs along its track, the gas becomes conducting, and sparks jump between the plates (and can be photographed).

spark counter A type of detector for counting alpha particles. It consists of a wire (or mesh) held a close distance above an earthed plate. The wire has a high positive potential, less than that required to cause a spark. When an alpha particle passes close to the wire the electric field increases and a spark occurs between the electrodes. The sparks can be counted by a suitable circuit.

special theory *See* relativity.

specific /spĕ-**sif**-ik/ Denoting a physical quantity per unit mass. For example, volume (V_m) per unit mass (m) is called specific volume: $V_m = V/m$. In certain physical quantities the term does not have this meaning: for example, the old term *specific gravity* is properly called relative density; likewise *specific resistance* is resistivity.

specific activity For a radioisotope, the number of disintegrations per unit mass per unit time.

specific gravity *See* relative density.

specific heat *See* specific thermal capacity.

specific humidity *See* humidity.

specific latent heat *See* latent heat.

specific latent thermal capacity *See* latent heat.

specific resistance *See* resistivity.

specific rotatory power Symbol: α_m The rotation of plane-polarized light in degrees produced by a 10 cm length of solution containing 1 g of a given substance per milliliter of stated solvent. The specific rotatory power is a measure of the optical activity of substances in solution. It is measured at 20°C using the D-line of sodium.

specific thermal capacity (**specific heat capacity**) Symbol: c The heat required to raise the temperature of unit mass of a given substance by unit temperature. Generally, it is given as the heat in joules that raises the temperature of one kilogram by one kelvin; the units are then joules per kilogram per kelvin (J kg^{-1} K^{-1}).

Note that specific THERMAL CAPACITY is thermal capacity per unit mass, and is a property of the material or substance rather than of a given sample. Water, for example, has a specific thermal capacity of around 4.2 kJ kg^{-1} K^{-1}.

Strictly speaking, specific thermal capacities depend on the way in which pressure and volume change during heat transfer. Two principal specific heat capacities are defined: one at constant pressure (c_p) and one at constant volume (c_v). For solids and liquids these differ by only a few percent. The quantity measured directly is always c_p, but c_v can then be calculated. For gases, the principal values differ much more and both can be measured directly. For an ideal gas,

$$c_p - c_v = R/M$$

where R is the molar gas constant and M the mass of a mole of gas.

The ratio c_p/c_v is given the symbol γ (*gamma*). According to classical theory, in the case of an ideal gas this depends on the number of degrees of freedom (F) of the molecules according to the relationship $\gamma = 1 + 2/F$.

See also molar thermal capacity.

spectral emissivity *See* emissivity.

spectral line A particular wavelength of light emitted or absorbed by an atom, ion, or molecule. *See* line spectrum.

spectral series A group of related lines in the absorption or emission spectrum of a substance. The lines in a spectral series occur when the transitions are all between the same energy level and a set of different levels. *See also* Bohr theory.

spectrograph /**spek**-trŏ-graf, -grahf/ An instrument for producing a photographic record of a spectrum.

spectrometer /spek-**trom**-ĕ-ter/ **1.** An instrument for examining the different wavelengths present in electromagnetic radiation. A simple type of spectrometer has a flat rotatable table on which a prism or diffraction grating is placed. Light is directed onto this through a collimator tube, and the transmitted light is observed through a telescope. The telescope can be moved on an angular scale, so that different wavelengths of radiation can be seen.

There are many other types of spectrometer for producing and investigating spectra over the whole range of the electromagnetic spectrum. Often spectrometers are called *spectroscopes*.

2. Any of various other instruments for analyzing the energies, masses, etc., of particles. *See* mass spectrometer.

spectrophotometer /spek-troh-foh-**tom**-ĕ-ter/ A form of spectrometer able to measure the intensity of radiation at different wavelengths in a spectrum. Spectrophotometers can be made to work with infrared and ultraviolet radiation as well as with visible radiation.

spectroscope /**spek**-trŏ-skohp/ An instrument for examining the different wavelengths present in electromagnetic radiation. *See also* spectrometer.

spectroscopy /spek-**tros**-kŏ-pee/ **1.** The production and analysis of spectra. There are many spectroscopic techniques designed for investigating the electromagnetic radiation emitted or absorbed by substances. Spectroscopy, in various forms, is used for analysis of mixtures, for identifying and determining the structures of chemical compounds, and for investigating energy levels in atoms, ions, and molecules. In astronomy it is used for determining the composition of celestial objects and for measuring red shifts.

2. Any of various techniques for analyzing the energy spectra of beams of particles or for determining mass spectra.

spectrum (*pl.* **spectra**) /**spek**-trŭm/ **1.** A range of electromagnetic radiation emitted or absorbed by a substance under particular circumstances. In an *emission spectrum*, light or other radiation emitted by the body is analyzed to determine the particular wavelengths produced. The emission of radiation may be induced by a variety of methods; for example, by high temperature, bombardment by electrons, absorption of higher-frequency radiation, etc. In an *absorption spectrum* radiation with a continuous spectrum is passed through the sample. The radiation is then analyzed to determine which wavelengths are absorbed. *See also* band spectrum; continuous spectrum; line spectrum.

2. In general, any distribution of a property. For instance, a beam of particles may have a spectrum of energies. A beam of ions may have a mass spectrum (the distribution of masses of ions). *See* mass spectrometer.

spectrum, electromagnetic *See* electromagnetic spectrum.

specular reflection /**spek**-yŭ-ler/ *See* reflection.

speed Symbol: c Distance moved in unit time: $c = d/t$. Speed is a scalar quantity; the equivalent vector is velocity, or displacement in unit time.

Usage can be confusing. It is common to meet the word velocity where speed is more correct. For instance c_0 is the speed of light in free space, not its velocity; temperature relates to the root-mean-square speed of particles.

speed of light A fundamental constant equal to $2.997\,924\,58 \times 10^8$ m s⁻1. *See* light.

speed of sound *See* sound.

spherical aberration *See* aberration.

spherical polar coordinates A method of defining the position of a point in space by its radial distance, r, from a fixed point, the origin O, and its angular position on the surface of a sphere centred at O. The angular position is given by two angles θ and ϕ. θ is the angle that the radius vector makes with a vertical axis through O (from the south pole to the north pole). It is called the *co-latitude*. For points on the vertical axis above O, $\theta = 0$. For points lying in the 'equatorial' horizontal plane, $\theta = 90°$. For points on the vertical axis below O, $\theta = 180°$. ϕ is the angle that the radius vector makes with an axis in the equatorial plane. It is called the *azimuth*. For all points lying in the axial plane, that is, on, vertically above, or vertically below this axis, $\phi = 0$ on the positive side of O and $\phi = 180°$ on the negative side. This plane corresponds to the plane $y = 0$ in rectangular Cartesian coordinates. For points in the vertical plane at 90° to this, ($x = 0$ in rectangular Cartesian coordinates) $\phi = 90°$ in the positive half-plane and 270° in the negative half-plane. For a point $P(r,\theta,\phi)$, the corresponding rectangular Cartesian coordinates (x,y,z) are:

$$x = r\cos\phi\sin\theta$$
$$y = r\sin\phi\sin\theta$$
$$z = r\cos\theta$$

Compare cylindrical polar coordinates.

spherometer /sfi-**rom**-ĕ-ter/ An instrument for measuring the curvature of a spherical surface.

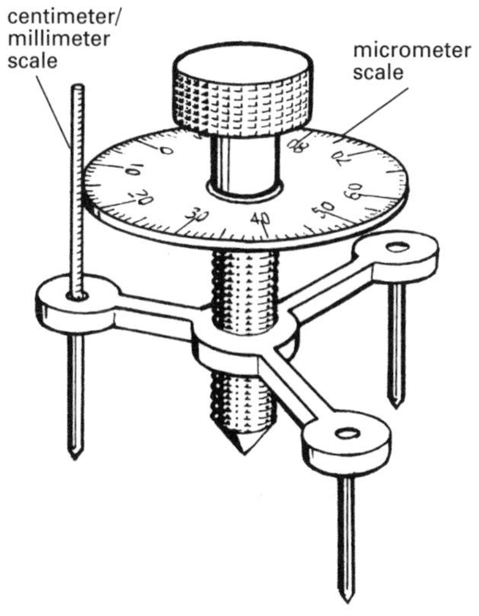

Spherometer

spin A property of certain elementary particles whereby the particle acts as if it were spinning on an axis; i.e. it has an angular momentum. Such particles also have a magnetic moment. In a magnetic field the spins line up at an angle to the field direction and precess around this direction. Certain definite orientations to the field direction occur such that $m_s h/2\pi$ is the component of angular momentum along this direction. m_s is the spin quantum number and h is the Planck constant. For an electron it has values +½ and –½.

spinthariscope /spin-**tha**-ră-skohp/ An instrument for observing the scintillations produced by ionizing radiation when it strikes a zinc sulfide screen.

S-pole *See* poles; magnetic.

square wave A stream of regular on-off pulses, usually in an electrical signal. The pulses are of equal height and the duration of each pulse is the same as the interval between them.

stability A measure of how hard it is to displace an object or system from EQUILIBRIUM.

Three cases are met in statics (*see* equilibrium). They differ in the effect on the center of mass of a small displacement. An object's stability is improved by: (a) lowering the center of mass; or (b) increasing the area of support; or by both. *See also* virtual work.

stable equilibrium Equilibrium such that if the system is disturbed a little, there is a tendency for it to return to its original state. *See* equilibrium; stability.

standard cell A voltaic cell whose e.m.f. is used as a standard. *See* Clark cell; Weston cadmium cell.

standard electrode A half cell used for measuring electrode potentials. The hydrogen electrode is the basic standard but, in practice, calomel electrodes are usually used.

standard model The combination of QUANTUM CHROMODYNAMICS and the WEINBERG–SALAM MODEL. In spite of its great success in describing all the non-gravitational interactions it is not a complete theory of elementary particles, firstly because it does not include gravity and secondly because many quantities of interest such as the mass of electrons, are empirical parameters rather than calculated from first principles. There are theories of elementary particles that attempt to go beyond the standard model including GRAND UNIFIED THEORIES and SUPERSTRING THEORY. *See* quark.

standard near-point distance *See* near-point.

standard pressure An internationally agreed value; a barometric height of 760 mmHg at 0°C; 101 325 Pa (approximately 100 kPa).

This is sometimes called the *atmosphere* (used as a unit of pressure). The *millibar*, used mainly in meteorology, is 100 Pa exactly.

standard temperature An internationally agreed value for which many measurements are quoted. It is the melting temperature of water, 0°C (273.15 K). *See also* STP.

standard temperature and pressure *See* STP.

standing wave *See* stationary wave.

star A heavenly body that generates its own light. The Sun is the most familiar example of a star. The heat and light given off by stars is due to nuclear fusion reactions inside the star. Stars gather together in galaxies.

Stark effect The splitting and displacement of atomic spectral lines in a strong transverse electric field. The effect is named for the German physicist Johannes Stark (1874–1957). *See also* Zeeman effect.

stat- A prefix used with a practical c.g.s. electrical unit to name the corresponding electrostatic unit. For example, the electrostatic unit of charge is called the *statcoulomb*. *Compare* ab-.

states of matter The three states – solid, liquid, and gas – in which matter normally exists.

Solids have fixed shapes and volumes – i.e. they do not flow, like liquids and gases, and they are difficult to compress. In solids the atoms or molecules occupy fixed positions in space. In most cases there is a regular pattern of atoms – the solid is crystalline.

Liquids have fixed volumes (i.e. low compressibility) but flow to take up the shape of the container. The atoms or molecules move about at random, but they are quite close to one another and the motion is hindered.

Gases have no fixed shape or volume. They expand spontaneously to fill the container and are easily compressed. The molecules have almost free random motion.

A PLASMA is sometimes considered to be a fourth state of matter.

static electricity 1. *See* electrostatics. 2. Electricity characteristic of charges that are at rest, as opposed to current electricity.

static friction *See* friction.

static pressure The pressure exerted by a fluid at rest, or flowing in such a way that the pressure at the point of interest is unaffected by the flow.

statics A branch of mechanics dealing with the forces on an object or in a system in equilibrium. In such cases there is no resultant force or torque and therefore no resultant acceleration.

stationary wave (**standing wave**) The interference effect resulting from two waves of the same type and equal frequency and intensity moving in opposite directions in the same region. The effect is most often caused when a WAVE is reflected back along its original path. At certain places the amplitude of one type of disturbance is a maximum. These places, which are called *antinodes* for this disturbance, occur at a separation of half a wavelength. Between these are places where this disturbance is always zero; these are called NODES. The nodes for one kind of disturbance are antinodes for the other. For example, in acoustics the closed end of a resonating pipe is a pressure antinode, which is a particle-velocity node. Just outside the open end there is a pressure node, which is an antinode for particle velocity. *See also* interference.

statistical mechanics The derivation of thermodynamics arising from the consideration of the distribution of energy (or certain other quantities) among the constituent particles of a system in equilibrium.

See Bose–Einstein statistics; entropy; Fermi–Dirac statistics; Maxwell–Boltzmann statistics

stator /**stay**-ter/ The fixed part of an electric motor or dynamo; or some similar device. *See also* rotor.

steam Water in the gaseous state. Note that the white 'steam' seen when boiling water contains tiny droplets of liquid water.

steam engine A type of HEAT ENGINE in which mechanical energy is generated from the pressure of steam. Steam engines were used extensively in the second half of the 19th century and the scientific analysis of their performance was one of the factors contributing to the development of thermodynamics as a subject. The steam engine is an example of an external combustion engine. *See also* internal combustion engine.

steam point *See* International Practical Temperature Scale.

steelyard An ancient weighing device, still widely used, as shown or in more complex arrangements. In use, the movable rider is placed so that the device rests horizontally. The weight of the load is then read from the scale directly.

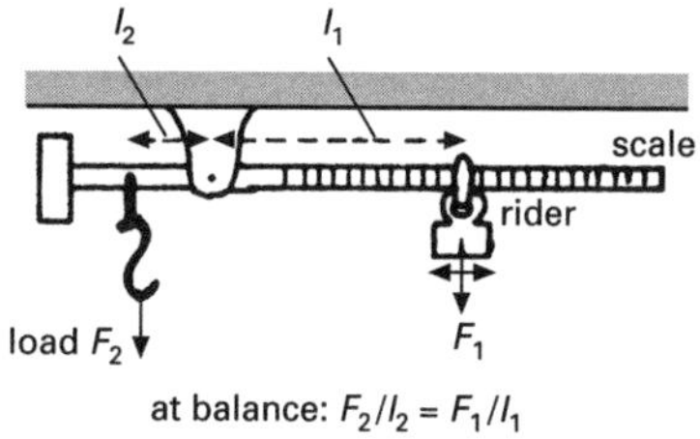

Steelyard

Stefan–Boltzmann constant /**stef**-ănz/ *See* Stefan's law.

Stefan's law The principle that the total energy, W, radiated by the surface of a BLACK BODY is proportional to the fourth power of the thermodynamic temperature (T). For a surface, $M = \sigma T^4$, where M is the energy radiated per unit area per second. The constant σ is known as the *Stefan constant* and is equal to 5.7×10^{-8} watt metre^{-1} kelvin^{-4} (W m^{-1} K^{-4}). The law was proposed by the Austrian physicist Josef Stefan (1835–93) in 1879 on experimental evidence and derived by Boltzmann in 1884 by thermodynamic theory. It was later shown to be a consequence of Planck's radiation law.

stellar dynamics The dynamics of stars insofar as they interact together in clusters or galaxies. Stellar dynamics is described by Newtonian gravity. For systems with a large number of stars Newtonian gravity is sometimes combined with the methods of statistical mechanics.

step-down transformer A transformer for reducing an alternating voltage.

step-up transformer A transformer for increasing an alternating voltage.

steradian /sti-**ray**-dee-ăn/ Symbol: sr The SI unit of solid angle, equal to that subtended at the center of a sphere by the area of a square on the sphere's surface, each of whose sides are equal in length to that of the sphere's radius.

stereophony /ste-ree-**off**-ŏ-nee, steer-ee-/ A type of sound reproduction that uses two or more channels and loudspeakers to produce sounds more like the original than can be reproduced by monaural techniques.

stereoscope /**ste**-ree-ŏ-skohp, **steer**-ee-/ A device that shows different views of the same scene to each eye in such a way as to give the impression of depth. Stereoscopes were very popular in the last century. Although they can be extremely effective, they are now rare.

stereoscopic vision /ste-ree-ŏ-**skop**-ik, steer-ee-/ *See* binocular vision.

stilb An m.k.s.a. unit of luminance, equal to the luminance produced by one candela (cd) per square centimeter (cm^{-2}). 1st = 1cd cm^{-2}.

stimulated emission If an atom or molecule is in an excited state it may make a transition to a lower-energy state spontaneously, with emission of a photon. Alternatively, radiation may be incident on the excited particle having the same frequency as would be emitted spontaneously. In this case the particle may undergo stimulated emission. The emitted radiation has the same direction and phase as the incident radiation, so the intensity is doubled. *See* laser.

stokes /stohks/ Symbol: St A former unit of kinematic viscosity used in the c.g.s. system. It is equal to 10^{-4} m^2 s^{-1}. The unit is named for the British mathematician and physicist Sir George Gabriel Stokes (1819–1903).

Stokes' law **1.** The law giving the viscous force *F* opposing the motion of a sphere radius *r* at a low speed *c* in a fluid of viscosity η

$$F = 6\pi r\eta c$$

When a sphere falls under gravity the speed increases until the downward force on it equals *F* when *c* has its maximum value, the *terminal speed*. Observation of the terminal speed can be used to determine viscosity.

For speeds such that REYNOLDS' NUMBER ($\rho rc/\eta$) is greater than about unity (ρ is the fluid density) the flow becomes turbulent and the resulting force is greater than that given by Stokes' law. The fall of a small raindrop is resisted by viscosity while that of a large raindrop is opposed by turbulent resistance. Stokes' law does not apply to large rapidly moving bodies such as tennis balls and pendulum bobs. *See also* turbulent flow.

2. In fluorescence, the rule that the emitted radiation has longer wavelength than the incident radiation.

stop An aperture, such as an iris or diaphragm, in an optical system.

storage battery *See* accumulator.

storage ring A torroidal evacuated ring that forms a part of some particle accelerators. Resembling a synchrotron in design, the ring has a stationary magnetic field to contain subatomic particles, which may be collided at high energies using two interlaced rings or by having particles circling in each direction in a single ring. *See* accelerator.

STP (**NTP**) Standard temperature and pressure. Conditions used internationally when measuring quantities that vary with both pressure and temperature (such as the density of a gas). The values are 101 325 Pa (approximately 100 kPa) and 0°C (273.15 K). *See also* standard pressure; standard temperature.

straight conductor An electrical conductor in the form of a long straight line. The magnetic flux density for the magnetic field associated with the electric current in a long straight conductor can easily be calculated and shown to be inversely proportional to the perpendicular distance from the conductor.

strain The fractional change in dimension produced by a STRESS applied to a body.

Tensile strain applies to the stretching of a body. It is the change in length divided by the original length ($\Delta l/l$). *Bulk strain* occurs when a body is subjected to a change of pressure. It is the change in volume divided by the original volume. *Shear strain* occurs when an angular deformation occurs, and is equal to the angular displacement produced. *See also* elasticity; elastic modulus.

strain gauge A device for measuring the strain or distortion on a surface. Some strain gauges work by changes in the electrical resistance of a wire attached to the surface. Others use magnetic induction in two coils, one fixed to the surface. Capacitance changes can be used in a similar way.

strange attractor A path in PHASE SPACE that is not closed. Strange ATTRACTORS are characteristic of chaotic behavior.

strangeness *See* quark.

streamline A line following the direction of the fluid in LAMINAR FLOW or *streamline flow*. Where the speed increases, as it does in a narrower section of a pipe, the streamlines are closer together.

stress When a system of opposing forces acts on a body the material is subject to some form of stress. This is expressed in terms of one of the forces divided by the area on which it acts. *Bulk stress* is a change of pressure applied to a fluid, or applied by a fluid to a solid. *Tensile stress* is a stress that stretches a body. *Shear stress* causes a deformation without any change of volume. *See also* elasticity; elastic modulus; strain.

string A one-dimensional object that appears in elementary-particle theory. Strings sometimes appear as solutions of the equations of quantum field theories. For example, cosmic strings are solutions for certain grand unified theories. Strings are the fundamental entities in *string theories*, i.e. theories in which pointlike elementary particles are replaced by strings of a finite length (*open strings*) or loops (*closed strings*). SUPERSTRING THEORY is the combination of string theory and SUPERSYMMETRY.

strobe /strohb/ *See* stroboscope.

stroboscope /**stroh**-bŏ-skohp/ (**strobe**) A device that allows something to be viewed as a series of short separate scenes, rather than continuously. It may be a light giving regular short intense flashes, or a rotating disk with slots through which the system is viewed. Stroboscopes are used in physics to investigate periodic motions (rotation or vibration). If the frequency of the illumination is equal to the frequency of vibration, the system is seen at the same point in its cycle at each view, and appears stationary.

strong interaction *See* interaction.

strontium /**stron**-shee-ŭm, -tee-/ A soft low-melting reactive metal. The electronic configuration is that of krypton with two additional outer 5s electrons.

Symbol: Sr; m.p. 769°C; b.p. 1384°C; r.d. 2.54 (20°C); p.n. 38; r.a.m. 87.62.

strontium unit Symbol S.U. A measure of the concentration of the radioisotope strontium-90 in bones, milk, or soil in relation to their calcium content.

subatomic particles /sub-ă-**tom**-ik/ *See* elementary particles.

subcritical /sub-**krit**-ă-kăl/ Describing an arrangement of fissile material that does not permit a sustained chain reaction because too many neutrons are absorbed without causing fission or otherwise lost. *Compare* supercritical. *See also* multiplication factor.

sublimation /sub-lă-**may**-shŏn/ A direct change of state from solid to vapor without melting.

subsonic /sub-**sonn**-ik/ Describing a speed that is less than the speed of sound in the medium concerned. *See* sound; supersonic.

substance /**sub**-stăns/ The substance of a body is expressed in terms of the number of constituent particles. *Compare* matter. *See also* mole.

subtractive process /sŭb-**trak**-tiv/ A process of color mixing by subtraction. *See* color.

sulfur /**sul**-fer/ A low melting non-metallic solid, yellow colored in its common forms. It has the electronic configuration $[Ne]3s^2 3p^4$. The element exhibits allotropy and its structure in all phases is quite complex. The common crystalline modification, *rhombic* sulfur, is in equilibrium with a *triclinic* modification above 96°C. Both have structures based on S_8-rings but the crystals are quite different. If molten sulfur is poured into water a dark red 'plastic' form is obtained in a semielastic form. The structure appears to be a helical chain of S atoms.

Symbol: S; m.p. 112.8°C; b.p. 444.6°C; r.d. 2.07; p.n. 16; r.a.m. 32.066.

Sun The star that is nearest to the Earth. It is nearly 150 million kilometers from the Earth and has a mass of about 1.9×10^{30} kilograms and a radius of nearly 700 000 kilometers. The Sun is about five billion years old and is about half-way through its life. The heat and light from the Sun, which sustains life on Earth, come from nuclear fusion reactions inside the Sun, which convert hydrogen into helium.

superatom /**soo**-per-at-ŏm/ *See* Bose-Einstein condensation.

superconductivity /soo-per-kon-duk-**tiv**-ă-tee/ Certain substances when cooled below a characteristic transition temperature become superconducting; that is, they lose all electrical resistance. At the same time they exhibit the Meissner effect – the exclusion of magnetic flux. Currents induced in circuits of materials in the superconducting state persist indefinitely without measurable change.

Most known superconductors have transition temperatures of only a few kelvins, so experiments have to be done with specimens cooled in liquid helium. Some superconductors with transition temperatures up to about 18 K have been known for many years. In the 1980s a number of ceramic materials were discovered that show superconductivity at much higher temperatures. Substances are known that are superconducting at liquid-nitrogen temperatures, and it is possible that materials may be produced that are superconducting at normal laboratory temperatures. Superconductors have important practical applications. For example, very strong magnetic fields can be produced by using electromagnets with superconducting coils (ordinary electromagnets are limited by the saturation of iron cores at about 2 tesla). Other possible uses may be in high-speed computers and in cheap long-distance power transmission. The theory of ordinary superconductivity is well established in terms of pairs of electrons, with the pairing being due to interaction between the electrons and the lattice vibrations. The theory of high-temperature superconductivity is not yet well established.

supercooling /soo-per-**koo**-ling/ The cooling of a liquid at a given pressure to a temperature below its melting temperature at that pressure without solidifying it. The liquid particles lose energy but do not spontaneously fall into the regular geometrical pattern of the solid. A supercooled liquid is in a metastable state and will usually solidify if a small crystal of the solid is introduced to act as a 'seed' for the formation of crystals. As soon as this happens, the temperature returns to the melting temperature until the substance has completely solidified.

supercritical /soo-per-**krit**-i-kăl/ Describing an arrangement of fissile material that sustains a branching chain reaction; i.e. more neutrons are being produced than are wasted and escape. *Compare* subcritical. *See also* multiplication factor.

superelastic collision /soo-per-i-**las**-tik/ A collision for which the restitution coefficient is greater than one. In effect the relative velocity of the colliding objects after the interaction is greater than that before.

The apparent energy gain is the result of transfer from potential energy. For example, if a collision between two trolleys causes a compressed spring in one to be released against the other, the collision may be superelastic. *See also* restitution; coefficient of.

superficial expansivity *See* expansivity.

superfluidity /soo-per-floo-**id**-ă-tee/ A property of liquid helium at very low temperatures. At 2.186 K liquid helium makes a transition to a superfluid state, which has a high thermal conductivity and flows without friction. The isotope helium-3 also becomes a superfluid but at much lower temperatures than helium-40.

superheating /soo-per-**hee**-ting/ **1.** The raising of the temperature of a liquid above the normal boiling temperature at the applied pressure. *See also* bumping.
2. The raising of the temperature of a vapor (usually steam) so that it is above the liquid–vapour equilibrium temperature at the applied pressure. The vapor is then said to be superheated or dry. In modern steam turbines the steam is normally passed from the boiler into a *superheater* before entering the turbine. This increases the efficiency and reduces damage to the turbine blades. *See also* efficiency; heat engine.

superheterodyne /soo-per-**het**-ĕ-roh-dÿn/ *See* heterodyne.

superhigh frequency /**soo**-per-hÿ/ (SHF) A radio frequency in the range between 30 GHz and 3 GHz (wavelength 1–10 cm).

supermembrane theory A theory that combines SUPERSYMMETRY with the basic entities being two-dimensional extended objects (called *membranes*). There are serious difficulties with supermembrane theory as a starting point for a unified theory of forces and particles but it may emerge from M THEORY.

supermodel A model that describes elementary particles that incorporates SUPERSYMMETRY. Supersymmetric GUTs and the supersymmetric standard model are examples of supermodels.

supernova The sudden explosive brightening of a star that occurs when a large star collapses to a NEUTRON STAR. A supernova is brighter than the whole of the rest of the galaxy it occupies. Supernova explosions are important in NUCLEOSYNTHESIS because many of the elements heavier than iron are created only in supernova explosions. In addition, supernova explosions scatter other elements made inside stars into space.

superposition, principle of /soo-per-pŏ-**zish**-ŏn/ *See* principle of superposition.

supersaturated solution /soo-per-**sach**-ŭ-ray-tit/ A solution that contains more solute than a saturated solution at the same temperature. It is unstable and readily crystallizes to form a saturated solution.

supersonic /soo-per-**sonn**-ik/ Describing a speed that is greater than the speed of SOUND in the medium concerned. *See* sound; subsonic.

superstring theory /**soo**-per-string/ A unified theory of the fundamental interactions that combines STRING THEORY with SUPERSYMMETRY. In superstring theory the basic entities are one-dimensional objects called *superstrings* which are about 10–35 m long. This very short distance is equivalent to energies of about 1019 GeV.

A purely *bosonic string*, i.e. a string that is associated with bosons and does not contain fermions is only consistent as a quantum theory in 26-dimensional space–time. A superstring contains both bosons and fermions and is only consistent as a quantum theory in 10-dimensional space–time. There is some evidence that superstring theory is free of the infinities that plague attempts to construct a quantum field theory of gravity. However, a complete proof of this has not yet been established.

Superstrings can either be open strings or closed strings. By the mid 1980s it was found that there were five superstring theories which are consistent. In the mid-

1990s it was found that the five types of superstring theory and also 11-dimensional supergravity theory are related to each other by a set of dualities. This has led to the postulate that there may be a theory in 11-dimensional space–time called *M-theory* which contains all the five superstring theories and supergravity theory as limiting cases.

supersymmetry /soo-per-**sim**-ĕ-tree/ A type of symmetry that can unify fermions and bosons into one mathematical structure. Supersymmetry is thought to be a fundamental principle of elementary particle theory, with supersymmetry being a broken symmetry. It is hoped that supersymmetric 'partners' of all the known elementary particles will be discovered by the end of the first decade of the twenty-first century.

supertransuranics (superheavy elements) The very heavy elements with atomic numbers of about 114, which are predicted to be relatively stable by nuclear shell theory, with 114 being a magic number. There is some evidence that some of these elements have been observed.

supplementary units The dimensionless units – the radian and the steradian – used along with base units to form derived units. *See* SI units.

suppressor grid *See* pentode.

surface charge density *See* charge density.

surface tension Symbol: γ or σ The attraction between molecules (*cohesion*) in the plane of the surface of a liquid, which thus acts a bit like an elastic skin containing the liquid. Surface tension explains why water can drip slowly from a tap and why mercury gathers into globules on a flat surface. Molecules that are surrounded by others are, on average, attracted equally in all directions, since the liquid is under pressure. At the surface, however, the intermolecular spacings are larger in the plane of the surface, and the molecules attract each other. Consequently, the surface layer is in tension.

γ is defined as the force per unit length acting normally in the plane of the surface on either side of a line in the surface. It is measured in units of newton metre^{-1}. Sometimes called *free surface energy*, it may also be defined as the work done in increasing the surface area by a square meter. Both are defined at given temperature, as surface tension falls with temperature rise.

surfactant /ser-**fak**-tănt/ (**surface active agent**) A substance that reduces the SURFACE TENSION of a liquid and thereby improves its wetting properties. Surfactants are used in detergents, foaming agents and wetting agents.

susceptibility /sŭ-sep-tă-**bil**-ă-tee/ Symbol: χ The ratio, for a given substance, of the magnetization of a sample to the magnetic field strength applied. In SI it equals (μ_r – 1), where μ_r is the relative permeability. The value of χ determines whether a substance shows *paramagnetism*, *diamagnetism*, or *ferromagnetism*. A diamagnetic material has a negative susceptibility while paramagnetic and ferromagnetic materials have small and large positive susceptibilities respectively.

suspension A mixture that consists of a fluid in which small solid particles are suspended, i.e. the particles neither float nor sink but are distributed uniformly throughout the fluid.

switch A mechanical or solid-state device for opening and closing an electric circuit.

symmetry The set of *invariances* of a system, i.e. the set of operations, known as *symmetry operations*, on a system that leave the system unchanged. Examples of symmetry operations include rotations and reflections for molecules and translations for crystals. Symmetry is studied systematically by a branch of mathematics called GROUP THEORY. Symmetry and BROKEN SYMMETRY are two of the most important concepts throughout physics and chemistry.

synchrocyclotron /**sink**-rŏ-**sÿ**-klŏ-tron/ A device for accelerating particles to high energies; a development of the cyclotron in which the frequency applied to the 'dees' is decreased to compensate for the progressively increased period of revolution caused by the relativistic increase in mass at high energies, typically hundreds of MeV. For example, the 4.7-m machine at Berkeley in the United States can produce particles of energy up to 730 MeV. The synchrocyclotron is also sometimes called a frequency modulated (FM) cyclotron.

synchronous motor An electric motor than runs at a speed proportional to the frequency of the voltage supply. A rotating magnetic field is produced by separate fixed coils energized by an alternating current supply. This field pulls the rotating part round with it. The rotating part of the device consists of a permanent magnet or an electromagnet. When starting, the motor behaves like an induction motor until the synchronous speed is reached. *See also* electric motor; induction motor.

synchronous orbit *See* parking orbit.

synchrotron radiation /**sink**-rŏ-tron/ Electromagnetic radiation produced by electric charges that are moving very rapidly in a magnetic field. Synchrotron radiation is used extensively in the study of matter and occurs naturally in various astrophysical contexts, notably in pulsars.

Système International d'Unités /see-**stem** an-tair-nas-yo-**nal** doo-nee-**tay**/ *See* SI units.

T

tachyon /**tak**-ee-on/ A hypothetical subatomic particle that can travel faster than the speed of light.

tandem generator An arrangement of two linear accelerators end to end. The ends are grounded and there is a common anode at the center at a potential difference of up to 20 million volts. Negative ions are accelerated from ground potential at one end to the center, where the surplus electrons are stripped off to convert them into positive ions, which then continue to accelerate to the far end. They thus get the acceleration due to the high potential twice over. Protons can be given energies up to 40 MeV.

tangent galvanometer A galvanometer in which a magnetized needle is pivoted to rotate in a horizontal plane at the center of a large fixed vertical coil. Current flowing through the coil turns the needle. Tangent galvanometers are now rarely used.

tantalum /**tan**-tă-lŭm/ A silvery transition element. It is strong, highly resistant to corrosion, and is easily worked. Tantalum is used in turbine blades and cutting tools and in surgical and dental work.

Symbol: Ta; m.p. 2996°C; b.p. 5425 ± 100°C; r.d. 16.654 (20°C); p.n. 73; r.a.m. 180.9479.

tau-particle /taw, tow/ A type of lepton that has exceptionally high mass and the same negative charge as the electron and muon. *See* lepton.

technetium /tek-**nee**-shee-ŭm/ A transition metal that does not occur naturally on Earth. It is produced artificially by bombarding molybdenum with neutrons and also during the fission of uranium. It is radioactive.

Symbol: Tc; m.p. 2172°C; b.p. 4877°C; r.d. 11.5 (est.); p.n. 43; r.a.m. 98.9063 (^{99}Tc); most stable isotope ^{98}Tc (half-life 4.2×10^6 years).

telecommunications Any method of communicating at a distance, usually taken to mean those methods that use electricity or radio waves. They include telegraphy, telephony, radio, television, and radar.

telegraphy A form of telecommunications that employs pulses of electricity as coded signals carried along wires. The signals are send along radio beams in radio telegraphy.

telemetry A method of sending information over long distances, usually as a modulated radio signal.

telephony A form of telecommunications that employs a varying electric current to carry voice signals along wires. The signals are send along radio beams in radio telephony.

telephoto lens A lens that is used in cameras to obtain an image of a distant object without using a lens with a very long focal length. A telephoto lens has the combination of a converging lens and a weak diverging lens, which is nearer the film in the camera.

telescope An optical system for collecting radiation from a distant object and producing an enlarged image. *Optical telescopes* use visible radiation. There are two types: REFRACTORS (or *refracting telescopes*) in which the light is collected by a converging lens; and REFLECTORS (or *reflecting telescopes*) in which a converging mirror is used. The light collected by the objective forms an image, which is magnified by an eyepiece. A large objective is desirable in that it collects more light from faint distant objects. A large objective also minimizes diffraction effects.

Reflector systems have a number of advantages over refractors in theory and practice, and the largest *astronomical telescopes* are all reflectors. They do not have the same problems of chromatic aberration; large, comparatively cheap, robust instruments can give good images. Telescopes for non-astronomical use are refractors that produce an upright image (known as *terrestrial telescopes*).

The performance of telescopes is described in terms of:

(1) *magnifying power* – the angular magnification obtained (usually the ratio of focal distance of objective to that of eyepiece).

(2) *aperture* – determining the amount of light admitted, and thus the minimum magnitude of sources that can be seen.

(3) *resolving power* – the ability to separate images of close objects.

Telescopes for use in astronomy have been developed for other forms of electromagnetic radiation.

television A form of telecommunications that employs radio to transmit picture (video) and sound (audio) signals over a distance. Closed-circuit television (CCTV) generally operates over only small distances and the signals are carried along wires.

tellurium /te-**loor**-ee-ŭm/ A brittle silvery metalloid element belonging to group 16 of the periodic table. It is found native and in combination with metals. Tellurium is used mainly as an additive to improve the qualities of stainless steel and various metals.

Symbol: Te; m.p. 449.5°C; b.p. 989.8°C; r.d. 6.24 (20°C); p.n. 52; r.a.m. 127.6.

temperature A measure of the 'hotness' of a body. Heat flows from bodies of higher temperature to ones at lower temperature. It can be shown that the energies of the particles in a body are related in a complicated way to temperature, in all cases increasing as the temperature is raised. Formerly temperature was defined arbitrarily in terms of the variations in certain physical properties. It is now defined in absolute terms in relation to the laws of thermodynamics. *See also* temperature scale; thermometer; thermodynamic temperature.

temperature coefficient A coefficient determining how some physical property changes with temperature. For example, the temperature coefficient of resistance is the coefficient in the equation showing how resistance or resistivity varies with temperature. *See* resistance thermometer.

temperature scale A practical scale for use in measuring temperature. Formerly temperature scales were defined arbitrarily in terms of the changes of certain physical quantities, calibrated at two *fixed points*. For example, a mercury-in-glass scale was defined such that temperature intervals were proportional to the length of the thread of mercury in the capillary tube of a mercury thermometer. The fixed points were originally the *ice point* and the normal human body temperature. The latter was later replaced by the *steam point*. The difference between the fixed points was called the *fundamental interval*. This interval was divided into temperature units. In the case of the Celsius (centigrade) scale the interval between the ice and steam points was 100 degrees. Such arbitrary scales necessarily agree at the fixed points, but generally differ at other temperatures. Thus a certain temperature might be 200°C on a mercury-in-glass scale, 210°C on a certain resistance thermometer scale, 198°C on a certain constant-volume gas scale, and so on.

Modern thermometry is based upon Kelvin's *thermodynamic* (*ideal gas*) scale, which is independent of the properties of any real substance. It has one fixed point, the triple point of water. The unit (the kelvin, K) is defined so that the triple point is at 273.16 K measured from absolute zero. In practice, temperatures are measured on the *International Practical Temperature Scale* (1968), which is defined by eleven fixed points together with standard methods of measurement between these points, so as to reproduce the theoretical scale as nearly as possible.

On the absolute thermodynamic scale the ice point is 273.15 K and the steam point is 373.15 K. Hence it is possible to use a Celsius thermodynamic scale where the ice point is 0°C and the steam point is 100°C. The degrees have the same size on the Celsius and absolute scales.

temporary magnetism Substances show temporary magnetism if any magnetism disappears as soon as the inducing field is removed. They are easy to magnetize, and are often called 'soft', as in 'soft iron'.

tensile strain *See* strain.

tensile strength The tensile *stress* at which a material breaks apart or deforms permanently.

tensimeter /ten-**sim**-ĕ-ter/ A device that measures the difference in vapor pressures between two liquids.

tension A stretching force. *See* strain; stress; surface tension.

tensometer /ten-**som**-ĕ-ter/ A machine for measuring the tensile strength and other mechanical properties of materials.

tera- Symbol: T A prefix denoting 10^{12}. For example, 1 terawatt (TW) = 10^{12} watts (W).

terbium /**ter**-bee-ŭm/ A soft ductile malleable silvery rare element of the lanthanoid series of metals. It occurs in association with other lanthanoids. One of its few uses is as a dopant in solid-state devices.

Symbol: Tb; m.p. 1356°C; b.p. 3123°C; r.d. 8.229 (20°C); p.n. 65; r.a.m. 158.92534.

terminal A point in an electric circuit to which a connecting lead can be attached.

terminal speed The steady final speed reached by a body in a fluid when the resultant force on it is zero.

terrestrial magnetism *See* Earth's magnetism.

terrestrial telescope A refracting telescope for non-astronomical use: one producing an erect image. Terrestrial telescopes are either GALILEAN TELESCOPES or KEPLERIAN TELESCOPES with added inverting lenses. *See also* refractor.

tesla /**tess**-lă/ Symbol: T The SI unit of magnetic flux density, equal to a flux density of one weber per square meter. 1 T = 1 Wb m^{-2}. The unit is named for the Croatian–American physicist Nikola Tesla (1856–1943).

Tesla coil A device for producing high voltage oscillations from a low voltage direct current source. It consists of a transformer with a high ratio of secondary to primary turns and a make-and-break mechanism to interrupt the current in the secondary circuit.

tetrode /**tet**-rohd/ A THERMIONIC TUBE with one grid more than the TRIODE, called the *screen grid*. It is placed between the control grid and the anode, and is normally kept at a fixed positive potential. The purpose of the screen grid is to minimize the capacitance between the control grid and the anode and thus improve the high-frequency performance of the triode as an amplifier or oscillator.

A disadvantage of the tetrode is that, at high anode voltages, secondary electrons emitted from the anode are captured by the screen grid, causing a drop in the anode current. This is overcome in the PENTODE.

thallium /**thal**-ee-ŭm/ A soft malleable grayish metallic element belonging to group 13 of the periodic table. Thallium is highly toxic and was used previously as a rodent and insect poison. Various compounds are now used in photocells, infrared detectors, and low-melting glasses.

Symbol: Tl; m.p. 303.5°C; b.p. 1457°C; r.d. 11.85 (20°C); p.n. 81; r.a.m. 204.3833.

theodolite /th'ee-**od**-ŏ-lÿt/ An optical instrument (using prisms and lenses) for measuring horizontal and vertical angles with great accuracy. It is an essential tool in surveying.

theorem of parallel axes If I_0 is the moment of inertia of an object about an axis, the moment of inertia I about a parallel axis is given by:

$$I = I_0 + md^2$$

m is the mass of the object and d is the separation of the axes.

theory of everything (TOE) A unified theory of all the fundamental interactions, elementary particles, and cosmology. A theory of everything will involve a model that accounts for both quantum mechanics and gravity. It is hoped that SUPERSTRING THEORY will be developed into such a theory.

therm A unit of heat energy equal to 10^5 British thermal units (1.055 056 joules).

thermal capacity (**heat capacity**) Symbol: C The heat required to raise the temperature of a body by one unit. It is usually in joules per kelvin (J K^{-1}). The thermal capacity may be affected by volume changes. Two principal thermal capacities are defined – one at constant pressure (C_p) and one at constant volume (C_v). For liquids and solids, these may differ by only a few percent, but for gases the differences are much larger. *See also* molar thermal capacity; specific thermal capacity.

thermal conductivity *See* conductivity (thermal).

thermal diffusion A method of separating gas molecules of different masses by maintaining one part of the gas at a lower temperature than the other (i.e. producing a temperature gradient along a column of gas) – the more massive molecules tend to stay at the low-temperature end. It can be used for the separation of isotopes.

thermal equilibrium A state in which there is no net flow of heat. If two bodies are in thermal equilibrium, then they have the same temperature. *See also* equilibrium.

thermal expansion *See* expansion; thermal.

thermalization /ther-mă-li-**zay**-shŏn/ The slowing down of FAST NEUTRONS to thermal energies (converting them to SLOW NEUTRONS) so that they can initiate nuclear fission. In a nuclear reactor, it is carried out by a MODERATOR.

thermal neutron *See* moderator; slow neutron.

thermal radiation *See* radiant heat; radiation.

thermal reactor *See* nuclear reactor.

thermionic diode /ther-mee-**on**-ik/ *See* diode

thermionic emission The emission of electrons from the surface of a heated metal or metal oxide. The energy of the emitted electrons is relatively low, and their rate of emission increases with the temperature of the surface. The thermal energy of the electrons is sufficient to eject them against the counter-attraction of the surface atoms. Heated cathodes are used as sources of electrons in thermionic tubes and electron guns.

thermionics /ther-mee-**on**-iks/ The branch of physics that deals with the emission of electrons or ions from a heated substance. It is a part of electronics. *See* thermionic emission.

thermionic tube A device that uses the thermionic effect and exhibits the action of a valve, i.e. current flow in one direction only. In the simplest type, a diode valve, the electron-emitting cathode surface is contained in an evacuated glass envelope together with the second electrode, the anode. Electrons emitted from the cathode by THERMIONIC EMISSION, flow toward the anode if it is made positive with respect to the cathode. This constitutes a current. Reversal of the potential causes the current to fall to zero. There is no current, or very little, in the reverse direction because the anode does not emit electrons.

There are many other types of vacuum tube incorporating additional electrodes, called GRIDS. Some of these types may be filled with gas at low pressure, in which case the valve is called a thyratron. For many of its original uses the thermionic tube has been replaced by solid-state transistors, but it is still used in special cases such as high-power amplification (for radio transmission), cathode-ray tubes, and the klystron and magnetron oscillators. *See also* pentode; tetrode; triode.

thermistor /ther-**mist**-er/ An electrical component with a high temperature coefficient of resistance (i.e. the electrical resistance changes rapidly with temperature). Thermistors are usually semiconductor devices; in one type the resistance decreases with temperature rise. They can be used as sensitive resistance thermometers, which operate at various temperatures. Their thermal capacity is small, so they rapidly attain the temperature to be measured. Another use is in electronic circuits in which it is desirable that the current should increase to a maximum value slowly.

thermocouple /**ther**-moh-kup-ăl/ A pair of wires of different conductors or semiconductors, welded or soldered at one end, used to measure temperature. A small e.m.f. is produced at the junction between the metals, and this changes with temperature. The thermocouple (or *thermojunction*) can thus be used as a *thermoelectric thermometer*.

Thermocouples are convenient and have low thermal capacity. In inaccurate work a single junction can be used with a galvanometer in the circuit to measure the current produced by the thermoelectric e.m.f. In more accurate work two identical junctions are connected in series. One junction is maintained at a constant low temperature and the other used for temperature measurement. A potentiometer is used to measure the e.m.f.

See also International Practical Temperature Scale; Seebeck effect; thermometer; thermopile.

thermodynamic equilibrium /ther-moh-dÿ-**nam**-ik/ *See* equilibrium.

thermodynamics /ther-moh-dÿ-**nam**-iks/ The study of the internal energy of bodies and the effects on this of the processes heat and work. It has applications in engineering, physical chemistry, and many branches of physics.

The *first law of thermodynamics* states that the total energy in a closed system is conserved (constant). This means that energy cannot be created or destroyed, but in all processes is simply converted from one form to another, or transferred from one system to another, or both.

A mathematical statement of the first law is:

$$\delta Q = \delta U + \delta W$$

Here, δQ is the heat transferred to the system, δU the change in internal energy (resulting in a rise or fall of temperature), and δW is the external work done by the system. A gas for example may expand at constant pressure, in which case $\delta W = p\delta V$.

The *second law of thermodynamics* can be stated in a number of ways, all of which are equivalent. One is that heat cannot pass from a cooler to a hotter body without some other process occurring. Another is the statement that heat cannot be totally converted into mechanical work – i.e. heat engine cannot be 100% efficient.

The *third law of thermodynamics* states that the entropy of a pure substance tends to zero as its thermodynamic temperature approaches zero. This implies that absolute zero is an unattainable limit.

Often a *zeroth law of thermodynamics* is given: that if two bodies are each in thermal equilibrium with a third body, then they are in thermal equilibrium with each other. This is considered to be more fundamental than the other laws because they assume it. *See also* Carnot cycle; entropy; statistical mechanics.

thermodynamic temperature Temperature measured on a scale that is independent of any real substance. The usual scale is that proposed by Thomson (later Lord Kelvin) in 1848, slightly amended in later work. It is based on the concept of an ideal reversible heat engine. Since the EFFICIENCIES of all such engines working between two given temperatures are equal, Thomson proposed defining the scale in terms of these efficiencies. The thermodynamic temperatures T_1 and T_2 of the isothermal processes of a Carnot cycle are defined to be in the ratio of the heat intake to the heat output:

$$T_1/T_2 = Q_1/Q_2$$

To complete the definition the value of one standard temperature must be specified. This is now that of the triple point of water (273.16 K). The unit on this scale is called the kelvin (symbol: K). Kelvin showed that this scale was identical to one defined in terms of an ideal gas. In statistical mechanics it can be shown that the thermodynamic temperature determines the distribution of energies among the particles of a body.

See also Carnot's principle; gas laws; International Practical Temperature Scale; temperature scale.

thermoelectric effects /ther-moh-i-**lek**-trik/ Effects relating thermal energy and electricity. *See* Peltier effect; Seebeck effect; Thomson effect.

thermoelectric thermometer *See* thermocouple.

thermojunction /ther-moh-**junk**-shŏn/ *See* thermocouple.

thermoluminescent dating /ther-moh-loo-mă-**ness**-ĕnt/ A method for dating the firing of samples of pottery. As a result of absorbed alpha radiation, electrons become trapped at energy levels higher than normal. If the temperature is raised, they revert to the ground state with emission of photons (light). The amount of light emitted can be measured. This depends on the number of trapped electrons, which in turn depends on the lapse of time since the pottery was fired. It also depends on the amount of irradiation by alpha particles during that period and the type of material.

thermometer /ther-**mom**-ĕ-ter/ Any instrument or apparatus used for measuring temperature. Thermometers depend for their action on a *thermometric property* – i.e. a physical property that changes in a known way with temperature. For instance, a mercury thermometer depends on the expansion of a thin column of mercury from a bulb into a capillary tube. The material or substance used for its thermometric property is the *working substance* of the thermometer. *See also* pyrometer.

thermometer, clinical *See* clinical thermometer.

thermometer, maximum and minimum *See* maximum-and-minimum thermometer.

thermometric property /ther-moh-**met**-rik/ *See* thermometer.

thermometry /ther-**mom**-ĕ-tree/ The branch of physics concerned with methods of temperature measurement.

thermonuclear reaction /ther-moh-**new**-klee-er/ *See* fusion.

thermopile /**ther**-moh-pÿl/ A radiant energy detector or meter. It consists of a series of thermocouples, usually made of antimony and bismuth. Radiation strikes the blackened hot junctions, while the cold junctions are kept shielded. Measurement of the thermoelectric e.m.f. produced, enables the temperature of the hot junction to be calculated and hence the total radiation received from the source.

thermostat An instrument used for maintaining temperature within certain limits. It uses a device that cuts off the power when the required temperature is exceeded, then reconnects the supply when the temperature falls. A common method is to exploit the expansion and contraction of a substance with temperature. *See also* bimetallic strip.

thin-film interference The interference of light that occurs when light interacts with a thin film of material. It occurs because there is reflection and refraction of the light in each layer in the material, with interference occurring between reflected light from different layers. The colors observed in a thin film of oil is an example of thin-film interference.

thixotropy /thiks-**ot**-rŏ-pee/ The property of certain types of fluid whose viscosity decreases when the fluid is stressed. Thixotropic fluids are used to make nondrip paints.

Thomson effect The production of an electric potential gradient along a conductor as a result of a temperature gradient along it. Points at different temperatures in the conductor have different electrical potentials. The effect is named for the British physicist William Thomson (subsequently Lord Kelvin) (1824–1907).

thorium /**thor**-ee-ŭm, **thoh**-ree-/ A toxic radioactive element of the actinoid series that is a soft ductile silvery metal. It has several long-lived radioisotopes found in a variety of minerals including monazite. Thorium is used in magnesium alloys, incandescent gas mantles, and nuclear fuel elements.

Symbol: Th; m.p. 1780°C; b.p. 4790°C (approx.); r.d. 11.72 (20°C); p.n. 90; r.a.m. 232.0381.

thorium series *See* radioactive series.

thoron /**thor**-on, **thoh**-ron/ *See* emanation.

three-body problem The problem of understanding the motion of three bodies that interact in some specified way. The three-body problem cannot be solved exactly either in Newtonian mechanics or in quantum mechanics. Related to this, chaotic motion can occur in the three-body problem.

thulium /**thoo**-lee-ŭm/ A soft malleable ductile silvery element of the lanthanoid series of metals. It occurs in association with other lanthanoids.

Symbol: Tm; m.p. 1545°C; b.p. 1947°C; r.d. 9.321 (20°C); p.n. 69; r.a.m. 168.93421.

thyratron /**th'ÿ**-ră-tron/ A gas-filled thermionic tube, a type of triode, that can be used as a RELAY.

thyristor /th'ÿ-**ris**-ter/ (**silicon-controlled rectifier**) A semiconductor device, a type of diode, that can be used to control current supply, for example to an a.c. motor.

tidal power Power obtained by harnessing the tidal movement of water in the oceans. One method employs water turbines turned by the flow of tidal water. Other experimental methods use large floats that rise and fall with the tide.

tides The variation in water levels on the surface of the Earth due to the gravitational interactions between the Earth and the Moon and the Sun. When the Sun, Moon and Earth are all in a straight line *spring tides*, i.e. the tides with the biggest range, result. When the gravitational forces between the Earth and the Moon and between the Earth and the Sun are at right angles to each other *neap tides*, i.e. the tides with the least range, result. The tide-producing effect of the Moon is about twice that of the Sun.

timbre /**tim**-ber, **tam**-ber/ *See* quality of sound.

time Measurement of duration between events. The SI unit, the second, is based on vibration of radiation absorbed by cesium atoms. Time is based on astronomical measurements. *Solar time* is measured with respect to the Sun. *Sidereal time* is based on measurement with respect to the stars. *Lunar time* is based on the lunar month (the time between successive New Moons).

Fluctuations in the motion of astronomical bodies have made it necessary to adopt the atomic clock using cesium radiation as an absolute standard.

time constant Symbol T or τ. A time that characterizes the rate of discharge of a capacitor in a circuit consisting of the capacitor and a resistor. The time constant is given by CR seconds where C is in farads and R is in ohms.

tin A white lustrous metal of low melting point. Its electronic structure has outer s^2p^2 electrons ($[Kr]4d^{10}5s^25p^2$). The element is of low abundance in the Earth's crust (0.004%) but is widely distributed. Tin has three crystalline modifications or allotropes, α-tin or 'gray tin' (diamond structure), β-tin or 'white tin', and γ-tin; the latter two are metallic with close packed structures. Tin also has several isotopes. It is used in a large number of alloys including Babbit metal, bell metal, Britannia metal, bronze, gun metal, and pewter as well as several special solders.

Symbol: Sn; m.p. 232°C; b.p. 2270°C; r.d. 7.31 (20°C); p.n. 50; r.a.m. 118.710.

tint The color sensation of mixing a hue (a spectral color) with white. This reduces the saturation of the hue.

titanium /tÿ-**tay**-nee-ŭm, ti-/ A silvery transition metal. It is used in the aerospace industry as it is strong, resistant to corrosion, and has a low density. It forms compounds with oxidation states +4, +3, and +2, the +4 state being the most stable.

Symbol: Ti; m.p. 1660°C; b.p. 3287°C; r.d. 4.54 (20°C); p.n. 22; r.a.m. 47.867.

TOE *See* theory of everything.

tog A unit of thermal insulance used for textiles, equal to ten times the difference, in degrees Celsius, between two surfaces when the heat transfer between them is one watt per square meter.

Tokamak /**toh**-kă-mak/ A type of reactor with a toroidal (donut) shape, used in experiments to produce nuclear fusion. The plasma is guided round the toroid by strong magnetic fields in such a way that it does not touch the walls of the container.

ton /tun/ **1.** Either of two f.p.s. units of mass. The *short ton*, chiefly used in North America, is equal to 2000 pounds avoirdupois, or to 907.184 74 kilograms (kg) in SI units. The *long ton*, chiefly used in the UK, is equal to 2240 pounds, or to 1016.046 91 kg in SI units.

2. *See* tonne.

3. An f.p.s. unit of refrigeration power, defined as the heat needed to melt one short ton of ice in 24 hours, or 12 000 British thermal units per hour. In SI derived units a ton of refrigeration is equal to 3514 watts.

4. *See* megaton.

5. (displacement ton) An f.p.s. unit used chiefly in the U.S. to measure the displacement of a long ton of seawater by ships, considered to be 35 cubic feet. In SI derived units it is equal to 991.09 cubic decimeters.

6. (register ton) An f.p.s. unit used chiefly in the U.S. to measure the internal capacity of ships, equal to 100 cubic feet. In SI derived units it is equal to 2.8317 cubic meters.

7. (measurement ton) An f.p.s unit used chiefly in the U.S. to determine the cargo capacity of a ship, considered to be 40 cubic feet. In SI derived units it is equal to 1.1327 cubic meters.

tone *See* note.

tonne /tun/ (**metric ton**) Symbol: t A unit of mass equal to 10^3 kilograms.

torque /tork/ Symbol: T A turning moment or couple. The torque of a force F about an axis (or point) is Fs, where s is the perpendicular distance from the axis to the line of action of the force. The unit is the newton meter. Note that the unit of *work*, also the newton meter, is called the joule. Torque is *not* however measured in joules. The two physical quantities are not in fact the same. Work (a scalar) is the scalar product of force and displacement. Torque is the vector product and is a vector at 90° to the plane of the force and displacement.

See also couple; moment.

torr A unit of pressure equal to a pressure of 1/760 atmospheres (133.322 Pa). It is equal to the mmHg. The unit is named for the Italian physicist Evangelista Torricelli (1608–47)

torsion /**tor**-shŏn/ Angular strain (twist) caused by applying opposing torques (couples) to opposite ends of a body such as a rod or wire. If Hooke's law applies the angle of twist is proportional to either couple. For a cylinder of radius *a* and length *l*, the angle ϕ is given by

$$T = (\pi a^2 G\phi)/2l$$

where *T* is the torque and *G* is the shear modulus.

torsional wave /**tor**-shŏ-năl/ A wave motion in which the vibrations in the medium are rotatory simple harmonic motions around the direction of energy transfer.

torsion balance A device for measuring very small forces by the torsion they cause in a wire or fiber. The angle of twist is usually measured by the displacement of a beam of light reflected from an attached mirror (i.e. by an optical lever).

total eclipse *See* eclipse.

total internal reflection The total reflection of radiation that can occur in a medium at the boundary with another medium of lower refractive constant. In such cases, rays incident at small angles will be refracted 'away from the normal'; in other words, the angle of refraction (r) is greater than the angle of incidence (i). Because r is greater than i, it is possible to increase i to a value at which r is 90° (or just under). That value of i is called the *critical angle c*. The relation for the refractive constant for radiation entering this medium is:

$$_1n_2 = 1/\sin c$$

If i is increased still further, refraction cannot occur. The radiation must then all be reflected. Normally, when radiation meets the boundary between two media some will be reflected and some will be refracted. Total internal reflection is the only exception – here *all* the energy is reflected. Because of this, optical instruments often include totally internally reflecting prisms rather than mirrors. The critical angles of optical glasses at visible wavelengths are typically 40° or less.

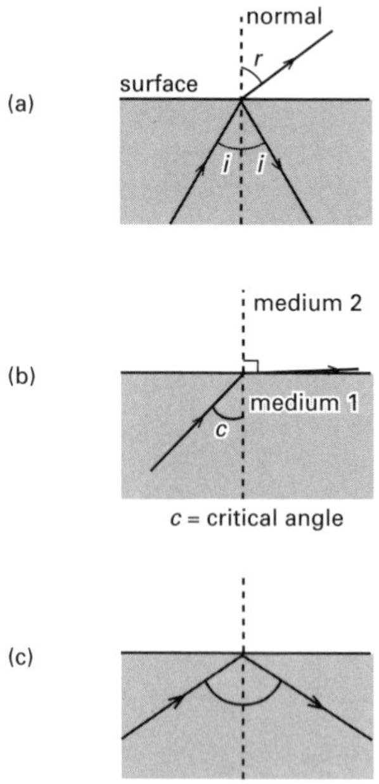

Total internal reflection

total radiation pyrometer An instrument used for measuring high temperatures. Radiation from the source is focused by a concave mirror onto a blackened foil to which is attached a thermocouple. The thermoelectric e.m.f. produced can be measured and hence the temperature of the foil. The temperature of the radiating source may then be calculated. *See also* optical pyrometer.

tourmaline /**toor**-mă-leen, -lin/ (**schorl**) A complex bluish or brownish mineral important for its polarizing effect on light. It is a dichroic material.

trajectory /tră-**jek**-tŏ-ree/ The path of a moving object – a projectile – acted upon by air resistance, gravity, and other forces.

transducer /trans-**dew**-ser, tranz-/ **1.** A device that converts one form of energy into another. For example, a microphone converts sound into an electric current, and a loudspeaker does the reverse.
2. A device for converting a physical effect into a voltage (so giving a measure of the physical effect).

transformer, voltage /trans-**for**-mer/ A device for changing the voltage of an alternating supply. Transformers have no moving parts and operate by the current in one coil, the *primary* winding, electromagnetically inducing a current in another, the *secondary* winding, which forms part of a separate electrical circuit. The ratio of the voltages in the primary and secondary circuits, V_1/V_2, is approximately equal to the ratio of the number of turns in the primary and secondary coils. In an ideal transformer there is no power loss and the primary to secondary current ratio, I_1/I_2, is the inverse of the voltage ratio; i.e. $V_1I_1 = V_2I_2$. In practice, some power is lost because of eddy currents, hysteresis, resistance of the coils, and incomplete linking of the magnetic flux between the coils. Usually the primary and secondary windings are wound around a laminated iron core designed to give maximum flux linkage. With careful design, transformers can be more than 98% efficient.

Step-up transformers are used to increase voltage and reduce the current of the output from power stations so that losses in transmission lines are minimized. The voltage is reduced in stages by *step-down transformers* nearer the user.

transient /**tran**-see-ĕnt, -zee-/ A sudden brief disturbance of a system.

transistor /tran-**zis**-ter/ A device incorporating two junctions between n-type and p-type SEMICONDUCTORS, and having the property that a current flowing across one junction modulates the current flowing across the other junction. Electrical contact is made by three electrodes attached one to each piece of semiconductor, called the emitter, base, and collector. There are two types of transistor, called p-n-p and n-p-n.

Both types may be regarded as two diodes connected back to back, so that when a voltage is applied across the device one diode is forward biased and the other reverse biased. Current flows across the forward biased junction, i.e. majority carriers (electrons in n-type and holes in p-type semiconductors) move from the emitter toward the collector. The base region is thin enough not to prevent this.

In the n-p-n transistor, electrons from the emitter cross into the base, and some holes from the base (p-type) cross into the emitter. Those electrons entering the base move slowly toward the collector-base junction, which is reverse biased. The base region is relatively thin, and once electrons enter the collector region they are swept through it by the applied voltage. In the opposite direction holes cross the base to the emitter junction, causing the hole concentration in the base to fall. This effect would reduce the forward voltage sufficiently to stop the forward current unless the base region was made to increase its hole concentration by losing electrons through the external connection to the base. Hence the base-emitter junction must always be forward biased. In such a case current amplification is possible; a small base current controls a larger collector current. The current gain is defined by the ratio of collector current to base current, given by the symbol β (beta). This use of the transistor is called the *common emitter* connection, in which an input signal is applied to the base. This is the most common way in which a transistor is used as an amplifier.

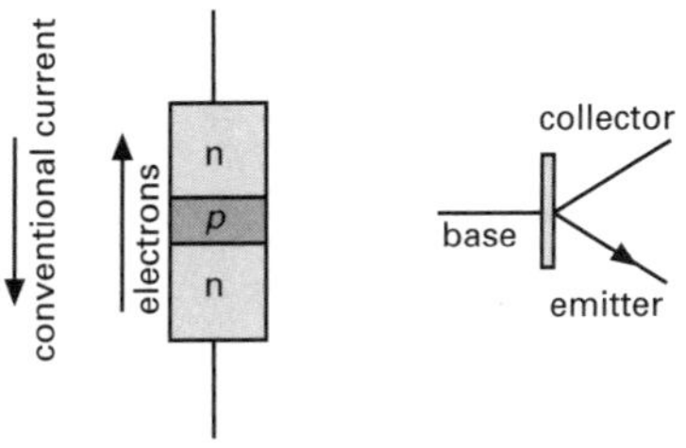

n-p-n transistor

transition temperature A temperature at which some definite physical change occurs in a substance. Examples of such transitions are change of state, change of crystal structure, and change of magnetic behavior.

translation /trans-**lay**-shŏn, tranz-/ *See* translatory motion.

translatory motion /**trans**-lă-tor-ee, -toh-ree, **tranz**-/ (**translation**) Motion involving change of position; it compares with rotatory motion (rotation) and vibratory motion (vibration). Each is associated with kinetic energy. Translatory motion is usually described in terms of (linear) speed or velocity and acceleration.

translucent /trans-**loo**-sĕnt, tranz-/ Able to pass radiation, but with much deviation and/or absorption. *Compare* transparent.

transmission electron microscope /trans-**mish**-ŏn, tranz-/ *See* electron microscope.

transmittance /trans-**mit**-ăns, tranz-/ (**transmission coefficient**) The ratio of the transmitted energy that a substance allows through to the energy incident upon it.

transmitter A device used in a telecommunication system to generate and propagate an electrical signal.

transparent Able to pass radiation without significant deviation or absorption. Note that a substance transparent to one radiation may be opaque to another. The divide between transparency and translucency is not well defined. Thus some people call a filter transparent (as it does not distort radiation): others would call it TRANSLUCENT (as it absorbs some of the radiation). *See also* opaque.

transport The transfer of matter or radiation. Transport phenomena can be analyzed from first principles using non-equilibrium statistical mechanics. In practice, it is convenient to use equations from kinetic theory. Electrical and thermal conductivity are examples of transport phenomena.

transport coefficients Quantities, such as electrical and thermal conductivity, that quantify the transport of matter or radiation. Transport coefficients may be calculated using non-equilibrium statistical mechanics or kinetic theory. An *inverse transport coefficient*, such as resistivity, quantifies the opposition to transport in a system.

transport number /**trans**-port (*n.*); trans-**port** (*vb.*)/ Symbol t The fraction of the total current passing through an electrolyte carried by a particular ion.

transuranic elements /trans-yû-**ran**-ik, tranz-/ Elements that have a proton number greater than 92 (i.e. greater than that of uranium). The transuranic elements are all unstable (radioactive) and are produced by nuclear reactions induced either by bombarding heavy elements with high-energy nuclei of light elements or by the capture of slow neutrons, followed by beta decay. In nuclear explosions, transuranic elements may be produced by numerous successive captures of neutrons by uranium nuclei.

transverse wave A wave motion in which the motion or change is perpendicular to the direction of energy transfer. Electromagnetic waves are examples of transverse waves. *See* polarization. *Compare* longitudinal waves.

traveling wave *See* wave

triangle of forces *See* triangle of vectors.

triangle of vectors A triangle describing three coplanar vectors acting at a point with zero resultant. When drawn to scale – shown correctly in size, direction, and sense, but not in position – they form a closed triangle. Thus three forces acting on an object at equilibrium form a *triangle of forces*. Similarly a *triangle of velocities* can be constructed.

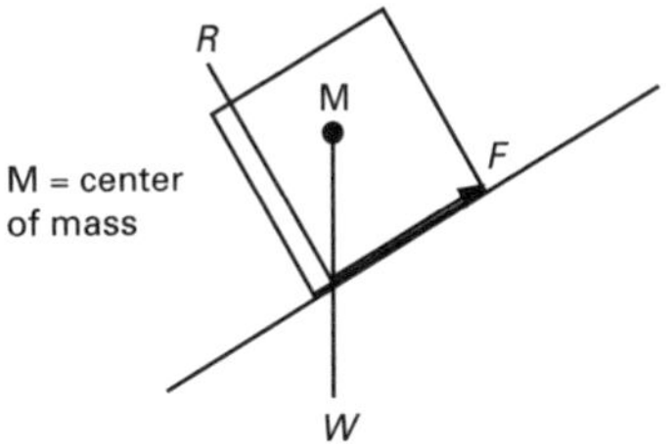

(a) force diagram for an object at rest on a rough slope

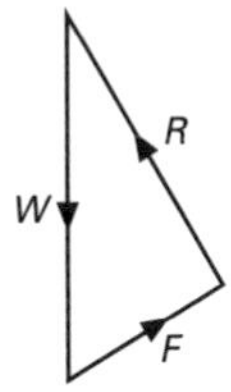

(b) corresponding triangle

Triangle of vectors

triangle of velocities *See* triangle of vectors.

triatomic /trȳ-ă-**tom**-ik/ Describing a molecule consisting of three atoms. Water (H_2O) and carbon dioxide (CO_2) are examples.

triboelectricity /trȳ-boh-ĕ-lek-**tris**-ă-tee/ (**frictional electricity**) Static electricity produced by friction.

tribology /trȳ-**bol**-ŏ-jee/ The study of friction, lubricants, and lubrication.

triboluminescence /trȳ-boh-loo-mă-**ness**-ĕns/ *See* luminescence.

triode /**trȳ**-ohd/ A THERMIONIC TUBE with three electrodes; i.e. one GRID, called the *control grid*, in addition to the anode and cathode. The grid is placed nearer to the cathode than the anode, and so is able to control the current by the application of relatively small voltages to this grid. In normal operation the grid is biased negatively, so that no current flows on to it; but a small voltage fluctuation, or input, superimposed on the GRID BIAS causes large anode current changes, or output. The ratio of output voltage change to input voltage change is called the voltage gain, or *amplification factor*.

triple point The only point at which the gas, solid, and liquid phases of a substance can coexist in equilibrium. The triple point of water (273.16 K) is used to define the kelvin. Some substances (e.g. carbon dioxide) have pressures at the triple point greater than atmospheric. Such materials can appear as liquids only when under pressure.

tritiated /**trit**-ee-ay-tid, **trish**-/ *See* tritium.

tritium /**trit**-ee-ŭm, **trish**-/ Symbol: T, ^3H A radioactive isotope of hydrogen of mass number 3. The nucleus contains 1 proton and 2 neutrons. Tritium decays with emission of low-energy beta radiation to give ^3He. The half-life is 12.3 years. Tritium is found in the atmosphere, possibly as a result of the bombardment of nitrogen by neutrons from cosmic rays:

$$^{14}N + {}^{1}n \rightarrow {}^{12}C + {}^{3}H$$

It is useful as a tracer in studies of chemical reactions. Compounds in which ^3H atoms replace the usual ^1H atoms are said to be *tritiated*.

triton /**trȳ**-tŏn/ A nucleus of a tritium atom, consisting of two neutrons and one proton.

true north Geographical north, the direction toward the North Pole (as opposed to MAGNETIC NORTH).

truth *See* quark.

tungsten /**tung**-stĕn/ A transition metal, formerly called *wolfram*. It is used as the filaments in electric lamps and in various alloys.

Symbol: W; m.p. 3410 ± 20°C; b.p. 5650°C; r.d. 19.3 (20°C); p.n. 74; r.a.m. 183.84.

tuning fork A metal fork designed to vibrate when struck gently to produce a pure sound tone.

tunnel diode A highly doped p-n junction DIODE that has a large reverse current, and, in the forward direction, a negative slope resistance over part of the voltage-current characteristic.

The 'tunnel effect' explains the shape of the characteristic. Electrons are able to tunnel through the conduction band, which is normally forbidden to them because they do not possess sufficient energy to cross the potential barrier of the junction. *See also* diode.

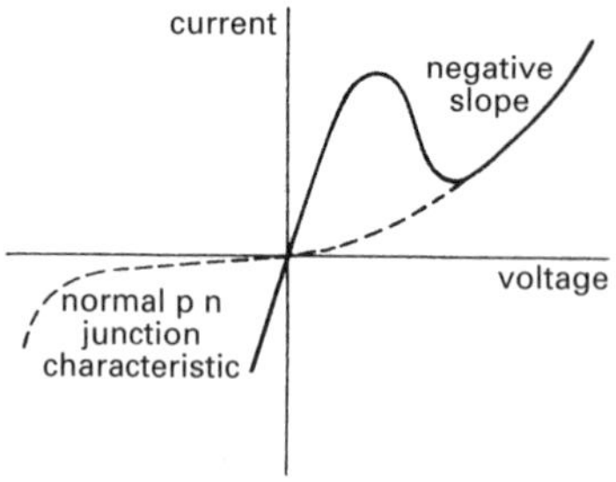

Tunnel diode

tunnel effect The passage of a particle through a potential barrier, even though it has not enough energy to pass the barrier on classical grounds. The tunnel effect can be explained by quantum mechanics. The process of alpha decay is an example of the tunnel effect. *See* tunnel diode.

turbulent flow Fluid flow in which the speed at any point varies rapidly in magnitude and direction. Fluid flow becomes turbulent when its speed increases beyond a *critical speed*. This corresponds to a critical value of the REYNOLDS NUMBER that depends on the geometry of the system. For highly turbulent flow, the resistance to motion is proportional to the product of the density of the fluid, the speed squared, and the square of the linear dimensions. It has proved very difficult to construct a mathematical theory of turbulence. *Compare* laminar flow.

turns ratio The ratio of the number of turns in the primary coil of a voltage transformer to the number in the secondary coil.

two-body problem The problem of understanding the motion of two bodies governed by some interaction. The two-body problem can be solved exactly in the case of two massive bodies interacting by Newtonian gravity and two electrically charged bodies in quantum mechanics. The problem of two interacting massive bodies described by general relativity theory cannot be solved exactly.

Tyndall effect /**tin**-dăl/ The scattering of light by suspended particles or a colloidal solution. The effect is named for the British physicist John Tyndall (1820–93).

U-V

UHF *See* ultrahigh frequency.

ultracentrifuge /ul-tră-**sen**-tră-fyooj/ A high-speed CENTRIFUGE used for separating out very small particles. The sedimentation rate depends on the particle size, and the ultracentrifuge can be used to measure the mass of colloidal particles and large molecules (e.g. proteins).

ultrahigh frequency /ul-tră-**hÿ**/ (UHF) A radio frequency in the range between 3 GHz and 0.3 GHz (wavelength 10 cm–1 m).

ultrahigh vacuum *See* vacuum.

ultramicroscope /ul-tră-**mÿ**-krŏ-skohp/ A modified compound microscope able to view the sub-microscopic particles in a colloidal suspension when this is suitably illuminated.

ultrasonic frequency /ul-tră-**sonn**-ik/ A sound frequency above the range normally audible by humans, i.e. greater than about 20 kHz.

ultrasonics /ul-tră-**sonn**-iks/ The study of sounds that are too high-pitched to be detected by the human ear; i.e. have a frequency greater than about 20 kHz. Ultrasound is used in a similar way to x-rays, for medical diagnosis and testing for flaws in metal. It is also used to clean surfaces and to break up particles in suspension. Ultrasound is often generated by the piezoelectric effect. *See also* sonar.

ultraviolet /ul-tră-**vÿ**-ŏ-lĕt/ (UV) A form of electromagnetic radiation, shorter in wavelength than visible light. Ultraviolet wavelengths range between about 1 nm and 400 nm. Ordinary glasses are not transparent to these waves; quartz is a much more effective material for making lenses and prisms for use with ultraviolet.

Like light, ultraviolet radiation is produced by electronic transitions between the outer energy levels of atoms. The distinction between the two types of radiation is in fact physiological rather than physical. However, having a higher frequency, ultraviolet photons carry more energy than those of light.

See also electromagnetic spectrum.

ultraviolet catastrophe In the late nineteenth century it was realized that the short-wavelength region of black-body radiation could not be explained by the theories of physics of the time (classical physics). The problem – sometimes called the ultraviolet catastrophe – was resolved by the concept of quantization of energy. *See* Planck's radiation law; Rayleigh–Jeans law.

umbra (*pl.* **umbrae** or **umbras**) /**um**-bră/ Total shadow. *See* shadow.

uncertainty principle *See* Heisenberg's uncertainty principle.

underdamping /un-der-**damp**-ing/ *See* damping.

uniaxial crystal /yoo-nee-**aks**-ee-ăl/ A type of birefringent crystal having only one axis along which the ordinary and extraordinary rays travel at the same speed. *See* optic axis.

unified field theory A single theory to account for the electromagnetic, gravitational, strong, and weak interactions by one set of equations. So far, attempts to find such a theory have been unsuccessful, although there has been some progress in unifying the weak and electromagnetic interactions.

uniform acceleration Acceleration that is constant in magnitude and direction.

uniform electric field An electric field in which the field strength is the same at all points in the region of the field. For example, a uniform electric field may be produced by two long parallel electrically charged plates, with the field strength having the same value between the plates.

uniform motion A vague phrase, usually taken to mean motion at constant velocity (constant speed in a straight line).

uniform speed Constant speed.

uniform velocity Constant velocity, describing motion in a straight line with zero acceleration.

unipolar transistor /yoo-nă-**poh**-ler/ *See* field-effect transistor.

unit A reference value of a quantity used to express other values of the same quantity. *See also* SI units.

unit cell The smallest unit of entities such as atoms, molecules or ions which are repeated in a periodic way in a crystal.

universal constants *See* fundamental constants.

unsaturated /un-**sach**-ŭ-ray-tid/ Describing a color that is a tint; i.e. a hue mixed with white. *See* saturation.

unstable equilibrium EQUILIBRIUM such that if the system is disturbed a little, there is a tendency for it to move farther from its original position rather than to return. *See* stability.

upthrust An upward force on an object in a fluid. In a fluid in a gravitational field the *pressure* increases with depth. The pressures at different points on the object will therefore differ; the resultant is vertically upward. *See* Archimedes' principle; buoyancy; flotation (law of).

uranium /yû-**ray**-nee-ŭm/ A toxic radioactive silvery element of the actinoid series of metals. Its three naturally occurring radioisotopes, ^{238}U (99.283% in abundance), ^{235}U (0.711%), and ^{234}U (0.005%), are found in numerous minerals including the uranium oxides pitchblende, uraninite, and carnotite. The readily fissionable ^{235}U is a major nuclear fuel and nuclear explosive, while ^{238}U is a source of fissionable ^{239}Pu.

Symbol: U; m.p. 1132.5°C; b.p. 3745°C; r.d. 18.95 (20°C); p.n. 92; r.a.m. 238.0289.

uranium–lead dating A method of radioactive DATING used for estimating the age of certain rocks. It is based on the decay of ^{238}U to ^{206}Pb (half-life 4.5×10^9 years) or ^{235}U to ^{207}Pb (half-life 0.7×10^9 years). The technique is useful for time periods from 10^7 years ago back to the age of the Earth (about 4.6×10^9 years).

uranium–radium series *See* radioactive series.

US Customary units The f.p.s. system of weights and measures established in the former American colonies prior to the American Revolutionary War. Today they remain in use only in the US, Liberia, and Myanmar (Burma).

US Customary units for length and, with the exception of the ton and stone, weight or mass, are essentially the same as those used in other f.p.s. systems, including the IMPERIAL UNITS system. Significant differences occur in units for volume and capacity, however. Thus the US gallon of 128 US fluid ounces, which is equal to 3.785 412 cubic decimeters (dm^3) in SI units, is considerably smaller than the Imperial gallon of 160 Imperial fluid ounces, which is equal to 4.546 09 dm^3 in SI units. Similarly, the Imperial bushel, equal in SI units to 36.367 72 dm^3, is about 3% larger than the US bushel, which is equal in SI units to 35.239 07 dm^3. Conversely the US fluid ounce, equal in SI units to 29.573 53 cubic centimeters (cm^3), is slightly larger than the Imperial fluid ounce, which in SI units is equal to 28.423 062 cm^3.

The US system further differs from the Imperial system in having separate dry and liquid units for certain volumetric measures. Whereas the Imperial quart, for example, equal in SI units to 1.136 522 dm^3, is used for both liquid and dry measure, there are separate US Customary quarts for dry goods, equal in SI units to 1.101 221 dm^3, and liquids, equal in SI units to 0.946 352 9 dm^3. The different US units arise from the fact that they are based on divisions of the US bushel (for dry goods) and the US gallon (for liquids).

UV *See* ultraviolet.

vacancy *See* defect.

vacuum (*pl.* **vacuums** or **vacua**) A space containing gas below atmospheric pressure. A perfect vacuum contains no matter at all, but for practical purposes *soft* (*low*) *vacuum* is usually defined as down to about 10^{-2} pascal, and *hard* (*high*) *vacuum* as below this. *Ultrahigh vacuum* is lower than 10^{-7} pascal.

vacuum distillation Distillation carried out at reduced pressure, which lowers the boiling point of the liquid.

vacuum flask *See* Dewar flask.

vacuum pump *See* diffusion pump.

vacuum state The lowest energy state in a system described by relativistic quantum field theory. The vacuum state is analogous to the ground state in a system described by quantum mechanics.

vacuum tube *See* thermionic tube.

valence band *See* energy bands.

valence electron In an atom, an electron in an incompletely filled (usually outer) shell, available for chemical bonding to form a molecule. *See* electron shell.

valley of stability A concept that is useful in discussing the stability of nuclei. If the number of protons in a nucleus is plotted on one horizontal axis, the number of neutrons in the nucleus is plotted on a horizontal axis perpendicular to the first axis, and the energy of the nucleus is plotted on a vertical axis for all nuclei the resulting surface is a valley. Stable nuclei are near the bottom of this valley of stability, with iron and nickel occupying positions close to the lowest points.

valve *See* thermionic tube.

valve voltmeter *See* electrometer.

vanadium /vă-**nay**-dee-ŭm/ A silvery transition element used in alloy steels.

Symbol: V; m.p. 1890°C; b.p. 3380°C; r.d. 6.1 (20°C); p.n. 23; r.a.m. 50.94.

Van de Graaff accelerator /**van**-dĕ-graf, -grahf/ A particle accelerator in which the high voltage from a Van de Graaff generator is used to accelerate charged particles. The accelerator is named for the American physicist Robert Jemison Van de Graaff (1901–67).

Van de Graaff generator An electrostatic generator capable of producing high p.d.s (up to millions of volts). It consists of a large smooth metal sphere on top of a hollow insulating cylinder. An endless insulating belt runs between pulleys at each end of the cylinder. Charge is sprayed from metal points connected to a high-voltage source on to the bottom of the belt. It is then carried up to the top of the belt where it is collected by other metal points and accumulated on the outside of the sphere, which becomes highly charged.

van der Waals' equation An equation of state for real gases. For one mole of gas the equation is

$$(p + a/V_m^2)(V_m - b) = RT,$$

where p is the pressure, V_m the molar volume, and T the thermodynamic temperature. *a* and *b* are constants for a given substance and *R* is the gas constant. The equation gives a better description of the behavior of real gases than the perfect gas equation ($pV_m = RT$).

The equation contains two corrections: *b* is a correction for the non-negligible size of the molecules; a/V_m^2 corrects for the fact that there are attractive forces between the molecules, thus slightly reducing the pressure from ideal. The equation is named for the Dutch physicist Johannes Diderik Van der Waals (1837–1923).

van der Waals' forces Attractive forces existing between molecules. These forces are the ones giving the pressure correction in the van der Waals' equation. They are much weaker than chemical bonds and act over short range (inversely proportional to the seventh power of distance). They are caused by attraction between dipoles of molecules. For atoms or molecules without permanent molecular dipole moments, the attractive forces result from attractions between nucleus-electron dipoles (called dispersion forces).

Van't Hoff's law At a given temperature, the osmotic pressure of a solution is the same as the pressure that the solute would exert if it were in the gaseous phase and occupying the same volume. The law is named for the Dutch chemist Jacobus Henricus Van't Hoff (1852–1911).

vapor A gas at any temperature at which it may be liquefied by pressure alone; i.e. a gas below its critical temperature. The critical temperatures for water, carbon dioxide, and nitrogen are 374°C, 31.1°C, and –147°C respectively. *See also* Andrews' experiments; critical state.

vapor density The mass of a volume of gas divided by the mass of an equal volume of hydrogen at the same temperature and pressure, equal to half its relative molecular mass.

vaporization A change of state from liquid or solid to vapor.

vapor pressure The pressure exerted at a particular temperature by a vapor. When a liquid or solid evaporates, molecules are continuously escaping from the surface at a rate that increases rapidly with temperature; they exert a vapor pressure. Those striking the surface tend to re-enter the liquid or solid, so that eventually a state of dynamic equilibrium is reached in which the number of molecules returning to the surface per second is the same as the number leaving it. The vapor now exerts its equilibrium pressure for that temperature, the *saturated vapor pressure* (SVP). The value of this depends only on the temperature, and is independent of the volume (unless all the vapor condenses or all the liquid evaporates).

variable capacitor A capacitor for which the capacitance can be changed. A common form has two sets of parallel semi-circular plates in which one set can swing into or out of the other set, thereby changing the area of overlap of the plates. Variable capacitors are used in radio receivers.

variable star A star in which the brightness changes with time. This variability can be regular or chaotic. There are many causes for the variability. For example, a SUPERNOVA, which is an extreme manifestation of the variability of a star, occurs when the core of a large star collapses to a NEUTRON STAR.

variation, magnetic *See* magnetic variation.

variometer /vair-ee-**om**-ĕ-ter/ A variable inductor, usually consisting of two coils connected in series and able to be moved with respect to each other, so that the total self-inductance can be changed.

vector /**vek**-ter/ A measure in which direction is important and must usually be specified. For instance, displacement is a vector quantity, whereas distance is a scalar. Weight, velocity, and magnetic field strength are other examples of vectors – they are each quoted as a number with a unit and a direction. The addition of vectors must take account of direction. The resultant may be found by the parallelogram of vectors.

There are various ways of multiplying vectors. A vector multiplied by a scalar gives another vector; for example mass (a scalar) multiplied by velocity (a vector) gives momentum (a vector). Two vectors can be multiplied together in two different ways. *See* scalar product; vector product.

For some purposes, it is useful to define two classes of vector. Those quantities defining something that happens or may be caused to happen in a particular direction (e.g. displacement, acceleration, momentum, force) are called *polar vectors*. Quantities using the orientation of an axis (e.g. angular velocity, moment of a force) are known as *axial vectors* or *pseudovectors*. The vector product of two polar vectors is a pseudovector.

vector calculus The branch of mathematics concerned with rates of change of vectors. There are three main operators in vector calculus: *curl*, *div*, and *grad*.
If V is a vector, curl V is a vector that results from the vector product of the operator ∇, where $\nabla = i\partial/\partial x + j\partial/\partial y + k\partial/\partial z$, and V, i.e. curl $V = \nabla \times V$.
The quantity div V is a scalar that results from the scalar product of ∇ and V, i.e. div $V = \nabla . V$.
The quantity grad ϕ, where ϕ is a scalar function, is a vector that results from the operation of ∇ on ϕ, i.e. grad $\phi = \nabla\phi$.
Vector calculus is used extensively in many branches of physics, particularly electromagnetism and fluid mechanics.

vector product A multiplication of two vectors to give a vector. The vector product of $\boldsymbol{A}$ and $\boldsymbol{B}$ is written $\boldsymbol{A} \times \boldsymbol{B}$. It is a vector of magnitude $AB\sin\theta$, where A and B are the magnitudes of $\boldsymbol{A}$ and $\boldsymbol{B}$ and θ is the angle between $\boldsymbol{A}$ and $\boldsymbol{B}$. The direction of the vector product is at right angles to $\boldsymbol{A}$ and $\boldsymbol{B}$. It points in the direction in which a right-handed screw would move turning from $\boldsymbol{A}$ toward $\boldsymbol{B}$.

Note, $\boldsymbol{A} \times \boldsymbol{B} = -\boldsymbol{B} \times \boldsymbol{A}$. An example of a vector product is the force $\boldsymbol{F}$ on a moving charge Q in a field $\boldsymbol{B}$ with velocity $\boldsymbol{v}$ (as in the motor effect). Here

$$\boldsymbol{F} = Q\boldsymbol{B} \times \boldsymbol{v}$$

The vector product is sometimes called the *cross product*.

See also scalar product.

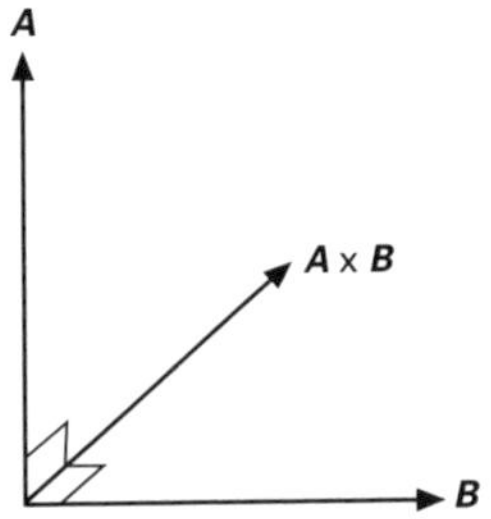

Vector product

vectors, parallelogram (law) of *See* parallelogram of vectors.

vectors, triangle (law) of *See* triangle of vectors.

velocity Symbol: v Displacement per unit time. The unit is the meter per second (m s^{-1}). Velocity is a vector, speed being the scalar form. If velocity is constant, it is given by the slope of a position/time graph, and by the displacement divided by the time taken. If it is not constant, the mean value is obtained. If x is the displacement, the instantaneous velocity is given by $v = dx/dt$. *See also* equations of motion.

velocity modulation A type of modulation in which the velocity of a beam of electrons is made to fluctuate at the frequency of a superimposed electric field, usually a radiofrequency. As faster electrons overtake slower electrons, bunching occurs in the beam. High-frequency oscillators, such as those used in radar, make use of velocity-modulated beams of electrons. *See also* modulation; radar.

velocity of light *See* speed of light.

velocity of sound *See* sound.

velocity ratio *See* distance ratio.

Venturi tube /ven-**toor**-ee/ A cylindrical pipe that has a constriction at its center. The rate of flow of a fluid through the tube increases at the constriction and its pressure decreases. The difference in pressure between the ends of the tube and at the constriction can be used to calculate the rate of flow. The tube is named for the Italian physicist Giovanni Battista Venturi (1746–1822).

vernier scale /**ver**-nee-er, **ver**-neer/ A small auxiliary scale in a measuring instrument which is put next to the main scale so that divisions in the main scale can be measured accurately. The scale is named for the French mathematician Pierre Vernier, who invented it in 1631.

very high frequency (**VHF**) A radio frequency in the range between 300 MHz and 30 MHz (wavelength 1 m–10 m).

very large scale integration (**VLSI**) The construction of electrical circuits in which a large number of transistors are connected together in a small region such as a *silicon chip*. The development of VLSI has been an essential ingredient in the success of the electronics industry, particularly computers.

very low frequency (**VLF**) A radio frequency in the range between 30 kHz and 3 kHz (wavelength 10 km–100 km).

VHF *See* very high frequency.

vibrating-reed electrometer *See* electrometer.

vibration (**oscillation**) Any regularly repeated to-and-fro motion or change. Examples are the swing of a pendulum, the vibration of a sound source, and the change with time of the electric and magnetic fields in an electromagnetic wave.

vibration magnetometer *See* magnetometer.

virial coefficient /**vi**-ree-ăl/ Quantities A, B, C, D, ... which occur in the *virial equation*:

$$PV = A + B/V + C/V^2 + D/V^4 + ...,$$

where P is the pressure and V is the volume. The virial coefficients B, C, D, etc. having nonzero values is an attempt to describe real gases rather than ideal gases.

virtual image *See* image.

virtual object *See* object.

virtual particle An elementary particle postulated to exist in interactions. For example, it is possible to explain the electromagnetic interaction between two charged particles by assuming that they are exchanging virtual photons.

virtual work The work done if a system is displaced infinitesimally from its position. The virtual work is zero if the system is in equilibrium.

viscometer /vis-**kom**-ĕ-ter/ An instrument for measuring fluid viscosity. Some types measure the rate of fluid flow through a narrow tube. Others measure the time a ball takes to fall through a known depth of fluid. Effects such as the damping of mechanical vibrations by the fluid and the friction forces it transmits between two moving parts (both of which are direct results of viscosity) can also be used in viscometers.

viscosity /vis-**koss**-ă-tee/ (**internal friction**) Symbol: η The resistance of a fluid to flow at low speeds. Molasses, for example, has a higher viscosity than water. It exerts a greater frictional force when it is stirred, the same object will fall through it more slowly than through water, and it flows more slowly.

In steady flow of a fluid over a plane surface of area A the surface will be subjected to a force F in its plane and in the direction of flow, given by

$$F = \eta A(dv/dx)$$

where (dv/dx) is the gradient in the direction normal to A of the velocity parallel to the surface. The quantity η is called viscosity, dynamic viscosity, or coefficient of viscosity. For most fluids viscosity is a constant at constant temperature and the equation expresses Newton's law of viscosity. For some fluids, such as pastes and paints, η depends upon the velocity gradient. These are called *non-Newtonian fluids*. The unit of viscosity is kg m^{-1} s^{-1}, or N m^{-2} s. *See also* Reynolds number; turbulent flow.

visible radiation *See* light.

visual acuity *See* acuity; visual.

vitreous humor The gelatinous substance behind the lens in the eye. *See* eye.

VLF *See* very low frequency.

VLSI *See* very large scale integration.

volt Symbol: V The SI unit of electrical potential, potential difference, and e.m.f., defined as the potential difference between two points in a circuit between which a constant current of one ampere flows when the power dissipated is one watt. 1 V = 1 J C^{-1}. The unit is named for the Italian physicist Count Alessandro Volta (1745–1827).

voltage Symbol: *V* E.m.f. or electrical potential difference measured in volts.

voltage divider (**potential divider**) A number of resistors, inductors, or capacitors connected in series with several terminals at intermediate points. The total voltage across the chain can be split into known fractions by making connections at the various points.

voltaic cell /vol-**tay**-ik/ *See* cell.

voltameter /vol-**tam**-ĕ-ter/ (**coulo(mb)-meter**) A device for determining electric charge or electric current using electrolysis. The mass *m* of material released is measured and this can be used to calculate the charge (*Q*) and the current (*I*) from the electrochemical equivalent (*z*) of the element, using $Q = m/z$ or $I = m/zt$. The term now often refers to any case in which two electrodes are immersed in an electrolyte.

voltmeter /**vohlt**-mee-ter/ A meter used to measure electrical voltage. The commonest types are moving-coil instruments with high series resistances to keep the current drawn to a low value. For alternating voltages a rectifier must be used. Moving-iron instruments with high series resistances can be used for both direct and alternating voltages. Electrostatic instruments are sometimes used, especially for measuring high potential differences. Electronic instruments can also be constructed, often with digital display (*digital voltmeters*). The cathode-ray oscilloscope can also be used as a voltmeter. *See also* electrometer; potentiometer.

volume Symbol: *V* A measure of the space occupied by an object or system.

volume charge density *See* charge density.

vortex /**vor**-teks/ (*pl.* **vortices**) A circulating eddy in a fluid. Vortices can be created in the rapid flow of fluid around a blunt body, with the streamlines forming vortices behind the body. The production of vortices in this way is associated with large drag forces on the body.

W

watt /wot/ Symbol: W The SI unit of power, defined as a power of one joule per second. 1 W = 1 J s^{-1} The unit is named for the Scottish instrument maker and inventor James Watt (1736–1819).

watt-hour A unit of electric power equal to one watt per hour consumed. *See* electric meter.

wattmeter /**wot**-mee-ter/ An instrument for measuring power in an electric circuit. The most common type, used in alternating-current circuits, consists of two coils, one fixed and one able to rotate. The fixed coil is connected in series with the circuit (like an ammeter). The rotating coil has a high resistance (or has an additional resistance in series) and is connected across the load like a voltmeter. The deflection of this coil is opposed by hair springs, and its value is proportional to the product of the current and the potential difference – hence, the instrument can be calibrated to give the power.

wave Waves may be *progressive* or *stationary*. A progressive (or *traveling* wave) is an oscillatory disturbance propagated through a medium or, in the case of electromagnetic waves, through space. In each type of wave there are two kinds of disturbance. Electromagnetic waves involve electric and magnetic fields oscillating in directions at right angles to each other and to the direction of propagation, equal amounts of energy being transferred by the electric and magnetic fields. Waves on the surface of a liquid involve motion of particles in a vertical plane, there being horizontal (longitudinal) and vertical (transverse) oscillations. In the case of regular waves such as ripples and ocean rollers the water particles move very nearly in circles so the longitudinal and transverse amplitudes are equal. Sound waves involve fluctuations of pressure and particle velocity. Waves in a stretched string involve oscillations of particles at right angles to the string and fluctuations of the tension.

A simple harmonic plane progressive wave can be expressed by the equation

$$y = a\sin 2\pi(ft - x/\lambda)$$

where y is the displacement at time t at a distance from the origin x. The maximum value of the displacement, a, is called the amplitude. f is the frequency and λ is the wavelength. The relationship can also be expressed in the forms

$$y = a\sin 2\pi(vt - x/\lambda)$$

or

$$y = a\sin 2\pi(t/T - x/\lambda)$$

where v is the wave speed and T is the period. Such waves can equally be expressed in terms of cosines. If the minus sign is replaced by a plus sign, these equations represent waves traveling in the opposite direction.

For electromagnetic waves two similar equations are required, the displacement being the electric field in one, and the magnetic field in the other. The two disturbances are in step. For waves on liquids two equations are needed, one representing the vertical displacement and the other the horizontal displacement. These are a quarter of a wavelength out of step. The wave profile has narrow peaks and broad troughs. In practice, waves never conform exactly to the ideal simple harmonic form since this imply that there was infinite duration and infinite space, without any attenuation.

A stationary (or *standing*) wave is an oscillation equivalent to the resultant of two equal progressive waves traveling in opposite directions. Combining the equations

$$y = a\sin 2\pi(ft - x/\lambda)$$

and

$$y = a\sin 2\pi(ft + x/\lambda)$$

we obtain a resultant displacement

$$Y = 2a\sin 2\pi ft\cos 2\pi x/\lambda$$

This represents a simple harmonic disturbance whose amplitude varies with x according to a cosine law. Any given type of wave requires two similar equations, one for each kind of disturbance such that where one has its maximum amplitude the other has its minimum.

See also antinode; interference; longitudinal wave; node; phase; transverse wave.

wave equation A partial differential equation that describes wave propagation. In three spatial dimension, x, y, and z, it has the form $\nabla^2 u = (1/c^2)\,\partial^2 u/\partial t^2$, where ∇^2 is the Laplace operator, i.e. $\nabla^2 = \partial^2/\partial x^2 + \partial^2/\partial y^2 + \partial^2/\partial z^2$, u is the displacement of the wave, c is the speed of propagation of the wave, and t is the time.

waveform *See* wave.

wavefront A continuous surface associated with a wave radiation, in which all the vibrations concerned are in phase. A parallel beam has plane wavefronts; the output of a point source has spherical wavefronts.

wave function A mathematical quantity, used in wave mechanics, analogous to the amplitude of ordinary waves. The square of the wave function of a particle at a particular point represents the probability of locating the particle in unit volume at that point.

waveguide *See* microwaves.

wavelength Symbol: λ The distance between the ends of one complete cycle of a wave. Wavelength relates to the wave speed (c) and its frequency (ν) thus:

$$c = \nu\lambda$$

wave mechanics *See* quantum theory.

wave motion Any form of energy transfer that may be described as a wave rather than as a stream of particles. The term is also sometimes used to mean any harmonic motion.

wave number Symbol: σ The reciprocal of the wavelength of a wave. It is the number of wave cycles in unit distance, and is often used in spectroscopy. The unit is the metre^{-1} (m^{-1}). The circular wave number (symbol: k) is given by:

$$k = 2\pi\sigma$$

wave theory of light The theory that light is propagated in the form of waves. Hooke and Huygens assumed this theory in the seventeenth century but it did not receive general acceptance until after the publication of researches on interference, diffraction, and polarization in the early nineteenth century, especially by Young, Fresnel, Arago and Fraunhofer. At first, light was believed to be a form of elastic wave in the *ether*, but after the publication of the theory of electromagnetism by Maxwell (1865) it was recognized as a form of electromagnetic wave. *See also* corpuscular theory; duality.

W-boson A spin-one, electrically charged (+1 or –1) particle that is the quantum of the field that mediates the weak interactions with a change of electric charge. The W-bosons were discovered at CERN in 1983, with their masses being about 80 GeV. *See also* Weinberg–Salam model; Z-boson.

weak interaction *See* interaction.

weber /**vay**-ber/ Symbol: Wb The SI unit of magnetic flux, equal to the magnetic flux that, linking a circuit of one turn, produces an e.m.f. of one volt when reduced to zero at a uniform rate in one second. 1 Wb = 1 V s.

The unit is named for the German physicist Wilhelm Eduard Weber (1804–91).

weight Symbol: W. Unit: joule. **1.** The force by which a mass is attracted by the Earth. This is usually expressed as $W = mg$, where g is the ACCELERATION OF FREE FALL, although there is a small error due to the rotation of the Earth.

2. The force which a body exerts on its support when at rest. This is equal to the force defined in (1) but acts in a different place, on a different body, by a different mechanism.

In problems in mechanics the two meanings (1) and (2) are often confused. Also, in everyday language 'weight' is often used to mean mass, so it is important to avoid misunderstanding.

weightlessness A condition experienced by persons and equipment in spacecraft when in free fall, that is in an orbit when no rockets are operating and there is no air resistance. All objects in the spacecraft have the same acceleration as the spacecraft itself in the gravitational field of the Earth, so there are no support forces and the objects are said to be weightless. It is often wrongly supposed that weightlessness means that there is no gravitational field present, and the misleading term 'zero gravity' is used for this condition.

Weinberg–Salam model /wȳn-berg să-lahm A quantum field theory involving broken symmetry which gives a unified theory of the weak and electromagnetic interactions. It predicts that weak interactions are mediated by fields, the quanta of which are electrically charged W-BOSONS and neutral Z-BOSONS. The masses of the W and Z-bosons and other experimental findings are in accord with the predictions of the model, which was put forward by Steven Weinberg in 1967 and independently by Abdus Salam in 1968.

Weiss constant /vȳss/ *See* Curie's law.

Weston cadmium cell A standard cell that produces a constant e.m.f. of 1.018 6 volts at 20°C. It consists of an H-shaped glass vessel containing a negative cadmium-mercury amalgam electrode in one leg and a positive mercury electrode in the other. The electrolyte – saturated cadmium sulfate solution – fills the horizontal bar of the vessel to connect the two electrodes. The e.m.f. of the cell varies very little with temperature, being given by the equation:

$$E = 1.018\,6 - 0.000\,037\,(T - 293)$$

where T is the absolute temperature.

wet and dry bulb hygrometer *See* hygrometer.

wet cell A type of cell, such as a car battery, in which the electrolyte is a liquid solution. *See also* Leclanché cell.

Wheatstone bridge /wheet-stohn/ A circuit for measuring electrical resistance. Four resistors, P, Q, R, and S, are connected in a loop (as if along four sides, or 'arms', of a square). A battery is connected across the junctions between P and S and between Q and R. A sensitive galvanometer is connected across the two opposite junctions (P–Q and R–S). If no current flows through the galvanometer, then $P/Q = S/R$. The bridge is then said to be *balanced*. If R is an unknown resistance and P, Q, and S are known, then R can be found. P and Q are called the *ratio arms* of the bridge (they determine the ratio of S to R).

There are various ways of using this circuit. In a meter bridge P and Q are a single length of uniform resistance wire with a sliding contact. S is a standard resistor. The position of the contact is moved until the galvanometer gives a zero reading. Then the ratio l_1/l_2 is equal to S/R. The circuit is

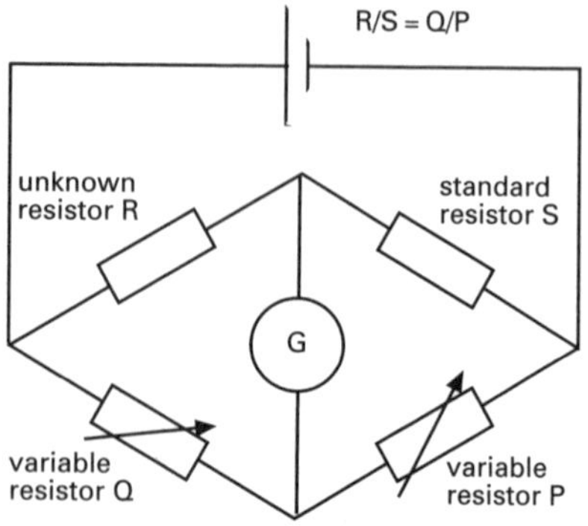

Wheatstone bridge

named for the British physicist Sir Charles Wheatstone (1802–75).

white dwarf A compact heavenly body formed when a star with a mass which is similar to that of the Sun exhausts its nuclear fuel. A white dwarf has a volume which is similar to that of the Earth. It is supported against further gravitational collapse to a NEUTRON STAR or BLACK HOLE by the DEGENERACY PRESSURE of electrons.

white light Visible radiation that gives a sensation of whiteness. The effect is very subjective and depends very much upon conditions and contrast. It is produced by a continuum over the whole visible spectrum. The sensation of whiteness can also be produced by suitable combinations of parts of the spectrum; for example, two complementary colors or three primary colors.

white noise *See* noise.

Wiedemann–Franz law /**vee**-dĕ-măn **frantz**/ An experimental finding about pure metals stating that, at a given temperature, the ratio of the thermal conductivity to the electrical conductivity, is approximately the same for all metals. This law is reasonably well obeyed except at temperatures close to absolute zero. It was discovered by Gustav Heinrich Wiedemann and Rudolph Franz in 1853.

Wien's displacement law /veenz/ For black-body radiation the rate of energy radiation per unit area per unit wavelength range at constant kelvin temperature T_1 can be plotted against wavelength. It is found that there is a peak at wavelength λ_1. For temperature T_2 the peak comes at λ_2, such that

$$\lambda_1 T_1 = \lambda_2 T_2 = \text{constant}$$

This rule was deduced by the German physicist Wilhelm Wien (1864–1928).

Wilson's cloud chamber *See* cloud chamber.

Wimshurst machine /**wimz**-herst/ A type of electrostatic generator. It consists of two circular disks of insulating material with radial strips of metal foil on their sides. The disks rotate in opposite directions and the charge is produced by friction and collected by metal points. The machine is named for the British physicist James Wimshurst (1832–1903).

Wollaston prism /**wûl**-ă-stŏn/ A type of polarizing prism named for the English scientist William Hyde Wollaston (1766–1828).

work When a force $\boldsymbol{F}$ acts on a body while it undergoes a displacement $\boldsymbol{s}$ the body exerting the force is said to do *work*, given by the scalar product of $\boldsymbol{F}$ and $\boldsymbol{s}$

$$W = \boldsymbol{F}.\boldsymbol{s}$$

that is

$$W = Fs\cos\theta$$

where F and s are the magnitudes of $\boldsymbol{F}$ and $\boldsymbol{s}$ and θ is the angle between them. This may be expressed as: work is the product of the force times the component of the displacement in the direction of the force; *or* work is the product of displacement times the component of the force in the direction of the displacement.

When a torque T acts on a body while it undergoes angular displacement θ about the same axis the work done $W = T\theta$ (θ measured in radians).

When a surface at pressure p displaces a volume ΔV the work done is $p\Delta V$.

The system of electric units is so defined that electric work can be expressed; work done electrically when a charge Q is displaced between two points with a potential difference V is given by:

$$W = QV$$

In all cases the SI unit of work is the joule (J).

See also conservation of energy; heat; thermodynamics.

work function A measure of the extent to which photoelectric or THERMIONIC EMISSION will take place, usually expressed as the energy needed to remove an electron. *See* photoelectric effect.

xenon /**zen**-on/ A colorless odorless monatomic element of the rare-gas group. It occurs in trace amounts in air. Xenon is used in thermionic tubes and strobe lighting.

Symbol: Xe; m.p. −111.9°C; b.p. −107.1°C; d. 5.8971 (0°C) kg m^{-3}; p.n. 54; r.a.m. 131.29.

x-radiation A form of electromagnetic radiation with short wavelength. The wavelength is commonly in the range 10^{-10} m to 10^{-11} m but much shorter or longer waves can be produced.

x-radiation is generated whenever high-energy electrons strike matter. In an x-ray tube electrons from a hot filament are accelerated through a large potential difference (typically 10^5 V) and focused upon a *target* or *anticathode* usually made of a high-melting-point metal. Radiation is emitted from the region where the electron beam strikes the target. The radiation is also caused by high-voltage cathode-ray tubes (such as are used in some television receivers and computer terminals) and some other electrical equipment. X-rays are also generated when beta radiation is absorbed in matter.

X-radiation is absorbed in matter mostly by the *photoelectric effect*, the probability of which is proportional to the fourth power of the proton number Z. Thus bone, which contains calcium (Z = 20) and phosphorus (Z = 15), absorbs far more x-radiation than does soft tissue, which contains few atoms with greater proton number than oxygen (Z = 8). Hence a body exposed to x-radiation casts a shadow on a photographic emulsion or fluorescent screen, which can be used in diagnosis. The radiation can be detected using ionization chambers, proportional counters, Geiger–Müller tubes, photography, fluorescence, and other methods.

The spectrum of x-radiation shows lines superimposed upon a continuum. The continuous spectrum is caused by the abrupt slowing-down of electrons as they pass through matter, and is called *bremsstrahlung* (German for 'braking radiation'). The short-wavelength limit λ_0 is produced by the whole kinetic energy of an electron going to generate a single quantum of radiation:

$$\lambda_0 = hc/qV$$

where h is the Planck constant, c is the speed of light, q is the electron charge, and V is the p.d. across the x-ray tube. The line spectrum is caused by atoms that have had inner electrons ejected by electron impact. As electrons from higher energy quantum states enter the empty inner states, radiation is emitted with wavelengths characteristic of the element. Hence these lines can be used in x-ray spectroscopy.

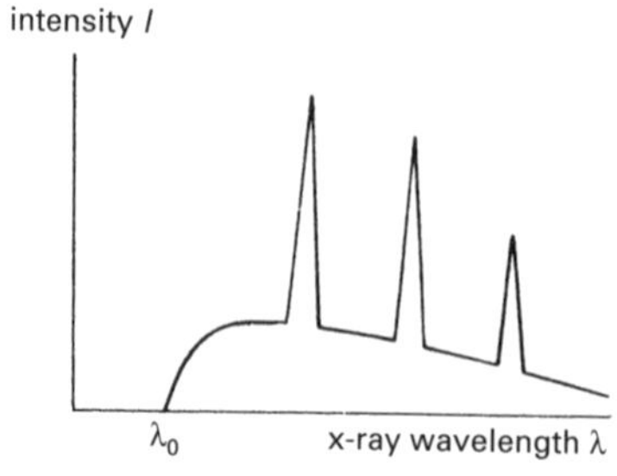

X-rays

x-ray crystallography The use of x-rays to study crystal structure. The way in which x-rays are diffracted gives information about the spacing between crystal planes. The structure of molecules of biological interest such as DNA and proteins have been determined in this way.

x-ray fluorescence Softer (i.e. less energetic) secondary x-rays emitted by a substance bombarded by high-energy electrons or primary x-rays. The secondary x-ray wavelengths are characteristic of the substance and can be used to identify it. *See* spectroscopy; x-radiation.

x-rays Streams of x-radiation.

x-ray tube A vacuum tube used for producing x-rays. A strong electrostatic field accelerates electrons emitted by a white-hot filament and directs them at a metal target, where x-rays are emitted.

year The time taken for the Earth to complete one orbit of the Sun. It is measured in various ways. The *solar year* is the time taken for the Sun to make two successive appearances at the first point of Aries. It is 365.242 19 mean solar days. The *calendar year* is regulated (using leap years) so that its average length is equal to that of the solar year. The *sidereal year* is measured with respect to the fixed stars. It is 365.256 36 mean solar days. The *lunar year*, which is 12 lunar months, is 365.3671 mean solar days.

yellow spot (**macula lutea**) An area a few millimeters across in the human retina. It has a high concentration of rods, giving high visual acuity and color vision but low sensitivity to dim light. *See* eye; fovea.

yield point *See* elasticity.

yocto- Symbol: y A prefix denoting 10^{-24}. For example, 1 yoctometer (ym) = 10^{-24} meter (m).

yoke A piece of ferromagnetic metal that connects two or more magnetic cores.

yotta- Symbol: Y A prefix denoting 10^{24}. For example, 1 yottameter (Ym) = 10^{24} meter (m).

Young modulus *See* elastic modulus.

Young's interference experiment An experiment first performed by the English physicist Thoms Young (1773–1829) in about 1801. Light passing through a small hole fell on a distant screen with two pinholes. The light spread out by diffraction on passing through the pinholes so that there was a region of overlap. Parallel colored fringes were observed in the overlap region as a result of interference.

Light is propagated in trains of waves lasting typically a few nanoseconds. Waves from the same original wavetrain are selected by the first hole and then pass through *both* pinholes. Hence in the overlap region there is a regular relationship between the phases of the waves coming by the two routes – the sources are said to be *coherent*. The appearance of the fringes is quite different from what is seen with just one pinhole.

From the results of such experiments, and comparison with such observations as *Newton's rings*, Young concluded that light was a form of wave motion. Similar experiments have been done since, replacing the pinholes by slits and using monochromatic light sources to give sharply defined light and dark fringes. The separation between two light fringes (or two dark fringes) is $\lambda D/2d$ where d is the separation of the pinholes (or slits) and D is their distance from the plane in which the fringes are observed.

See also biprism; Lloyd's mirror.

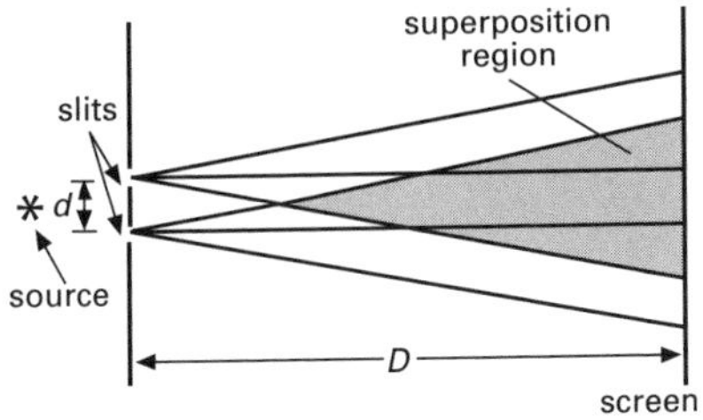

Young's slits

ytterbium /i-**ter**-bee-ŭm/ A soft malleable silvery element having two allotropes and belonging to the lanthanoid series of metals. It occurs in association with other lanthanoids. Ytterbium has been used to improve the mechanical properties of steel.

Symbol: Yb; m.p. 824°C; b.p. 1193°C; r.d. 6.965 (20°C); p.n. 70; r.a.m. 173.04.

yttrium /**it**-ree-ŭm/ A silvery metallic element. It is found in almost every lanthanoid mineral, particularly monazite. Yttrium is used in various alloys, in yttrium-aluminum garnets used in the electronics industry and as gemstones, as a catalyst, and in superconductors. A mixture of yttrium and europium oxides is widely used as the red phosphor on television screens.

Symbol: Y; m.p. 1522°C; b.p. 3338°C; r.d. 4.469 (20°C); p.n. 39; r.a.m. 88.90585.

Z-boson A spin-one, electrically neutral particle that is the quantum of the field that mediates the weak interactions that do not involve a change of electric charge. The Z-boson was discovered at CERN in 1983, with its mass being about 90 GeV. *See also* W-boson; Weinberg–Salam model.

Zeeman effect /**zay**-mahn/ The splitting of atomic spectral lines into two or more components in a transverse magnetic field. The effect is named for the Dutch physicist Pieter Zeeman (1865–1943) who discovered it in 1897. A full explanation of the Zeeman effect requires quantum mechanics. *See also* Stark effect.

Zener breakdown /**zee**-ner, **zen**-er/ *See* Zener diode.

Zener diode A semiconductor DIODE with high doping levels on each side of the junction. If the junction is reverse-biased, breakdown occurs at a well-defined potential, giving a sharp increase in current. The effect is called *Zener breakdown*; it occurs because electrons are excited directly from the valence band into the conduction band. Zener diodes are used as voltage regulators. The Zener diode and Zener breakdown are named for the American physicist Clarence Zener (1905–93). *See also* transistor.

zepto- Symbol: z A prefix denoting 10^{-21}. For example, 1 zeptometer (zm) = 10^{-21} meter (m).

zero-point energy The energy possessed by the atoms and molecules of a substance at absolute zero (0 K).

zetta- Symbol: Z A prefix denoting 10^{21}. For example, 1 zettameter (Zm) = 10^{21} meter (m).

zinc A bluish-white transition metal, applied as a coating (galvanizing) to protect steel from corrosion.

Symbol: Zn; m.p. 419.58°C; b.p. 907°C; r.d. 7.133 (20°C); p.n. 30; r.a.m. 65.39.

zirconium /zer-**koh**-nee-ŭm/ A hard lustrous silvery transition element that occurs in a gemstone, zircon ($ZrSiO_4$). It is used in some strong alloy steels.

Symbol: Zr; m.p. 1850°C; b.p. 4380°C; r.d. 6.506 (20°C); p.n. 40; r.a.m. 91.224.

zone refining A method used to purify solids, especially semiconductors. The material in the form of a bar is passed slowly through a localized heating region. A molten zone forms, which moves slowly along the bar. Impurities tend to concentrate in the molten zone and they can be localized at the end of the bar.

APPENDIXES

Appendix I

Periodic Table of the Elements - giving group, atomic number, and chemical symbol

Period	1	2	3	4	5	6	7	8	9	10	11	12	13	14	15	16	17	18
1	1 H																	2 He
2	3 Li	4 Be											5 B	6 C	7 N	8 O	9 F	10 Ne
3	11 Na	12 Mg											13 Al	14 Si	15 P	16 S	17 Cl	18 Ar
4	19 K	20 Ca	21 Sc	22 Ti	23 V	24 Cr	25 Mn	26 Fe	27 Co	28 Ni	29 Cu	30 Zn	31 Ga	32 Ge	33 As	34 Se	35 Br	36 Kr
5	37 Rb	38 Sr	39 Y	40 Zr	41 Nb	42 Mo	43 Tc	44 Ru	45 Rh	46 Pd	47 Ag	48 Cd	49 In	50 Sn	51 Sb	52 Te	53 I	54 Xe
6	55 Cs	56 Ba	57-71 La-Lu	72 Hf	73 Ta	74 W	75 Re	76 Os	77 Ir	78 Pt	79 Au	80 Hg	81 Tl	82 Pb	83 Bi	84 Po	85 At	86 Rn
7	87 Fr	88 Ra	89-103 Ac-Lr	104 Rf	105 Db	106 Sg	107 Bh	108 Hs	109 Mt	110 Ds	111 Rg	112 Uub	113 Uut	114 Uuq	115 Uup	116 Uuh		

6	Lanthanides	57 La	58 Ce	59 Pr	60 Nd	61 Pm	62 Sm	63 Eu	64 Gd	65 Tb	66 Dy	67 Ho	68 Er	69 Tm	70 Yb	71 Lu
7	Actinides	89 Ac	90 Th	91 Pa	92 U	93 Np	94 Pu	95 Am	96 Cm	97 Bk	98 Cf	99 Es	100 Fm	101 Md	102 No	103 Lr

The above is the modern recommended form of the table using 18 groups. Older group designations are shown below.

Modern form	1	2	3	4	5	6	7	8	9	10	11	12	13	14	15	16	17	18
European convention	IA	IIA	IIIA	IVA	VA	VIA	VIIA	VIII (or VIIIA)			IB	IIB	IIIB	IVB	VB	VIB	VIIB	0 (or VIIIB)
N. American convention	IA	IIA	IIIB	IVB	VB	VIB	VIIB	VIII (or VIIIB)			IB	IIB	IIIA	IVA	VA	VIA	VIIA	VIIIA (or 0)

Appendix II

The Chemical Elements

(* *indicates the nucleon number of the most stable isotope*)

Element	*Symbol*	*p.n.*	*r.a.m*	*Element*	*Symbol*	*p.n.*	*r.a.m*
actinium	Ac	89	227*	europium	Eu	63	151.965
aluminum	Al	13	26.982	fermium	Fm	100	257*
americium	Am	95	243*	fluorine	F	9	18.9984
antimony	Sb	51	112.76	francium	Fr	87	223*
argon	Ar	18	39.948	gadolinium	Gd	64	157.25
arsenic	As	33	74.92	gallium	Ga	31	69.723
astatine	At	85	210	germanium	Ge	32	72.61
barium	Ba	56	137.327	gold	Au	79	196.967
berkelium	Bk	97	247*	hafnium	Hf	72	178.49
beryllium	Be	4	9.012	hassium	Hs	108	265*
bismuth	Bi	83	208.98	helium	He	2	4.0026
bohrium	Bh	107	262*	holmium	Ho	67	164.93
boron	B	5	10.811	hydrogen	H	1	1.008
bromine	Br	35	79.904	indium	In	49	114.82
cadmium	Cd	48	112.411	iodine	I	53	126.904
calcium	Ca	20	40.078	iridium	Ir	77	192.217
californium	Cf	98	251*	iron	Fe	26	55.845
carbon	C	6	12.011	krypton	Kr	36	83.80
cerium	Ce	58	140.115	lanthanum	La	57	138.91
cesium	Cs	55	132.905	lawrencium	Lr	103	262*
chlorine	Cl	17	35.453	lead	Pb	82	207.19
chromium	Cr	24	51.996	lithium	Li	3	6.941
cobalt	Co	27	58.933	lutetium	Lu	71	174.967
copper	Cu	29	63.546	magnesium	Mg	12	24.305
curium	Cm	96	247*	manganese	Mn	25	54.938
darmstadtium	Ds	110	269*	meitnerium	Mt	109	266*
dubnium	Db	105	262*	mendelevium	Md	101	258*
dysprosium	Dy	66	162.50	mercury	Hg	80	200.59
einsteinium	Es	99	252*	molybdenum	Mo	42	95.94
erbium	Er	68	167.26	neodymium	Nd	60	144.24

The Chemical Elements

Element	*Symbol*	*p.n.*	*r.a.m*	*Element*	*Symbol*	*p.n.*	*r.a.m*
neon	Ne	10	20.179	selenium	Se	34	78.96
neptunium	Np	93	237.048	silicon	Si	14	28.086
nickel	Ni	28	58.69	silver	Ag	47	107.868
niobium	Nb	41	92.91	sodium	Na	11	22.9898
nitrogen	N	7	14.0067	strontium	Sr	38	87.62
nobelium	No	102	259*	sulfur	S	16	32.066
osmium	Os	76	190.23	tantalum	Ta	73	180.948
oxygen	O	8	15.9994	technetium	Tc	43	99*
palladium	Pd	46	106.42	tellurium	Te	52	127.60
phosphorus	P	15	30.9738	terbium	Tb	65	158.925
platinum	Pt	78	195.08	thallium	Tl	81	204.38
plutonium	Pu	94	244*	thorium	Th	90	232.038
polonium	Po	84	209*	thulium	Tm	69	168.934
potassium	K	19	39.098	tin	Sn	50	118.71
praseodymium	Pr	59	140.91	titanium	Ti	22	47.867
promethium	Pm	61	145*	tungsten	W	74	183.84
protactinium	Pa	91	231.036	ununbium	Uub	112	285*
radium	Ra	88	226.025	ununtrium	Uut	113	284*
radon	Rn	86	222*	ununquadium	Uuq	114	289*
rhenium	Re	75	186.21	ununpentium	Uup	115	288*
rhodium	Rh	45	102.91	ununhexium	Uuh	116	292*
roentgenium	Rg	111	272*	uranium	U	92	238.03
rubidium	Rb	37	85.47	vanadium	V	23	50.94
ruthenium	Ru	44	101.07	xenon	Xe	54	131.29
rutherfordium	Rf	104	261*	ytterbium	Yb	70	173.04
samarium	Sm	62	150.36	yttrium	Y	39	88.906
scandium	Sc	21	44.956	zinc	Zn	30	65.39
seaborgium	Sg	106	263*	zirconium	Zr	40	91.22

Appendix III

Symbols for Physical Quantities

acceleration	a	moment of inertia	I, J
angular acceleration	α	momentum	p
angular frequency, $2\pi f$	ω	period	T
angular momentum	L	plane angle	$\theta, \phi, \alpha, \beta$, etc.
angular velocity	ω		
area	A	potential energy	E_p, V
breadth	b	power	P
circular wavenumber	k	pressure	p
diameter	d, D	radius	r, R
distance	s, L	reduced mass	μ
dynamic viscosity	η	relative density	d
energy	E, W	solid angle	Ω, ω
force	F	thickness	d
frequency	f, ν	time	t
height	h	torque	T
kinetic energy	E_k, T	velocity	v
length	l	volume	V
mass	m	wavelength	λ
mass density	ρ	wavenumber	σ
moment of force	M	weight	W, F_g

Appendix IV

The Greek Alphabet

Α	α	alpha	Ν	ν	nu
Β	β	beta	Ξ	ξ	xi
Γ	γ	gamma	Ο	ο	omikron
Δ	δ	delta	Π	π	pi
Ε	ε	epsilon	Ρ	ρ	rho
Ζ	ζ	zeta	Σ	σ	sigma
Η	η	eta	Τ	τ	tau
Θ	θ	theta	Υ	υ	upsilon
Ι	ι	iota	Φ	φ	phi
Κ	κ	kappa	Χ	χ	chi
Λ	λ	lambda	Ψ	ψ	psi
Μ	μ	mu	Ω	ω	omega

Appendix V

Conversion Factors

Length

To convert	*into*	*multiply by*
inches	meters	0.0254
feet	meters	0.3048
yards	meters	0.9144
miles	kilometers	1.60934
nautical miles	kilometers	1.85200
nautical miles	miles	1.15078
kilometers	miles	0.621371
kilometers	nautical miles	0.539957
meters	inches	39.3701
meters	feet	3.28084
meters	yards	1.09361

Area

To convert	*into*	*multiply by*
square inches	square centimeters	6.4516
square inches	square meters	6.4516×10^{-4}
square feet	square meters	9.2903×10^{-2}
square yards	square meters	0.836127
square miles	square kilometers	2.58999
square miles	acres	640
acres	square meters	4046.86
acres	square miles	1.5625×10^{-3}
square centimeters	square inches	0.155
square meters	square feet	10.7639
square meters	square yards	1.19599
square meters	acres	2.47105×10^{-4}
square meters	square miles	3.86019×10^{-7}
square kilometers	square miles	0.386019

Volume

To convert	*into*	*multiply by*
cubic inches	liters	1.63871×10^{-2}
cubic inches	cubic meters	1.63871×10^{-5}
cubic feet	liters	28.3168
cubic feet	cubic meters	0.0283168
cubic yard	cubic meters	0.764555
gallon (US)	liters	3.785438
gallon (US)	cubic meters	3.785438×10^{-3}
gallon (US)	gallon (UK)	0.83268

Conversion Factors

Mass

To convert	*into*	*multiply by*
pounds	kilograms	0.453592
pounds	tonnes	4.53592×10^{-4}
hundredweight (short)	kilograms	45.3592
hundredweight (short)	tonnes	0.0453592
tons (short)	kilograms	907.18
tons (short)	tonnes	0.90718
kilograms	pounds	2.204623
kilograms	hundredweights (short)	0.022046
kilograms	tons (short)	1.1023×10^{-3}
tonnes	pounds	2204.623
tonnes	hundredweights (short)	22.0462
tonnes	tons (short)	0.90718

The short ton is used in the USA and is equal to 2000 pounds. The short hundredweight (also known as the cental) is 100 pounds.

The long ton, which is used in the UK, is equal to 2240 pounds (1016.047 kg). The long hundredweight is 112 pounds (50.802 kg). 1 long ton equals 20 long hundredweights.

Force

To convert	*into*	*multiply by*
pounds force	newtons	4.44822
pounds force	kilograms force	0.453592
pounds force	dynes	444822
pounds force	poundals	32.174
poundals	newtons	0.138255
poundals	kilograms force	0.031081
poundals	dynes	13825.5
poundals	pounds force	0.031081
dynes	newtons	10^{-5}
dynes	kilograms force	1.01972×10^{-6}
dynes	pounds force	2.24809×10^{-6}
dynes	poundals	7.2330×10^{-5}
kilograms force	newtons	9.80665
kilograms force	dynes	980665
kilograms force	pounds force	2.20462
kilograms force	poundals	70.9316
newtons	kilograms	0.101972
newtons	dynes	100000
newtons	pounds force	0.224809
newtons	poundals	7.2330

Conversion Factors

Work and energy

To convert	*into*	*multiply by*
British Thermal Units	joules	1055.06
British Thermal Units	calories	251.997
British Thermal Units	kilowatt-hours	2.93071×10^{-4}
kilowatt-hours	joules	3600000
kilowatt-hours	calories	859845
kilowatt-hours	British Thermal Units	3412.14
calories	joules	4.1868
calories	kilowatt-hours	1.16300×10^{-6}
calories	British Thermal Units	3.96831×10^{-3}
joules	calories	0.238846
joules	kilowatt hours	2.7777×10^{-7}
joules	British Thermal Units	9.47813×10^{-4}
joules	electron volts	6.2418×10^{18}
joules	ergs	10^{7}
electronvolts	joules	1.6021×10^{-19}
ergs	joules	10^{-7}

Pressure

To convert	*into*	*multiply by*
atmospheres	pascals*	101325
bars	pascals	100000
pounds per square inch	pascals	68894.76
pounds per square inch	kilograms per square meter	703.068
pounds per square inch	atmospheres	0.068046
kilograms per square meter	pascals	9.80661
kilograms per square meter	pounds per square inch	1.42234×10^{-3}
kilograms per square meter	atmospheres	9.67841×10^{-5}
pascals	kilograms per square meter	0.101972
pascals	pounds per square inch	1.45038×10^{-4}
pascals	atmospheres	9.86923×10^{-6}

*1 pascal = 1 newton per square meter

Appendix VI

Web Sites

General physics resources:

American Physical Society	www.physicscentral.com
American Association of Physics Teachers	www.psrc-online.org
American Institute of Physics	www.aip.org
Atomic, nuclear, and particle physics:	
Brookhaven National Laboratory (BNL)	www.bnl.gov
CERN (European Laboratory for Particle Physics)	www.cern.ch
Fermi National Accelerator Laboratory (Fermilab)	www.fnal.gov
JET (Joint European Torus)	www.jet.efd.org
JINR (Joint Institute for Nuclear Research in Dubna, Russia)	www.jinr.ru
Lawrence Berkeley National Laboratory (LBNL)	www.lbl.gov
Los Alamos National Laboratory	www.lanl.gov
Stanford Linear Accelerator Center (SLAC)	www.slac.stanford.edu
UCLA Particle Beam Physics Laboratory	http://pbpl.physics.ucla.edu
History of physics:	
Center for History of Physics	www.aip.org/history
Biographies:	
Nobel Prizes	http://nobelprize.org

Bibliography

Comprehensive texts covering physics:

Cutnell, John D. and Kenneth W. Johnson. *Physics.* 6th ed. New York: Wiley, 2003.

Giancoli, Douglas C. *Physics: Principles with Applications.* 5th ed. New York: Prentice Hall, 1997.

Other sources:

Barrow, John D. *The World within the World.* Oxford, U.K.: Oxford University Press, 1988.

Chown, Marcus. *The Magic Furnace: The Search for the Origins of Atoms.* London: Jonathan Cape, 1999.

Emsley, John. *Nature's Building Blocks: An A–Z Guide to the Elements.* Oxford, U.K.: Oxford University Press, 2001.

Feynman, Richard P. *QED: The Strange Theory of Light and Matter.* Princeton, N.J.: Princeton University Press, 1985.

Gribbin, John. *Companion to the Cosmos.* London: Weidenfeld & Nicolson, 1996.

Gribbin, John. *Q is for Quantum.* London: Weidenfeld & Nicolson, 1998.

Hawking, Stephen. *A Brief History of Time.* New York: Bantam, 1988.

Penrose, Sir Roger. *The Road to Reality: The Mathematics and Physics of the Universe.* London: Vintage, 2002.

Stewart, Ian. *Does God Play Dice? The New Mathematics of Chaos.* 2nd ed. London: Penguin, 1997.

't Hooft, Gerard. *In Search of the Ultimate Building Blocks.* Cambridge, U.K.: Cambridge University Press, 1997.

Weinberg, Steven. *The First Three Minutes: A Modern View of the Origin of the Universe.* London: André Deutsch, 1977.

Weinberg, Steven. *The Discovery of Subatomic Particles.* New York: Scientific American Library, 1983.